Lecture Notes in Computer Science　　11343

Commenced Publication in 1973
Founding and Former Series Editors:
Gerhard Goos, Juris Hartmanis, and Jan van Leeuwen

More information about this series at http://www.springer.com/series/7407

Shaojie Tang · Ding-Zhu Du
David Woodruff · Sergiy Butenko (Eds.)

Algorithmic Aspects in Information and Management

12th International Conference, AAIM 2018
Dallas, TX, USA, December 3–4, 2018
Proceedings

Editors
Shaojie Tang
The University of Texas at Dallas
Richardson, TX, USA

David Woodruff
University of California, Davis
Davis, CA, USA

Ding-Zhu Du
University of Texas at Dallas
Richardson, TX, USA

Sergiy Butenko
Texas A&M University
College Station, TX, USA

ISSN 0302-9743 ISSN 1611-3349 (electronic)
Lecture Notes in Computer Science
ISBN 978-3-030-04617-0 ISBN 978-3-030-04618-7 (eBook)
https://doi.org/10.1007/978-3-030-04618-7

Library of Congress Control Number: 2018963820

LNCS Sublibrary: SL1 – Theoretical Computer Science and General Issues

This Springer imprint is published by the registered company Springer Nature Switzerland AG
The registered company address is: Gewerbestrasse 11, 6330 Cham, Switzerland

Preface

AAIM 2018 Co-chairs' Message

On behalf of the Organizing Committees, it is our pleasure to welcome you to the proceedings of the AAIM 2018, the 12th International Conference on Big Data Computing and Communications, held in Dallas, USA, during December 3–4, 2018. All submissions to this conference were carefully peer-reviewed, and were ranked according to their original contribution, quality, presentation, and relevance to the conference. Based on the review, 25 high-quality papers were accepted for presentation and inclusion in the proceedings. We aim to attract and bring together researchers working on algorithms, data structures, and their applications to share their latest research results.

The organization of this conference would have been impossible without the valuable help of many dedicated people. First of all, we would like to take this opportunity to thank all the authors for their submissions. We also kindly thank all the members of the Technical Program Committee who shared their valuable time and made a great effort in providing us reviews in time. We hope you find the proceedings interesting.

September 2018

Shaojie Tang
Ding-Zhu Du
David Woodruff
Sergiy Butenko

Organization

General Chairs

Ding-Zhu Du   University of Texas at Dallas, USA
David Woodruff   University of California, Davis, USA

Program Chairs

Shaojie Tang   University of Texas at Dallas, USA
Sergiy Butenko   Texas A&M University, USA

Technical Program Committee

Ding-Zhu Du   University of Texas at Dallas, USA
Minghui Jiang   Utah State University, USA
Zhao Zhang   Zhejiang Normal University, China
Romeo Rizzi   University of Verona, Italy
Michael Fellows   University of Bergen, Norway
Jinhui Xu   State University of New York at Buffalo, USA
Marek Chrobak   University of California, Riverside, USA
My Thai   University of Florida, USA
Guohui Lin   University of Alberta, Canada
Binhai Zhu   Montana State University, USA
Bhaskar Dasgupta   University of Illinois at Chicago, USA
Anthony Bonato   Ryerson University, USA
Chuangyin Dang   City University of Hong Kong, SAR China
Kazuo Iwama   Kyoto University, Japan
Binay Bhattacharya   Simon Fraser University, Canada
Bo Chen   University of Warwick, UK
Peng-Jun Wan   Illinois Institute of Technology, USA
Leizhen Cai   Chinese University of Hong Kong, SAR China
Dennis Komm   ETH Zürich, Switzerland
Hans-Joachim   ETH Zürich, Switzerland.
  Boeckenhauer
Jun Pei   Hefei University of Technology, China
Dalila Fontes   University of Porto, Portugal
Minming Li   City University of Hong Kong, SAR China
Shaojie Tang   University of Texas at Dallas, USA

Contents

Minimum Diameter k-Steiner Forest

Wei Ding[1(✉)] and Ke Qiu[2]

[1] Zhejiang University of Water Resources and Electric Power, Hangzhou 310018,
Zhejiang, China
dingweicumt@163.com

[2] Department of Computer Science, Brock University, St. Catharines, Canada
kqiu@brocku.ca

Abstract. Given an edge-weighted undirected graph $G = (V, E, w)$ and
a subset $\mathcal{T} \subseteq V$ of p terminals, a k-*Steiner forest* spanning all the terminals in $\mathcal{T}$ includes k branches, where every branch is a Steiner tree.
The diameter of a k-Steiner forest is referred to as the maximum distance between two terminals of a branch. This paper studies the *minimum diameter k-Steiner forest problem (MDkSFP)* and establishes the
relationship between MDkSFP and the *absolute Steiner k-center problem
(ASkCP)*. We first obtain a 2-approximation to ASkCP by a dual approximation algorithm and then achieve a 2-approximation to MDkSFP. Further, we achieve a (better) 2ρ-approximation to MDkSFP, where $\rho < 1$
in general, by modifying the sites of centers and re-clustering all the
terminals.

Keywords: k-Steiner forest · Diameter · Modification

1 Introduction

Given an undirected graph $G = (V, E)$ and a subset $\mathcal{T} \subseteq V$ of *terminals*, a
Steiner tree is referred to as a connected subgraph of G that spans all the terminals in $\mathcal{T}$ [6]. A *Steiner forest* is a group of *disjoint* connected branches spanning
$\mathcal{T}$, where every *branch* is a Steiner tree [1,2,5,7,11–14]. They have a wide variety
of applications in communication networks, computational biology, and etc. [6].

1.1 Related Results

Let $G = (V, E, w)$ be an edge-weighted undirected graph, where V is the set of n
vertices, E is the set of m edges, and $w(\cdot)$ is an edge weight function $w : E \to \mathbb{R}^+$.
The *cost* of a Steiner forest is equal to the sum of all the weights of edges used
by the Steiner forest. Given a subset $\mathcal{T} \subseteq V$ of p terminals and an integer,
$1 \leq k \leq p$, the k-**Steiner forest problem** (k**SFP**) asks to find a Steiner forest
of minimum cost which contains at most k branches. In [11], Ravi presented a
primal-dual approximation algorithm with a factor of $2(1 - \frac{1}{p-k+1})$ for kSFP.

The **generalized Steiner tree problem (GSTP)** (also called the **Steiner
forest problem (SFP)**) is another version of Steiner forest problem. Given a

© Springer Nature Switzerland AG 2018
S. Tang et al. (Eds.): AAIM 2018, LNCS 11343, pp. 1–11, 2018.
https://doi.org/10.1007/978-3-030-04618-7_1

collection $\mathcal{D} = \{(s_i, t_i) : 1 \leq i \leq K\}$ of K terminal pairs, the aim of GSTP is to find an edge subset $F \subseteq E$ of minimum cost such that F contains an s_i-t_i path. Agrawal *et al.* [1] gave the first approximation algorithm with a factor of $2(1 - \frac{1}{p})$, where p is the number of terminals. Refer readers to [2,13,14] for related results in special graphs. In [7], Hajiaghayi and Jain proposed the **Steiner k-forest problem (SkFP)**, which aims to find a subgraph of minimum cost such that at least $k \leq K$ terminal pairs of $\mathcal{D}$ are connected. In [12], Segev and Segev gave an $O(n^{2/3} \log K)$-approximation algorithm by using Lagrangian relaxation technique. Very recently, Dinitz [5] developed the best known approximation algorithm with a factor of $n^{\frac{1}{3}(7-4\sqrt{2})+\epsilon}$, where $\frac{1}{3}(7 - 4\sqrt{2}) < 0.44772$ and $\epsilon > 0$ is an arbitrary constant.

The *diameter* of a tree or forest is referred to as the longest distance between vertices (leaves) on the tree or forest. Hassin and Tamir [8] sought a **minimum diameter spanning tree problem (MDSTP)** of G as well as the **minimum diameter k-forest problem (MDkFP)**. In [3], Bui *et al.* presented a distributed algorithm for MDSTP, with time complexity of $O(n)$ and message complexity of $O(mn)$. In [4], Ding and Qiu proposed the **minimum diameter Steiner tree (MDSTT)**, and gave an $O(n \log p)$-time 2-approximation algorithm and $O(mn + n^2 \log n)$-time exact algorithm.

1.2 Our Results

In the real-world problems, we are sometimes required to find a Steiner forest spanning a subset of terminals with the aim of minimizing the maximum diameter of all the branches. For instance, in a communication network with every link having a delay, a set of clients are clustered into k groups arbitrarily and every group of clients are interconnected by a Steiner tree, resulting in a k-Steiner forest. The goal is to make the maximum delay between two clients as small as possible. This problem can be modelled as the **minimum diameter k-Steiner forest problem (MDkSFP)**, which is formally defined in Sect. 2.1. To the best of our knowledge, this paper is the first one to propose MDkSFP formally.

We first establish the relationship between MDkSFP and the **absolute Steiner k-center problem (ASkCP)**. ASkCP is NP-hard. We use the dual approximation framework [9,10] to design a 2-approximation algorithm for ASkCP. By clustering all the terminals into k groups with the 2-approximation as the centers, we design an $O(mp^2 \log m)$-time 2-approximation algorithm for MDkSFP. Further, we obtain a better approximation to ASkCP by modification. By re-clustering all the terminals with the new approximation as the centers, we develop an $O(mp^2 \log m + n^2 p)$-time 2ρ-approximation algorithm, where ρ is the ratio of the diameter of the latter approximation to MDkSFP over that of the former.

Organization. The rest of this paper is organized as follows. In Sect. 2, we define ASkCP and MDkSFP formally, and also show the relationship between ASkCP and MDkSFP. In Sect. 3, we give a dual approximation algorithm for ASkCP. In Sect. 4, we first design a 2-approximation algorithm for MDkSFP and further develop a better approximation algorithm by modification. In Sect. 5, we conclude the paper.

2 Preliminaries

2.1 Definitions and Notations

Let $G = (V, E, w)$ be an edge-weighted undirected graph, where $V = \{v_1, v_2, \ldots, v_n\}$ is the set of n vertices, E is the set of m edges, and $w : E \to \mathbb{R}^+$ is a weight function on edges. Let $\mathcal{T} = \{t_1, t_2, \ldots, t_p\} \subseteq V$ be a subset of p terminals in V. For any $e \in E$, we use $\mathcal{P}(e)$ to denote the set of continuum points on e. Let $\mathcal{P}$ be the set of all the continuum points on edges of G. So, $\mathcal{P} = \bigcup_{e \in E} \mathcal{P}(e)$. For any pair of vertices, $v_i, v_j \in V$, we use $d(v_i, v_j)$ to denote the v_i-v_j shortest path distance (SPD). Obviously, $d(\cdot)$ is a metric, i.e., $d(v_i, v_i) = 0$ and $d(v_i, v_k) \leq d(v_i, v_j) + d(v_j, v_k), \forall i, j, k$. Let $M = (d)_{n \times n}$ be the $n \times n$ distance matrix of shortest paths in G. We assume that M is available in this paper. Note that we also use $d(x, y)$ to denote the x-y SPD, for any two points, x and y. A set of cardinality of k is called a k-set.

Given a k-set, $\mathcal{X} = \{x_1, x_2, \ldots, x_k\} \subset \mathcal{P}$, the distance from terminal t_i to $\mathcal{X}$, $d(t_i, \mathcal{X})$, is referred to as the distance from t_i to the closest facility in $\mathcal{X}$, for any $1 \leq i \leq p$. So, $d(t_i, \mathcal{X}) = \min_{1 \leq j \leq k} d(t_i, x_j)$. The k-set, $\mathcal{X}$, is called a absolute Steiner k-center (ASkC). The maximum distance from terminals to $\mathcal{X}$ (ASkC) is called the k-radius from $\mathcal{T}$ to $\mathcal{X}$, denoted by $r(\mathcal{T}, \mathcal{X})$. Let $\mathcal{X}^* = \{x_1^*, x_2^*, \ldots, x_k^*\}$ denote an optimal ASkC, which minimizes the k-radius, and $\mathcal{C}_j$ be the subset of terminals are assigned to x_j^* (i.e., x_j^* is the closest facility). We have

$$r(\mathcal{T}, \mathcal{X}) = \max_{1 \leq i \leq p} d(t_i, \mathcal{X}). \tag{1}$$

and

$$r(\mathcal{T}, \mathcal{X}^*) = \min_{\mathcal{X} \subset \mathcal{P}} r(\mathcal{T}, \mathcal{X}). \tag{2}$$

This paper first deals with the following problem.

Definition 1. *Given an edge-weighted undirected graph* $G = (V, E, w)$, *a subset* $\mathcal{T} \subseteq V$ *of* p *terminals and a positive integer* $k \geq 1$, *the goal of* **absolute Steiner k-center problem (ASkCP)** *is to find an optimal k-set,* $\mathcal{X}^* = \{x_1^*, x_2^*, \ldots, x_k^*\} \subset \mathcal{P}$, *to minimize the k-radius.*

The **discrete (vertex) Steiner k-center problem (DSkCP)** is the special case of ASkCP, where the facilities are restricted to the vertices. Similarly, we can define the *discrete Steiner k-center (DSkC)*. Obviously, the ordinary **absolute k-center problem (AkCP)** is the special case of ASkCP with $\mathcal{T} = V$. Besides, DS1C and AS1C are the special cases of DSkC and ASkC with $k = 1$, respectively.

Let $\mathcal{F}$ denote a k-*Steiner forest* spanning all the terminals in $\mathcal{T}$, and $\mathcal{F}_j$ denote the j-th branch (Steiner tree). So, $\mathcal{F} = \bigcup_{j=1}^k \mathcal{F}_j$. Note that the k branches are both *edge-disjoint* and *vertex-disjoint*. One is called a *pseudo* Steiner forest if it has two branches having at least a common edge or vertex.

The *diameter* of branch $\mathcal{F}_j$, diam$(\mathcal{F}_j)$, is referred to as the longest distance between leaves of $\mathcal{F}_j$. As every branch $\mathcal{F}_j$ is a Steiner tree, its diameter is also the longest distance between terminals. Let $d_j(t_{i_1}, t_{i_2})$ denote the distance on branch $\mathcal{F}_j$ between t_{i_1} and t_{i_2}. The *diameter* of a k-Steiner forest $\mathcal{F}$, diam$(\mathcal{F})$, is referred to as the maximum diameter of branches. We have

$$\text{diam}(\mathcal{F}_j) = \max_{t_{i_1}, t_{i_2} \in \mathcal{F}_j, t_{i_1} \neq t_{i_2}} d_j(t_{i_1}, t_{i_2}). \tag{3}$$

and

$$\text{diam}(\mathcal{F}) = \max_{1 \leq j \leq k} \text{diam}(\mathcal{F}_j). \tag{4}$$

Since ASkCP and MDkSFP are trivial when $p = 2$, we focus on the cases with $p \geq 3$. On the other hand, the optimum is obvious (every branch contains one terminal) when $k = p$. Therefore, we focus on the cases where k is (quite) smaller than p in general.

Definition 2. *Given an edge-weighted undirected graph $G = (V, E, w)$, a subset $\mathcal{T} \subseteq V$ of p terminals and a positive integer $1 \leq k \leq p$, the **minimum diameter k-Steiner forest problem (MDkSFP)** asks to find an optimal k-Steiner forest, $\mathcal{F}^*$, such that the diameter is minimized.*

2.2 Fundamental Properties

Theorem 1 shows a sufficient condition of k-Steiner forest, which gives a method of constructing a k-Steiner forest while avoiding a pseudo forest.

Theorem 1. *Given a k-set with no duplicates, $\mathcal{X} = \{x_1, x_2, \ldots, x_k\} \subset \mathcal{P}$, we let $T_j \subseteq \mathcal{T}$ be the set of terminals, which are closest to x_j, for all $1 \leq j \leq k$. One **shortest path tree (SPT)** spanning T_{j_1} with x_{j_1} as the origin and the other SPT spanning T_{j_2} with x_{j_2} as the origin are both edge-disjoint and vertex-disjoint, for any $j_1 \neq j_2$.*

Theorem 2 is a necessary condition of minimum diameter k-Steiner forest, which shows the relationship between MDkSFP and ASkCP. The proof of Theorem 2 implies we can obtain a minimum diameter k-Steiner forest by first finding an optimal ASkC and then computing a collection of SPT's with the optimal ASkC as the origins. First, we give Lemma 1.

Lemma 1. *There are surely at least one facility $x_{j^*}^* \in \mathcal{X}^*$ and at least two terminals, $t_{i_1^*}$ and $t_{i_2^*}$, such that $d(t_{i_1^*}, x_{j^*}^*) = d(t_{i_2^*}, x_{j^*}^*) = r(\mathcal{T}, \mathcal{X}^*)$.*

Theorem 2. diam$(\mathcal{F}^*) = 2 \cdot r(\mathcal{T}, \mathcal{X}^*)$.

3   Approximation to ASkCP

In this section, we study ASkCP in edge-weighted graph $G = (V, E, w)$, and use the dual approximation framework [9, 10] to devise an approximation algorithm with a factor of 2. Let opt be the optimal value (of k-radius), $\mathcal{X}^*$ and $\mathcal{X}^A$ denote an optimal solution and an algorithm solution with no duplicate, respectively.

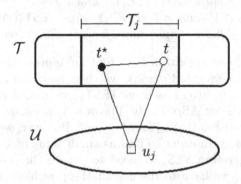

Fig. 1. Illustration for the proof of Theorem 3.

Procedure TEST$_1$ plays an important role, where k terminals are selected from $\mathcal{T}$ as a possible solution. Let $\mathcal{X}(\Delta)$ denote the solution generated by TEST$_1(\Delta)$, where $\Delta > 0$ is a given real number, and T record the unlabelled terminals. Initially, set $\mathcal{X}(\Delta) = \emptyset$ and $T = \mathcal{T}$. Every time one terminal t^* is selected arbitrarily as a facility, all unlabelled terminals $t \in T$ satisfying that

Procedure TEST$_1(\Delta)$:
Input: a graph $G = (V, E, w)$ and distance matrix M, a set $\mathcal{T} \subset V$ of terminals, a positive real number $\Delta > 0$;
Output: NO or YES with $\mathcal{X}(\Delta)$.
Step 1: $\mathcal{X}(\Delta) \leftarrow \emptyset$; $T \leftarrow \mathcal{T}$;
Step 2: while $\|\mathcal{X}(\Delta)\| < k$ and $T \neq \emptyset$ **do** 　　　　Select a terminal t^* arbitrarily from T; 　　　　$L(t^*) \leftarrow \{t \in T : d(t, t^*) \leq 2\Delta\}$; 　　　　$T \leftarrow T \setminus L(t^*)$; $\mathcal{X}(\Delta) \leftarrow \mathcal{X}(\Delta) \cup \{t^*\}$; 　　**end**

Step 3: if $T \neq \emptyset$ **then** Return NO;
　　else
　　　　if $\|\mathcal{X}(\Delta)\| \neq k$ **then**
　　　　　　Let $\mathcal{X}_0$ be a set of $k - \|\mathcal{X}(\Delta)\|$ terminals,
　　　　　　selected arbitrarily from $\mathcal{T} \setminus \mathcal{X}(\Delta)$;
　　　　　　$\mathcal{X}(\Delta) \leftarrow \mathcal{X}(\Delta) \cup \mathcal{X}_0$;
　　　　endif Return YES and $\mathcal{X}(\Delta)$;
　　endif

$d(t, t^*) \leq 2\Delta$ are labelled. This operation is repeated until $|\mathcal{X}(\Delta)| = k$ or $T = \emptyset$. If TEST_1 ends with $T \neq \emptyset$, then TEST_1 fails and returns NO. Otherwise, TEST_1 can find a solution in the following way, and so returns YES. Since the solution is a k-set, TEST_1 needs to judge whether $|\mathcal{X}(\Delta)| \neq k$ or not. If the former occurs, then a set $\mathcal{X}_0$ of $k - |\mathcal{X}(\Delta)|$ terminals are selected arbitrarily from $T \setminus \mathcal{X}(\Delta)$, and the union of $\mathcal{X}_0$ and $\mathcal{X}(\Delta)$ is output as a solution. Otherwise, it returns $\mathcal{X}(\Delta)$.

Theorem 3. *Given a graph $G = (V, E, w)$ and a subset $T \subset V$ of terminals, if there is a k-set $\mathcal{U} \subset V$ with $r(T, \mathcal{U}) \leq \Delta$, then $\mathsf{TEST}_1(\Delta)$ can find a k-set $\mathcal{X} \subset V$ with $r(T, \mathcal{X}) \leq 2\Delta$ in $O(pk)$ time, for any $\Delta > 0$ (see Fig. 1).*

The main idea of our algorithm is to find an approximation, a k-set of T, to ASkCP in vertex-unweighted graphs, which is based on the fact that every terminal can act as a facility. Based on TEST_1, we design our approximation algorithm ASkCP-ALG for ASpCP. By Theorem 3, we know that $\mathsf{TEST}_1(opt)$ returns YES as $\mathcal{X}^*$ satisfies that $r(T, \mathcal{X}^*) \leq opt$. However, we do not know the value of opt beforehand. In order to find as small value of k-radius as possible which makes TEST_1 return YES, we need to get all the possible values of k-radius. By Lemma 1, we have to discuss all the possible combinations of the edge containing x_{j*}^* and two terminals, $t_{i_1^*}$ and $t_{i_2^*}$. Given an edge $e = \{v', v''\}$ and two terminals, t_{i_1} and t_{i_2}, the corresponding two possible values of k-radius are, see Fig. 2,

$$r' = \frac{1}{2} \left(d(t_{i_1}, v') + w(v', v'') + d(t_{i_2}, v'') \right), \tag{5}$$

and

$$r'' = \frac{1}{2} \left(d(t_{i_1}, v'') + w(v', v'') + d(t_{i_2}, v') \right). \tag{6}$$

So, the number of possible values of k-radius is $2 \cdot m \binom{p}{2} = mp(p-1)$. By deleting duplicates of all the possible values and then arranging the values left into an increasing sequence, $l_1 < l_2 < \cdots < l_\lambda$, where $\lambda \leq mp(p-1)$, Step 1 of ASkCP-ALG can be done in $O(mp^2 \log m)$ time by sorting.

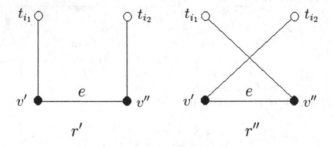

Fig. 2. Two possible values of k-radius for given t_{i_1}, t_{i_2} and edge $e = \{v', v''\}$.

Obviously, $\mathsf{TEST}_1(l_\lambda)$ returns YES. Our task is to find the smallest one, l_{s^*}, from all the possible values $l_s, 1 \le s \le \lambda$ which make $\mathsf{TEST}_1(l_s)$ return YES. In Step 2, we find l_{s^*} by applying a *binary search* to $l_s, 1 \le s \le \lambda$. So, it is sufficient to apply TEST_1 $O(\log \lambda)$ times. Recall that applying TEST_1 once takes at most $O(pk)$ time. Since $\lambda \le mp(p-1)$ and $p \le n \le m$, Step 2 takes at most $O(pk \log m)$ time. Therefore, ASkCP-ALG takes at most $O(mp^2 \log m)$ time. The details of Step 2 are presented as follows. Our search begins with l_1. If $\mathsf{TEST}_1(l_1)$ outputs YES, then we claim that $l_{s^*} = l_1$ and get an approximation $\mathcal{X}^A = \mathcal{X}(l_1)$. Otherwise, we find l_{s^*} by applying a binary search to $l_1, l_2, \ldots, l_\lambda$. Let LB and UB be the lower bound and upper bound on the index s^*, respectively, and let $M = \lfloor \frac{\mathrm{LB+UB}}{2} \rfloor$. Initially, set $\mathrm{LB} = 1$ and $\mathrm{UB} = \lambda$. If $\mathsf{TEST}_1(l_M)$ outputs NO, then M becomes a new lower bound on s^* and UB is also its upper bound. If $\mathsf{TEST}_1(l_M)$ outputs YES, then M becomes a new upper bound on s^* and LB is also its lower bound. Above operation is repeated until LB and UB become two consecutive integer numbers. During the whole operation, $\mathsf{TEST}_1(l_{\mathrm{LB}})$ always outputs NO while $\mathsf{TEST}_1(l_{\mathrm{UB}})$ always outputs YES. As a result, the final UB is s^*. The binary search ends with an approximation, $\mathcal{X}^A = \mathcal{X}(l_{s^*})$, to ASkCP in a vertex-unweighted graph G. According to above discussions, we obtain the following theorem.

Algorithm ASkCP-ALG:
Input: a graph $G = (V, E, w)$ and distance matrix M, a set $\mathcal{T} \subset V$ of terminals;
Output: a k-set $\mathcal{X}^A$.
Step 1: Delete duplicates of all the possible values of k-radius, and then arrange the values left into an increasing sequence, $l_1 < l_2 < \cdots < l_\lambda$.

Step 2: if $\mathsf{TEST}_1(l_1) = $ YES then Return $\mathcal{X}(l_1)$;
      else LB $\leftarrow 1$; UB $\leftarrow \lambda$;
        while UB $-$ LB $\ne 1$ **do**
          M $\leftarrow \lfloor \frac{\mathrm{LB+UB}}{2} \rfloor$;
          if $\mathsf{TEST}_1(l_M) = $ NO **then** LB $\leftarrow$ M;
          else UB $\leftarrow$ M; **endif**
        end Return $\mathcal{X}(l_{\mathrm{UB}})$;
      endif

Theorem 4. *Given a graph $G = (V, E, w)$ with n vertices and m edges, and a set $\mathcal{T} \subset V$ of p terminals, ASkCP-ALG is a 2-approximation algorithm for ASkCP in G, with time complexity of $O(mp^2 \log m)$.*

4   Approximations to MDkSFP

In this section, we design a 2-approximation algorithm for MDkSFP, and further develop an improved approximation algorithm with a factor of 2ρ, where $\rho < 1$ in general, based on the approximation to ASkCP.

4.1   A 2-Approximation Algorithm

In this subsection, we first get a better solution to the vertex-unweighted ASkCP by modifying the algorithm solution of ASkCP-ALG, and then design an approximation algorithm for MDkSFP based on the modification.

The algorithm solution, $\mathcal{X}^A = \{t_1^\star, t_2^\star, \ldots, t_k^\star\}$, produced by AS$k$CP-ALG is a set of k terminals. Let $C_j, 1 \leq j \leq k$ be the set of terminals which are labelled when $t_j^\star$ is selected as a facility. In details, we cluster all the terminals with $t_1^\star, t_2^\star, \ldots, t_k^\star$ as centers. Let C_j^A denote the *cluster* of terminals to which $t_j^\star$ is the closest facility. Note that C_j^A includes $t_j^\star$. The maximum distance to $t_j^\star$ in C_j^A (*resp.* C_j) is denoted by $r(C_j^A)$ (*resp.* $r(C_j)$). Lemma 2 implies that the multiset $\{C_j^A : 1 \leq j \leq k\}$ is better than the multiset $\{C_j : 1 \leq j \leq k\}$ with respect to the value of k-radius.

Lemma 2. $\max_{1 \leq j \leq k} r(C_j^A) \leq \max_{1 \leq j \leq k} r(C_j)$.

Let $\mathcal{F}_j^A$ be the SPT (branch) spanning C_j^A with $t_i^\star$ as the origin. By Theorem 1 and the definition of C_j^A, we claim that all $\mathcal{F}_j^A, \forall j$ are both edge-disjoint and vertex-disjoint. Hence, the collection of $\bigcup_{j=1}^k \mathcal{F}_j^A$ forms a Steiner forest (algorithm solution), denoted by $\mathcal{F}^A$. This leads to an approximation algorithm MDkSFP-ALG for MDkSFP. Its performance analysis is shown in Theorem 5. Let j_i be the index such that $d(t_i, t_{j_i}^\star) = \min_{1 \leq j \leq k} d(t_i, t_j^\star)$, for any $1 \leq i \leq p$.

Algorithm MDkSFP-ALG:
Input: a graph $G = (V, E, w)$ and distance matrix M, a set $\mathcal{T} \subset V$ of terminals;
Output: a Steiner forest $\mathcal{F}^A$.
Step 1: Call ASkCP-ALG to get $\mathcal{X}^A = \{t_1^\star, t_2^\star, \ldots, t_k^\star\}$; $\quad\quad\quad C_j^A \leftarrow \emptyset, 1 \leq j \leq k$; $\quad\quad\quad$ **for** $i := 1$ to p **do** $\quad\quad\quad\quad\quad j_i \leftarrow \arg\min_{1 \leq j \leq k} d(t_i, t_j^\star)$; $\quad\quad\quad\quad\quad C_{j_i}^A \leftarrow C_{j_i}^A \cup \{i\}$; $\quad\quad\quad$ **end**
Step 2: **for** $j := 1$ to k **do** $\quad\quad\quad\quad\quad$ Compute a SPT (branch) $\mathcal{F}_j^A$ spanning C_j^A; $\quad\quad\quad$ **end** $\quad\quad\quad \mathcal{F}^A \leftarrow \bigcup_{j=1}^k \mathcal{F}_j^A$; Return $\mathcal{F}^A$;

Theorem 5. *Given an edge-weighted graph $G = (V, E, w)$ with n vertices and m edges, and a set $T \subset V$ of p terminals, MDkSFP-ALG can compute a 2-approximation to MDkSFP in G within $O(mp^2 \log m)$ time.*

4.2 A Better Approximation Algorithm

The 2-approximation algorithm MDkSFP-ALG clusters all the terminals with the algorithm solution of ASkCP-ALG as centers. The algorithm solution is a set of k terminals, $\mathcal{X}^A = \{t_1^\star, t_2^\star, \ldots, t_k^\star\}$. However, the optimal solution, $\mathcal{X}^* = \{x_1^*, x_2^*, \ldots, x_k^*\}$, is a set of k points. In this subsection, we improve the performance of the solution to some extent by modifying k terminals to k points.

Let $|C_j^A| = p_j, \forall j$. So, $T = \bigcup_{j=1}^k C_j^A$ and $p = \sum_{j=1}^k p_j$. For any C_j^A, we use Ding and Qiu's algorithm [4] to compute the MDST of C_j^A, where C_j^A is the set of terminals. By Lemma 3, we claim that the SPT spanning C_j^A with the AS1C, $x_j^\star$, as the origin is just the MDST of C_j^A, while the SPT spanning C_j^A with the DS1C, $t_j^\star$, as the origin is a 2-approximate MDST of C_j^A.

Lemma 3 (see [4]). *Given a graph $G = (V, E, w)$ and a subset $T \subset V$ of terminals, the MDST is the SPT spanning T with the AS1C as the origin. In addition, the SPT spanning T with the DS1C as the origin is a 2-approximate MDST where the factor of 2 is tight.*

The center of C_j^A is perturbed from $t_j^\star$ to $x_j^\star$, for all $1 \le j \le k$, resulting in a better solution to ASkCP, denoted by $\mathcal{X}^B = \{x_1^\star, x_2^\star, \ldots, x_k^\star\}$. It is certain that $t_j^\star$ is closest to all the terminals in C_j^A, but it is uncertain that $x_j^\star$ is closest to all of them. Therefore, if the MDST (a.k.a., the SPT with $x_j^\star$ as the origin) spanning C_j^A is taken as one branch, then two different branches may have at least a common edge. In other words, the resulting solution may be a pseudo Steiner forest. Let $\mathcal{F}_j^P$ denote the MDST spanning C_j^A, and $\mathcal{F}^P$ denote the union of all $\mathcal{F}_j^P, \forall j$.

In order to avoid a pseudo solution, we re-cluster all the terminals with $\mathcal{X}^B$ as centers. Let C_j^B be the cluster of terminals to which $x_j^\star$ is the closest facility, and let $r(C_j^B)$ denote the maximum distance to $x_j^\star$ in C_j^B. The SPT spanning C_j^B with $x_j^\star$ as the origin is taken as one branch, denoted by $\mathcal{F}_j^B$. By Theorem 1, we conclude from the definition of C_j^B that all $\mathcal{F}_j^B, 1 \le j \le k$ are both edge-disjoint and vertex-disjoint. As a consequence, all the branches $\mathcal{F}_j^B, \forall j$ form a Steiner forest, denoted by $\mathcal{F}^B$. Based on above discussions, we devise an improved approximation algorithm MDkSFP-IMP for MDkSFP. Its performance analysis is shown in Theorem 6. For any $1 \le s \le p$, we let j_s be the index such that $d(t_s, x_{j_s}^\star) = \min_{1 \le j \le k} d(t_s, x_j^\star)$.

```
Algorithm MDkSFP-IMP:
Input: a graph G = (V, E, w) and distance matrix M,
a set T ⊂ V of terminals;
Output: a Steiner forest F^B.
Step 1: The same as MDkSFP-ALG;
Step 2: for j := 1 to k do
            Use the Ding and Qiu's algorithm in [4] to
            compute the AS1C, x*_j, of C^A_j;
        end
        C^B_j ← ∅, 1 ≤ j ≤ k;
        for s := 1 to p do
            j_s ← arg min_{1≤j≤k} d(t_s, x*_j);
            C^B_{j_s} ← C^B_{j_s} ∪ {s};
        end
Step 3: for j := 1 to k do
            Compute a SPT (branch) F^B_j spanning C^B_j;
        end
        F^B ← ∪^k_{j=1} F^B_j; Return F^B;
```

Theorem 6. *Given a graph $G = (V, E, w)$ with n vertices and m edges, and a set $T \subset V$ of p terminals, MDkSFP-IMP produces a possible better approximation than MDkSFP-ALG within $O(mp^2 \log m + n^2 p)$ time.*

The proof of Theorem 6 shows $\mathrm{diam}(\mathcal{F}^B) \leq \mathrm{diam}(\mathcal{F}^P) \leq \mathrm{diam}(\mathcal{F}^A)$. MDkSFP-IMP can produce the exact values of $\mathrm{diam}(\mathcal{F}^A_j)$ and $\mathrm{diam}(\mathcal{F}^B_j)$, for $\forall j$, and further the values of $\mathrm{diam}(\mathcal{F}^A)$ and $\mathrm{diam}(\mathcal{F}^B)$. Let

$$\rho = \frac{\mathrm{diam}(\mathcal{F}^B)}{\mathrm{diam}(\mathcal{F}^A)}. \tag{7}$$

Clearly, $\rho \leq 1$. Recall that $\mathrm{diam}(\mathcal{F}^A) \leq 2 \cdot \mathrm{diam}(\mathcal{F}^*)$. Therefore,

$$\mathrm{diam}(\mathcal{F}^B) = \rho \cdot \mathrm{diam}(\mathcal{F}^A) \leq 2\rho \cdot \mathrm{diam}(\mathcal{F}^*).$$

Corollary 1. *The approximation factor of MDkSFP-IMP is 2ρ, where ρ is the ratio of $\mathrm{diam}(\mathcal{F}^B)$ over $\mathrm{diam}(\mathcal{F}^A)$.*

In most cases, $\rho < 1$. This is because $\rho = 1$ if and only if the center (point) corresponding to the k-radius is at a terminal (vertex). Therefore, the approximation of MDkSFP-IMP improves that of MDkSFP-ALG by a factor of $2(1 - \rho)$.

5   Conclusions

This paper studied MDkSFP in undirected graphs and established the relationship between MDkSFP and ASkCP. First, we obtained an approximate ASkC and then designed a 2-approximation algorithm for MDkSFP. Further, we achieved a better approximate ASkC by modification, and then developed a 2ρ-approximation algorithm for MDkSFP, where $\rho < 1$ in general. Our algorithm can be adapted to the version where the number of the branches of Steiner forest is at most k instead of equal to k.

The performance ratio of the approximate ASkC is one of the major factors that influence the performance ratio of our algorithm solution of MDkSFP. We suggest to improve the approximation to MDkSFP by achieving a better approximation to ASkCP, and conjecture that ASkCP could admit a ρ-approximation algorithm with $\rho < 2$.

References

1. Agrawal, A., Klein, P., Ravi, R.: When trees collide: an approximation algorithm for the generalized Steiner problem in networks. SIAM J. Comput. **24**(3), 440–456 (1995)
2. Bateni, M., Hajiaghayi, M., Marx, D.: Approximation schemes for Steiner forest on planar graphs and graphs of bounded treewidth. J. ACM **58**(5), 1–21 (2011)
3. Bui, M., Butelle, F., Lavault, C.: A distributed algorithm for constructing a minimum diameter spanning tree. J. Parallel Distrib. Comput. **64**(5), 571–577 (2004)
4. Ding, W., Qiu, K.: Algorithms for the minimum diameter terminal Steiner tree problem. J. Comb. Optim. **28**(4), 837–853 (2014)
5. Dinitz, M., Kortsarz, G., Nutov, Z.: Improved approximation algorithm for Steiner k-forest with nearly uniform weights. In: 17th APPROX, pp. 115–127 (2014)
6. Du, D., Hu, X.: Steiner Tree Problems in Computer Communication Networks. World Scientific Publishing Co., Pte. Ltd., Singapore (2008)
7. Hajiaghayi, M., Jain, K.: The prize-collecting generalized steiner tree problem via a new approach of primal-dual schema. In: Proceedings of the 17th SODA, pp. 631–640 (2006)
8. Hassin, R., Tamir, A.: On the minimum diameter spanning tree problem. Inf. Process. Lett. **53**, 109–111 (1995)
9. Hochbaum, D.S., Shmoys, D.B.: A best possible heuristic for the k-center problem. Math. Oper. Res. **10**(2), 180–184 (1985)
10. Plesník, J.: A heuristic for the p-center problem in graphs. Discrete Appl. Math. **17**, 263–268 (1987)
11. Ravi, R.: A primal-dual approximation algorithm for the Steiner forest problem. Inf. Process. Lett. **50**(4), 185–189 (1994)
12. Segev, D., Segev, G.: Approximate k-Steiner forests via the Lagrangian relaxation technique with internal preprocessing. Algorithmica **56**, 529–549 (2010)
13. Winter, P.: Generalized Steiner problem in outerplanar graphs. BIT Numer. Math. **25**(3), 485–496 (1985)
14. Winter, P.: Generalized Steiner problem in series-parallel networks. J. Algorithms **7**(4), 549–566 (1986)

Factors Impacting the Label Denoising of Neural Relation Extraction

Tingting Sun[✉], Chunhong Zhang, and Yang Ji

Beijing University of Posts and Telecommunications, Beijing 100876, China
{suntingting,zhangch,jiyang}@bupt.edu.cn

Abstract. The goal of relation extraction is to obtain relational facts from plain text, which can benefit a variety of natural language processing tasks. To address the challenge of automatically labeling large-scale training data, a distant supervision strategy is introduced to relation extraction by heuristically aligning entity pairs in plain text with the knowledge base. Unfortunately, the method is vulnerable to the noisy label problem due to the incompletion of the exploited knowledge base. Existing works focus on the specific algorithms, but few works summarize the commonalities between different methods and the influencing factors of these denoising mechanisms. In this paper, we propose three main factors that impact the label denoising of distantly supervised relation extraction, including labeling assumption, prior knowledge and confidence level. In order to analyze how these factors influence the denoising effectiveness, we build a unified neural framework with word, sentence and label denoising modules for relation extraction. Then we conduct experiments to evaluate and compare these factors according to ten neural schemes. In addition, we discuss the typical cases of these factors and find that influential word-level prior knowledge and partial confidence for distantly supervised labels can significantly affect the denoising performance. These implicational findings can provide researchers with more insight of distantly supervised relation extraction.

Keywords: Relation extraction · Distant supervision · Label denoising

1 Introduction

Relation extraction is an active research task in information extraction [7] and natural language processing, which aims to predict the relation between two entities mentioned in plain text. It can benefit a variety of artificial intelligence tasks such as question answering [1], information search [16] and knowledge base construction [22]. However, supervised learning for relation extraction requires a large amount of annotated data, which is costly to obtain.

Therefore, a distant supervision [12] strategy for relation extraction is proposed to automatically label large-scale training data by heuristically aligning entity pairs in plain text with the Knowledge Base (KB). However, distant supervision tends to have the noisy labeling problem due to the incompletion

© Springer Nature Switzerland AG 2018
S. Tang et al. (Eds.): AAIM 2018, LNCS 11343, pp. 12–23, 2018.
https://doi.org/10.1007/978-3-030-04618-7_2

of the exploited KB. As shown in Fig. 1, for a relational triple *born_in(Donald Trump, United States)* in KB, all the sentences including two entities *Donald Trump* and *United States* are labeled as the instances of relation *born_in*, despite the sentence *"Donald Trump is the 45th president of the United States"* fails to express the relation. A similar noisy labeling problem exists in the entity pair *(Donald Trump, New York)*, thus it is a great challenge for distantly supervised relation extraction.

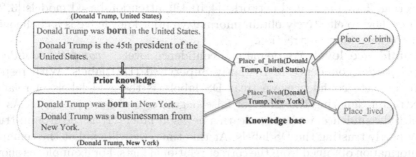

Fig. 1. A case of distantly supervised relation extraction.

To handle the noisy labeling issue, many researchers have made great efforts to distantly supervised relation extraction. Recently, deep learning and neural networks have achieved an excellent performance on this task. Zeng et al. [19] selected the most likely sentence for each entity pair as a valid instance to train the neural model. Then Lin et al. [10] selected multiple valid sentences to predict the relation by assigning higher weight for more informative sentences. Subsequently, Liu et al. [11] used a soft label as the ground-truth to correct the noisy labels, which directly achieved label denoising. Further, we also investigate some common methods that may affect the label denoising in relation extraction such as word-level prior knowledge [3,21] and class imbalance [6,9].

On the whole, existing works focus on specific algorithms but fail to explore the factors impacting the label denoising of neural relation extraction. Hence, we did a lot of research and experiments and found there are some commonalities between different denoising methods. In summary, we propose three fundamental and influential factors as follows:

(1) **Labeling assumption**. We mainly consider two labeling assumptions: *one-valid-instance* assumption and *multiple-valid-instance* assumption. Specifically, one-valid-instance assumption [18,19] indicates that we only choose the most effective sentence within all the sentences that mention an entity pair to predict the relation. Multiple-valid-instance assumption [4,10] assumes that one-valid-instance assumption may lose a lot of information from the neglected sentences, thus we select multiple informative sentences to benefit the relation prediction. Based on these assumptions, we can achieve sentence denoising at different levels with respect to the distantly supervised labels.

(2) **Prior knowledge.** We regard prior knowledge as all the information about the target in training instances, which are objective facts and independent of the knowledge base. Generally speaking, distant supervision mainly depends on informative prior knowledge in plain text. For instance, as shown in Fig. 1, in addition to entity mentions *Donald Trump*, *United States* and *New York*, the contextual words *born* and *president* are more important to relation extraction than the other words such as *the* and *was*. There are a variety of methods to capture influential prior knowledge, including feature engineering [2,13,14], neural networks [4,10,19], attention-based models [3,21]. Thus, how to effectively obtain informative prior knowledge is a challenging factor for relation extraction.

(3) **Confidence level.** We define two confidence levels – *complete* confidence and *partial* confidence in Distantly Supervised (DS) labels. With respect to complete confidence, we treat DS labels as the gold labels of relation extractor, ignoring the effect of false negatives and false positives. As for partial confidence, we hope to learn a new soft label [11] as the ground-truth by partly trusting the DS labels. At this time, we will exploit the semantic information obtained from the correct relational facts. For example, as shown in Fig. 1, the sentence *"Donald Trump was born in New York"* will be labeled as the instance of relation *Place_lived* between two entities *Donald Trump* and *New York* because the triple exists in KB, even it fails to describe the relation. However, if we choose partial confidence in DS labels, we can correct the noisy labels according to the similar semantic patterns (blue fonts) in our corpora, and find the right relation *Place_of_birth*. Therefore, choosing different confidence levels for DS labels is an important factor that impacts the denoising work.

The main contributions of this paper are to:

(1) propose three main factors that impact the label denoising of distantly supervised relation extraction, including labeling assumption, prior knowledge and confidence level.

(2) analyze these factors in a unified neural framework with word, sentence and label denoising modules. Through the combination of different methods in each module, we can test the denoising effectiveness of the proposed factors.

(3) conduct experiments to evaluate and compare these factors with ten neural schemes, discuss the typical cases and discover that the important word-level prior knowledge and partial confidence for distantly supervised labels can remarkably improve the denoising effect.

The remainder of this paper is structured as follows. Section 2 describes our proposed methodology in detail. Section 3 reports our experimental results and analysis. Section 4 gives a conclusion of the whole paper.

2   Methodology

As shown in Fig. 2, in order to analyze the denoising impact of our proposed factors for relation extraction, we design a unified neural model and employ some typical methods to model the factors. With respect to two labeling assumptions, we use the corresponding labeling schemes to reduce the influence of noisy sentences for each entity pair. To evaluate the prior knowledge factor, we also use two methods to verify the word-level denoising performance. As for different confidence levels, we respectively apply DS label (complete confidence) and soft label (partial confidence) as the ground-truth of relation extractor. Besides, based on the partial confidence idea, we propose a dynamic soft-label (DySoft) method to gradually learn the corrected labels during training. By combining three modules – word, sentence and label denoising, we build a neural relation extraction framework in an end-to-end training way.

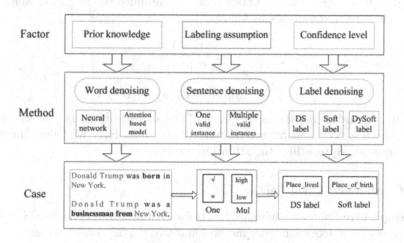

Fig. 2. An overview of our denoising framework for the proposed factors.

2.1 Labeling Assumption Factor

To analyze the labeling assumption factor, we consider two typical assumptions - one-valid-instance and multiple-valid-instance. Following these assumptions, we will achieve sentence denoising at different levels for each entity pair. For this reason, we utilize a sentence denoising module to model the two assumptions.

One-valid-instance assumption requires our model to only extract the most likely sentence for each entity pair to predict the relation. We follow the work of Zeng et al. [19] to realize this assumption, which chooses the sentence of the highest relation probability as the valid instance.

Multiple-valid-instance assumption will select multiple informative sentences for each entity pair as the valid candidates. For this assumption, different algorithms [4,5,10] are introduced to achieve sentence-level denoising. Our model uses the effective selective attention mechanism [10], which can learn high weight for the valid sentences and low weight for the noisy sentences.

2.2 Prior Knowledge Factor

The denoising impact of the prior knowledge factor depends on the design of network architecture. CNN [8,20], PCNN [10,19] and LSTM [17] are common architectures to learn sentence representation, which can obtain the discriminative word sequence information of a sentence. Further, an attention mechanism [3,21] provides us with the insight to achieve word-level denoising. Therefore, we define a word denoising module to analyze the prior knowledge factor.

Concretely, the widely used PCNN in relation extraction is used to learn the sentence representation. Besides, we introduce a word attention method to learn the importance of each word in an instance. To obtain the relatedness between each word in the sentence and entity pair, we compute the attention weight α_i of the i-th word in a sentence as follows:

$$\alpha_i = \frac{\exp(\mathbf{u}_i)}{\sum_k \exp(\mathbf{u}_k)} \tag{1}$$

$$\mathbf{u}_i = \mathbf{W}^w \tanh([\mathbf{w}_i, \mathbf{e}_1, \mathbf{e}_2]) + \mathbf{b}^w \tag{2}$$

where $\mathbf{w}_i$, $\mathbf{e}_1$ and $\mathbf{e}_2$ are the embeddings of the i-th word, entity pair e_1 and e_2 in the sentence. $\mathbf{W}^w$ and $\mathbf{b}^w$ are training parameters. Here we follow the work of Huang et al. [3] to utilize the attention mechanism.

2.3 Confidence Level Factor

With respect to different confidence levels for Distantly Supervised (DS) labels, we employ a complete confidence method and two partial confidence methods to test the impact of this factor. We model the factor as the label denoising module so that we can correct DS labels directly according to the partial confidence schemes. Firstly, for complete confidence, we use DS labels as the gold labels of relation extraction task. While for partial confidence, we reproduce a soft-label [11] method to learn a new label, which combines the predicted relation score and the DS labels via a static confidence vector.

In order to resolve the problem of statically presetting the confidence vector, we propose a Dynamic Soft-label (DySoft) denoising method. Let $\mathbf{c}_i^{(\tau)}$ denote the i-th label confidence at the τ-th training step, then the soft label $\hat{y}^{(\tau)}$ at the τ-th training step is iteratively calculated as follows:

$$\hat{y}^{(\tau)} = \arg\max_i \left\{ \frac{\mathbf{c}_i^{(\tau-1)} + \mathbf{p}_i^{(\tau)}}{\sum_k \mathbf{c}_k^{(\tau-1)} + \mathbf{p}_k^{(\tau)}} \right\} \tag{3}$$

where $\mathbf{p}^{(\tau)}$ represents the relation score predicted by relation extraction model at τ-th training step. The initial label confidence $\mathbf{c}^{(0)}$ is fed with one-hot vector $\mathbf{y}$ of the DS label y.

We simply learn the label confidence of the current training step by a linear combination of previous label confidence and current relation score. The iteration operation can model the correlation between label confidences at different training steps, and use preceding obtained soft labels to learn current new soft labels.

Hence, our proposed dynamic soft-label method remains soft-label consistency learning and gradually achieves label denoising during training.

3   Experiments

In this section, we conduct experiments to evaluate and compare the proposed factors. By combining different denoising modules, we implement 10 relation extraction schemes in Table 1, including 4 reproduction and 6 original schemes.

Table 1. Our denoising schemes for the proposed factors.

Methods	One	One+WA	Multiple	Multiple+WA
Complete confidence	ONE [19]	ONE+WA	MUL [4,10,17]	MUL+WA
Partial confidence	+Soft [11]	–	+Soft [11]	–
	+DySoft	+WA+DySoft	+DySoft	+WA+DySoft

For each column in Table 1, one-valid-instance and multiple-valid-instance assumptions respectively correspond to ONE and MUL methods. With respect to the prior knowledge factor, we use PCNN as the baseline to extract informative prior knowledge, and Word Attention (WA) module is a selective operation for word denoising. For each row in Table 1, complete confidence uses DS labels as the ground-truth, and partial confidence uses Soft-label method [11] and our proposed Dynamic Soft-label method (DySoft) as the gold labels.

3.1   Dataset and Evaluation

Our proposed factors are compared and evaluated on a widely used distantly supervised relation extraction benchmark dataset, which was developed by Riedel et al. [13] via aligning the plain text of the New York Times corpus with relational triples of Freebase. The dataset has 52 relation classes and a non-relation NA class. The training set includes 522,611 sentences, 281,270 entity pairs and 18,252 relational facts. The test set includes 172,448 sentences, 96,678 entity pairs and 1,950 relational facts. Similar to previous work [5,10,11,19], we adopt the held-out evaluation with aggregate precision-recall curves and top N Precision (P@N). The held-out evaluation compares the predicted relations of entity pairs with the distantly supervised relations without requiring human evaluation. As for parameter settings, we follow the settings of Liu et al. [11]. Specially, our proposed soft label is utilized after 3 training epochs.

3.2   Result and Analysis

As shown in Fig. 3, the neural models significantly performs better than traditional feature-based models Mintz [12], MultiR [2] and MIMLRE [15]. All the methods in Fig. 3 have compete confidence in distantly supervised labels. Obviously, adding a word attention (WA) module is beneficial to both one-valid-instance (ONE) and multiple-valid-instance (MUL) assumptions.

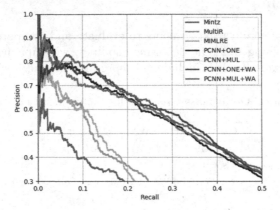

Fig. 3. Performance comparison of neural methods with feature-based methods.

From Fig. 4, we present the impact of different denoising modules. For both ONE and MUL assumptions, the Soft-label [11] method performs well when the recall is less than 0.3, but it declines notably when the recall is greater than 0.4. Our proposed DySoft-label method performs stable and achieves higher precision in the MUL assumption. In particular, combining word denoising (WA) with label denoising (Soft-label) module fails to improve the baseline. Consider that correcting labels may have a bad impact on the attention of prior knowledge, we get that the combination effect of WA+DySoft scheme may be not as good as we expect.

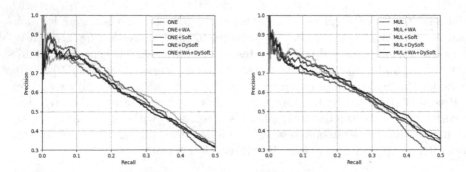

Fig. 4. Performance comparison of different denoising schemes.

For a clear comparison, Table 2 presents the precisions of a variety of models for different recalls $R = 0.1/0.2/0.3/0.4/0.5$ and Average Precision (AP) which corresponds to the area under the precision-recall curve. Compared with the neural baselines PCNN+ONE and PCNN+MUL which use DS labels as the ground-truth, We can observe that adding WA or adding DySoft can achieve the better performance than the other neural schemes. Next we will analyze the effect of each factor in detail.

Table 2. Precisions of a variety of models for different recalls.

Method	$R\_0.1$	$R\_0.2$	$R\_0.3$	$R\_0.4$	$R\_0.5$	AP
Mintz [12]	0.399	0.286	0.168	-	-	0.106
MultiR [2]	0.609	0.364	-	-	-	0.126
MIMLRE [15]	0.607	0.338	-	-	-	0.120
PCNN+ONE [19]	0.752	0.645	0.539	0.416	0.315	0.352
+WA	0.756	0.661	0.583	0.449	0.328	0.369
+Soft-label [11]	0.775	0.673	0.562	0.414	0.241	0.332
+DySoft-label	**0.783**	0.638	0.534	0.419	0.313	0.360
+WA+DySoft-label	0.768	0.646	0.522	0.430	0.318	0.358
PCNN+MUL [10]	0.699	0.641	0.548	0.439	0.338	0.365
+WA	**0.783**	0.672	0.554	0.444	0.346	**0.382**
+Soft-label [11]	0.777	**0.682**	0.553	0.390	0.225	0.331
+DySoft-label	0.762	0.668	**0.584**	**0.481**	**0.358**	0.370
+WA+DySoft-label	0.728	0.675	0.571	0.457	0.332	0.360

3.3 Effect of Labeling Assumption Factor

To test the effect of labeling assumption factor, we compare and evaluate two typical assumptions – one-valid-instance assumption [19] and multiple-valid-instance assumption [10]. Figure 4 shows that both assumptions bring an excellent precision-recall performance, and multiple-valid-instance assumption performs slightly better than one-valid-instance assumption due to the utilization of more sentences. In particular, we find that the performance of both assumptions is better than the report of Liu et al. [10] and the gap between two assumptions is not as large as theirs. We owe the discriminative results to the CPU/GPU hardware environment and code structure.

As multiple-valid-instance assumption has the effect on the entity pairs which have multiple sentences, we compare and analyze the performance of two assumptions on these entity pairs. Following the evaluation of Liu et al. [10], we randomly select one, two and all sentences from the entity pairs which have more than one sentence to conduct relation extraction. Table 3 presents top N precision (P@N) of different settings for top 100/200/300 relational facts. We discover that the whole performance of one-valid-instance and multiple-valid-instance assumptions has 8% and 5% improvements compared with the reported

Table 3. Top N precision (P@N) for relation extraction in the entity pairs with different number of sentences.

Settings	One				Two				All			
P@N(%)	100	200	300	Mean	100	200	300	Mean	100	200	300	Mean
ONE	76.0	69.5	65.3	70.3	77.0	71.5	68.3	72.3	83.0	72.5	68.0	74.5
+WA	77.0	72.0	**68.7**	72.6	78.0	**77.5**	70.0	75.2	79.0	77.5	71.3	75.9
SATT	75.0	75.5	68.0	72.8	74.0	75.0	**70.7**	73.2	76.0	77.0	74.3	75.8
+WA	**83.0**	**76.0**	67.0	**75.3**	**83.0**	**77.5**	**70.7**	**77.1**	**85.0**	**78.0**	**74.7**	**79.2**

results of Liu et al. [10]. The average precision for different settings of multiple-valid-instance assumption has 1% to 4% improvements compared with one-valid-instance assumption, which is also less than the gap of 5% to 9% in previous work [10]. Specially, the case study of labeling assumption for sentence-level denoising has been reported [10], thus we omit it in our paper.

3.4 Effect of Prior Knowledge Factor

Since the denoising impact of the prior knowledge factor in neural networks is usually invisible and directly reflected in model performance, we evaluate this factor in the precision-recall curve and visualization of word attention. From Fig. 4 and Table 2, we can find that adding a word-level denoising WA mechanism can greatly benefit both ONE and MUL assumptions. Particularly, MUL+WA model achieves the best AP performance compared with all the other models. As shown in Table 2, combining WA mechanism makes the average precision improve 3% and 4% for ONE and MUL baselines. These results demonstrate the effectiveness of adding the word-level importance for the prior knowledge factor.

Table 4. Visualization of word attention.

Relation	Instance
place_of_birth	*Roscoe Lee Browne* was **born** on May 2, **1925**, in *Woodbury*.
place_of_death	His most recent **book** is "Murder in *Amsterdam*: the **death** of *Theo Van Gogh* and the limits of tolerance."
company	"I like **Eli Manning**," the *NBC* football **analyst** *John Madden* said in a conference call Thursday.
founder	Still , some early **investors** like *Peter Munk*, the **founder** and **chairman** of *Barrick Gold*, were **skeptical** at first.

Table 4 presents the visualization of word attention mechanism, the bold font and size of a word denote its importance, the italic words are entity mentions. For example, with respect to the instance of relation *place_of_birth*, the word *born* is assigned the highest weight, entity mentions *Roscoe Lee Browne*, *Woodbury* and temporal word *1925* are also given higher weights than the other words in the sentence. It is not difficult to find that the informative prior knowledge can provide us with the clues to predict the relation of entity pairs. Therefore, influential word-level prior knowledge is effective to improve the performance of relation extraction.

3.5   Effect of Confidence Level Factor

As shown in Fig. 4 and Table 2, the different confidence levels can impact the label denoising. Concretely, with respect to the partial confidence in DS labels, both the Soft-label [11] and our proposed DySoft-label methods can improve the performance of baselines. In addition, we give some typical cases of soft-label corrections to check the label denoising of our proposed DySoft method in Table 5.

Table 5. Case study of label denoising.

DS label	Soft label	Instance
NA	**nationality**	..., said *Gani Fawehinmi*, one of *Nigeria*'s most prominent awyers and a longtime campaigner for good governance.
place_lived	**place_of_birth**	And surely we must recognize *Alton B. Parker*, born in *Cortland*, N.Y., who lost the 1904 election to another ...
company	**NA**	*George W. Bush* offered paeans to boyhood home in Midland, *Texas*.
place_lived	place_of_death $\rightarrow$ **NA**	He was *Antonello da Messina*, commonly called Antonello, not da *Messina*.

Firstly, we find that the noisy labels of some false negatives can be corrected by soft labels. For example, the relation between *Gani Fawehinmi* and *Nigeria* is missed in Freebase, but we can correctly recognize their relation *nationality*. For the instance of entity pair (*Alton B. Parker, Cortland*) which is labeled as relation *place_lived*, we can correct it with *place_of_birth* due to the high label confidence of this relation. Furthermore, false positives can be identified by label denoising method. For example, the instance of entity pair (*George W. Bush, Texas*) fails to express the relation *company*, we will correct it with NA relation. In particular, with respect to the last case, the instance also fails to express the DS label *place_lived* between entity pair (*Antonello da Messina, Messina*). However, we observe that their label can be gradually corrected based on our proposed DySoft-label method. At the beginning, the relation is firstly corrected as *place_of_death*, then it is corrected as *NA*. The results show that partial confidence in DS labels can achieve the label-level denoising of relation extraction.

4   Conclusions

In this paper, we propose three main factors – labeling assumption, prior knowledge and confidence level that impact the label denoising of distantly supervised relation extraction. Moreover, we build a unified neural framework with word, sentence and label denoising modules to analyze the effect of these factors. At last, we·empirically evaluate and compare these factors based on ten neural schemes. The experimental results demonstrate these factors can effectively impact the label denoising of relation extraction. In summary, we find that both typical labeling assumptions are effective for relation extraction. Furthermore, the prior knowledge factor can affect word-level denoising and improve the model performance. Besides, we discover that partial confidence in distantly supervised labels can reliably correct noisy labels. These findings are beneficial to the further research of relation extraction.

In the future, we will explore the feasibility and interpretability of relation extraction based on reinforcement learning and generative adversarial network via the proposed factors.

Acknowledgements. This work is supported by National Natural Science Foundation of China, 61602048, 61520106007, BUPT-SICE Excellent Graduate Students Innovation Funds, 2016.

References

1. Banko, M., Etzioni, O.: The tradeoffs between open and traditional relation extraction. In: Proceedings of ACL 2008, HLT, pp. 28–36 (2008)
2. Hoffmann, R., Zhang, C., Ling, X., Zettlemoyer, L., Weld, D.S.: Knowledge-based weak supervision for information extraction of overlapping relations. In: Proceedings of the 49th Annual Meeting of the Association for Computational Linguistics: Human Language Technologies, vol. 1, pp. 541–550. Association for Computational Linguistics (2011)
3. Huang, X., et al.: Attention-based convolutional neural network for semantic relation extraction. In: Proceedings of COLING 2016, the 26th International Conference on Computational Linguistics: Technical Papers, pp. 2526–2536 (2016)
4. Ji, G., Liu, K., He, S., Zhao, J.: Distant supervision for relation extraction with sentence-level attention and entity descriptions. In: AAAI, pp. 3060–3066 (2017)
5. Jiang, X., Wang, Q., Li, P., Wang, B.: Relation extraction with multi-instance multi-label convolutional neural networks. In: COLING, pp. 1471–1480 (2016)
6. Khan, S.H., Hayat, M., Bennamoun, M., Sohel, F.A., Togneri, R.: Cost-sensitive learning of deep feature representations from imbalanced data. IEEE Trans. Neural Netw. Learn. Syst. **29**, 3573–3587 (2017)
7. Kim, J.T., Moldovan, D.I.: Acquisition of semantic patterns for information extraction from corpora. In: Ninth Conference on Artificial Intelligence for Applications, Proceedings, pp. 171–176. IEEE (1993)
8. Kim, Y.: Convolutional neural networks for sentence classification. In: EMNLP, pp. 1746–1751 (2014)
9. Kukar, M., Kononenko, I., et al.: Cost-sensitive learning with neural networks. In: ECAI, pp. 445–449 (1998)

10. Lin, Y., Shen, S., Liu, Z., Luan, H., Sun, M.: Neural relation extraction with selective attention over instances. In: ACL (2016)
11. Liu, T., Wang, K., Chang, B., Sui, Z.: A soft-label method for noise-tolerant distantly supervised relation extraction. In: Proceedings of the 2017 Conference on Empirical Methods in Natural Language Processing, pp. 1791–1796 (2017)
12. Mintz, M., Bills, S., Snow, R., Jurafsky, D.: Distant supervision for relation extraction without labeled data. In: Proceedings of the Joint Conference of the 47th Annual Meeting of the ACL and the 4th International Joint Conference on Natural Language Processing of the AFNLP, vol. 2, pp. 1003–1011. Association for Computational Linguistics (2009)
13. Riedel, S., Yao, L., McCallum, A.: Modeling relations and their mentions without labeled text. In: Balcázar, J.L., Bonchi, F., Gionis, A., Sebag, M. (eds.) ECML PKDD 2010. LNCS (LNAI), vol. 6323, pp. 148–163. Springer, Heidelberg (2010). https://doi.org/10.1007/978-3-642-15939-8_10
14. Ritter, A., Zettlemoyer, L., Etzioni, O., et al.: Modeling missing data in distant supervision for information extraction. Trans. Assoc. Comput. Linguist. 1, 367–378 (2013)
15. Surdeanu, M., Tibshirani, J., Nallapati, R., Manning, C.D.: Multi-instance multi-label learning for relation extraction. In: Proceedings of the 2012 Joint Conference on Empirical Methods in Natural Language Processing and Computational Natural Language Learning, pp. 455–465. Association for Computational Linguistics (2012)
16. Wu, F., Weld, D.S.: Open information extraction using wikipedia. In: Proceedings of the 48th Annual Meeting of the Association for Computational Linguistics, pp. 118–127. Association for Computational Linguistics (2010)
17. Wu, Y., Bamman, D., Russell, S.: Adversarial training for relation extraction. In: Proceedings of the 2017 Conference on Empirical Methods in Natural Language Processing, pp. 1779–1784 (2017)
18. Xiangrong, Z., Kang, L., Shizhu, H., Jun, Z., et al.: Large scaled relation extraction with reinforcement learning. In: Proceedings of the 15th AAAI Conference on Artificial Intelligence (2018)
19. Zeng, D., Liu, K., Chen, Y., Zhao, J.: Distant supervision for relation extraction via piecewise convolutional neural networks. In: EMNLP, pp. 1753–1762 (2015)
20. Zeng, D., Liu, K., Lai, S., Zhou, G., Zhao, J., et al.: Relation classification via convolutional deep neural network. In: COLING, pp. 2335–2344 (2014)
21. Zhang, T., Huang, M., Zhao, L.: Learning structured representation for text classification via reinforcement learning. In: AAAI (2018)
22. Zhu, J., Nie, Z., Liu, X., Zhang, B., Wen, J.R.: Statsnowball: a statistical approach to extracting entity relationships. In: Proceedings of the 18th International Conference on World Wide Web, pp. 101–110. ACM (2009)

Makespan Minimization on Unrelated Parallel Machines with a Few Bags

Daniel R. Page[✉][iD] and Roberto Solis-Oba

Department of Computer Science, Western University, London, Canada
dpage6@uwo.ca, solis@csd.uwo.ca

Abstract. Let there be a set M of m parallel machines and a set J of n jobs, where each job j takes $p_{i,j}$ time units on machine M_i. In makespan minimization the goal is to schedule each job non-preemptively on a machine such that the length of the schedule, the makespan, is minimum. We investigate a generalization of makespan minimization on unrelated parallel machines ($R||C_{max}$) where J is partitioned into b bags $B = (B_1, \ldots, B_b)$, and no two jobs belonging to the same bag can be scheduled on the same machine. First we present a simple b-approximation algorithm for $R||C_{max}$ with bags ($R|bag|C_{max}$). Two machines M_i and $M_{i'}$ have the same machine type if $p_{i,j} = p_{i',j}$ for all $j \in J$. We give a polynomial-time approximation scheme (PTAS) for $R|bag|C_{max}$ with machine types where both the number of machine types and bags are constant. This result infers the existence of a PTAS for uniform parallel machines when the number of machine speeds and number of bags are both constant. Then, we present a $b/2$-approximation algorithm for the graph balancing problem with $b \geq 2$ bags; the approximation ratio is tight for $b = 3$ unless P = NP and this algorithm solves the graph balancing problem with $b = 2$ bags in polynomial time. In addition, we present a polynomial-time algorithm for the restricted assignment problem on uniform parallel machines when all the jobs have unit length. To complement our algorithmic results, we show that when the jobs have lengths 1 or 2 it is NP-hard to approximate the makespan with approximation ratio less than $3/2$ for both the restricted assignment and graph balancing problems with $b = 2$ bags and $b = 3$ bags, respectively. We also prove that makespan minimization on uniform parallel machines with $b = 2$ bags is strongly NP-hard.

Keywords: Makespan minimization · Unrelated parallel machines Approximation algorithms · Scheduling · Bag constraints

1 Introduction

Let M be a set of m *unrelated parallel machines* and J be a set of n *jobs*, where job j has length or processing time $p_{i,j} \in \mathbb{Z}^+$ on machine M_i. In makespan min-

Daniel Page is supported by an Ontario Graduate Scholarship. Roberto Solis-Oba was partially supported by the Natural Sciences and Engineering Research Council of Canada, grant 04667-2015 RGPIN.

S. Tang et al. (Eds.): AAIM 2018, LNCS 11343, pp. 24–35, 2018.
https://doi.org/10.1007/978-3-030-04618-7_3

imization, the goal is to schedule all the jobs on the machines so as to minimize the length of the schedule—the *makespan*. Makespan minimization on unrelated parallel machines is denoted as $R||C_{max}$ in the notation of Graham *et al.* [11]. Two extensively studied machine environments that are special cases of the unrelated parallel machine environment are the identical and uniform parallel machine environments: if the machines are *identical* then job j has the same *length* $p_j \in \mathbb{Z}^+$ on any machine; and if the machines are *uniform* then each machine M_i has a *speed* $s_i \in \mathbb{Z}^+$ and job j has processing time p_j/s_i on M_i.

We consider a generalization of $R||C_{max}$ where the jobs J are partitioned into b sets $B = (B_1, B_2, \ldots, B_b)$ called *bags*, and any feasible solution must satisfy the *bag constraints*: no two jobs from the same bag can be scheduled on the same machine. This problem is called *makespan minimization on unrelated parallel machines with bags*, and we denote it as $R|bag|C_{max}$. Notice that if $b = |J|$, then every job is in a distinct bag, and we get the classic setting $R||C_{max}$. As discussed in [7], the bag constraints appear in settings such as in the scheduling of tasks for on-board computers in airplanes. That is, these systems have multiple processors and it is required for some of the tasks to be scheduled on different processors so that the airplane continues to operate safely even if one of the processors were to fail. Thus, parallel machine scheduling problems where the bag constraints are imposed are a kind of fault-tolerant scheduling that finds applications in complex parallel systems where system stability is desired [4].

The best-known approximation algorithms for $R||C_{max}$ have approximation ratio 2 [8,16,19], and it is NP-hard to approximate the makespan with approximation ratio less than $3/2$ [16]. Two special cases of $R||C_{max}$ have become of recent interest to try to understand the $3/2$-to-2 approximation gap for $R||C_{max}$:

- *Restricted Assignment Problem* $(P|\mathcal{M}_j|C_{max})$. This is makespan minimization on identical parallel machines (i.e., $P||C_{max}$) with the constraint that some jobs are not eligible to be scheduled on some of the machines. That is, for each job $j \in J$, there is a set $\mathcal{M}_j$ of machines where job j can be scheduled. A schedule that assigns each job j to an eligible machine in $\mathcal{M}_j$ is said to satisfy the *eligibility constraints*.
- *Graph Balancing Problem* $(P|\mathcal{M}_j, |\mathcal{M}_j| \leq 2|C_{max})$: This is a special case of the restricted assignment problem where the number of eligible machines for each job is at most 2. An alternate way to interpret an instance of this problem is as a weighted multigraph where the jobs are edges and the machines are vertices, and every edge must be directed to one of its endpoints so as to minimize the maximum sum of the edge lengths directed toward a vertex.

We study variants of the above two problems with bag constraints. In addition to this, we investigate $R|bag|C_{max}$ in the setting with so-called machine types. As discussed by Gehrke *et al.* [10], a natural scenario in parallel machine scheduling is where the machines are clusters of processors where each processor in a cluster is of the same type, e.g. clusters of CPUs and/or GPUs. More formally, two machines M_i and $M_{i'}$ have the same *machine type* if $p_{i,j} = p_{i',j}$ for all $j \in J$. We study $R|bag|C_{max}$ with machine types, where the number δ of machine types is constant.

2   Related Work

For the restricted assignment problem with two different job lengths $p_j \in \{\alpha, \beta\}$ for each job j and $\alpha < \beta$, there are approximation algorithms with approximation ratio slightly less than 2 [3,15], most notably a $(2 - \gamma)$-approximation algorithm for some small value $\gamma > 0$ and a $(2 - \alpha/\beta)$-approximation algorithm, both by Chakrabarty et al. [3]. Jansen and Rohwedder [14] showed that in quasi-polynomial time the restricted assignment problem can be approximated within a factor $11/6 + \epsilon$ of the optimum for any $\epsilon > 0$. Ebenlendr et al. [6] presented a $7/4$-approximation algorithm for the graph balancing problem. For the graph balancing problem with two job lengths there are $3/2$-approximation algorithms [12,18], and there is no p-approximation algorithm with $p < 3/2$, unless P = NP [1,6]. Jansen and Maack [13] presented an efficient PTAS for $R||C_{max}$ with machine types when the number of machine types is constant. For more literature on makespan minimization with machine types see [10,13].

Makespan minimization with bags is a type of conflict scheduling problem, where two jobs conflict if two jobs from the same bag are scheduled on the same machine. A natural way to model this type of conflict is with an incompatibility graph: there is a vertex for each job and an edge $\{j, j'\}$ if jobs j and j' cannot be scheduled on the same machine. Then, makespan minimization with bags is when the incompatibility graph consists of b disjoint cliques. Bodlaender et al. [2] developed several results for $P||C_{max}$ with incompatibility graphs. In addition, Dokka et al. [5] considered a related, but generalized version of $P|bag|C_{max}$ called the multi-level bottleneck assignment problem, and gave a 2-approximation algorithm for three bags. For further discussion on related variants of conflict scheduling refer to Sect. 1.3 of [4].

In 2017, Das and Wiese [4] presented a PTAS for $P|bag|C_{max}$, and an 8-approximation algorithm for the restricted assignment problem with bags in the special case when for each bag B_k all the jobs $j \in B_k$ have the same eligibility constraints, i.e. each set of machines on which a job in B_k can be scheduled is the same. For any $\epsilon > 0$, Das and Wiese proved there is no $((\log n)^{1/4 - \epsilon})$-approximation algorithm for the restricted assignment problem with bags, unless NP $\subseteq$ ZPTIME($2^{(\log n)^{O(1)}}$).

3   Preliminaries

First, we give a couple of basic properties for $R|bag|C_{max}$. If the number b of bags is one, at most one job can be scheduled on each machine. Hence, we can solve $R||C_{max}$ with one bag in polynomial time as follows: build a weighted bipartite graph $G = (J \cup M, E)$, where $E = \{(j, i) \mid \text{job } j \text{ can be scheduled on machine } M_i\}$, and $w(j, i) = p_{i,j}$ for every $(j, i) \in E$. Compute a maximum cardinality bottleneck matching $\mathcal{M}$ of G and for each arc $(j, i) \in \mathcal{M}$, schedule job j on machine M_i; there is no feasible solution if any job is not scheduled. Thus, in the sequel we focus on $R|bag|C_{max}$ when there are $b > 1$ bags.

Property 1. For any schedule that satisfies the bag constraints with b bags, there are at most b jobs scheduled on a machine.

For some of our algorithmic results, we employ a *p-relaxed decision procedure* as given in Lenstra *et al.* [16]. Let $U \in \mathbb{Z}^+$ be an upper bound on the optimal makespan for some scheduling problem. We use binary search over the interval $[0, U]$ to determine the smallest value $d \in \mathbb{Z}^+$ for which the p-relaxed decision algorithm produces a schedule, it either: computes a schedule with makespan at most pd; or returns FAIL if there is no solution with value at most d. In the binary search, if the p-relaxed decision algorithm returns FAIL then the value d is increased, and if a schedule is returned the value d is decreased. If we keep track of the schedule with minimum makespan found, after $O(\log U)$ iterations the binary search guarantees that $d \leq OPT$ and a schedule with makespan at most $p \cdot OPT$ is found. Therefore, if the overall p-relaxed decision procedure takes polynomial time, this is a p-approximation algorithm.

4   Our Results

In Sect. 5 we provide a simple b-approximation algorithm for $R|bag|C_{max}$. In Sect. 6 we present a PTAS for $R|bag|C_{max}$ with machine types when there is a fixed number of machine types and bags. As we will explain, this implies the existence of a PTAS for $Q|bag|C_{max}$ when both the number of machine speeds and the number of bags are constant. Then, in Sect. 7 we give a $b/2$-approximation algorithm for the graph balancing problem with $b \geq 2$ bags, the approximation ratio for this algorithm is tight for $b = 3$ unless P = NP and implies that the graph balancing problem with $b = 2$ bags is solvable in polynomial time. Finally, in Sect. 8 we show that the restricted assignment problem with bags when the machines are uniform and every job has unit length ($Q|bag, p_j = 1, \mathcal{M}_j|C_{max}$) is polynomial-time solvable. As a note, we designed a $O(m \log m)$-time algorithm for $P|bag|C_{max}$ with $b = 2$ bags.

To complement our algorithmic results, we present a series of inapproximability and strong NP-hardness results in Sect. 9. We first show how to extend the classic 3/2-inapproximability lower bound of Lenstra *et al.* [16] to the restricted assignment problem with job lengths $p_j \in \{1, 2\}$ when there are $b = 2$ bags. Then, we prove that there is no approximation algorithm with approximation ratio less than 3/2 for the graph balancing problem with $b = 3$ bags and job lengths $p_j \in \{1, 2\}$, unless P = NP. Finally we show that $Q|bag|C_{max}$ with $b = 2$ bags is strongly NP-hard.

5   A b-Approximation Algorithm for $R|bag|C_{max}$

Let $p_{max} = \max_{1 \leq j \leq n, 1 \leq i \leq m}(p_{i,j})$. Our approximation algorithm uses binary search and a b-relaxed decision procedure with $U = p_{max}n$. For makespan estimate $d \leq U$, the idea is to treat each bag independently and simply schedule the jobs j in each bag $B_k \in B$ on machines M_i where $p_{i,j} \leq d$ so that the bag constraints are satisfied. We do this by using a bipartite flow network $N(B, P, d)$ where there is a source s, a *job node* for each $j \in J$, a *bag-machine node* $M_{B_k,i}$ for each $M_i \in M$ and $B_k \in B$, a *machine node* for every $M_i \in M$, and a sink t. Do the following for each bag $B_k \in B$: for each $j \in B_k$, add an arc from s to each job node j and set its capacity to 1. Next, for each $j \in B_k$ and $M_i \in M$, if $p_{i,j} \leq d$, then add an arc from job node j to $M_{B_k,i}$ with capacity 1. For each $M_i \in M$ and $B_k \in B$, add an arc from each bag-machine node $M_{B_k,i}$ to machine node M_i with capacity 1. Finally, from each machine node $M_i \in M$, add an arc from M_i to t with capacity b. The b-relaxed decision algorithm is as follows.

1. Build flow network $N(B, P, d)$ and compute an integral maximum flow f.
2. **For each** $j \in B_k$, if $f(s, j) = 0$ **then return** FAIL.
3. **For each** job $j \in B_k$, schedule j on machine M_i **if** $f(j, M_{k,i}) = 1$, and **return** this schedule.

If an integral maximum flow is computed and all the arcs incident on s are saturated, the jobs in bag B_k can be scheduled on the machines so as to satisfy the bag constraints, and such that each job takes at most d time units.

Theorem 1. *There is b-approximation algorithm for $R|bag|C_{max}$.*

6   A PTAS for $R|bag|C_{max}$ with a Constant Number of Machine Types and Bags

Recall that two machines M_i and $M_{i'}$ have the same machine type if, for every $j \in J$, $p_{i,j} = p_{i',j}$. Let $N_t(v)$ be the number of machines of machine type v, and let δ be the number of machine types. Now we describe the PTAS. Compute a b-approximate value ρ to the optimal makespan, so that $\rho/b \leq OPT \leq \rho$; value ρ can be computed using the b-approximation algorithm in Sect. 5. Then use binary search over the interval $[\rho/b, \rho]$ to find the smallest value τ for which the algorithm given below computes a schedule. Consider all possible schedules of length τ. We can simplify the structure of these schedules so that there is only a constant number of different load configurations for the machines (defined below), while increasing the length of the schedule by a factor of at most $(1 + \epsilon)$ for any constant $\epsilon > 0$. This simplification will allow us to design a PTAS.

First subdivide the timeline of a schedule into units of length $(\tau\epsilon)/b^2$ for some constant $\epsilon > 0$. Let the *load configuration* $(\ell_{i,1}, \ell_{i,2}, \ldots, \ell_{i,b})$ of machine M_i be such that if job j from bag B_k is scheduled on M_i then $(\ell_{i,k} - 1)(\tau\epsilon)/b^2 < p_{i,j} \leq \ell_{i,j} \cdot (\tau\epsilon)/b^2$. The number of possible values each $\ell_{i,k}$ can take is at most $1 + \lceil \frac{\tau}{\frac{\tau\epsilon}{b^2}} \rceil = 1 + \lceil \frac{b^2}{\epsilon} \rceil$, and so $\ell_{i,k} \in \{0, 1, 2, \ldots, \lceil b^2/\epsilon \rceil\}$. As there are b values in any load configuration, the number of possible load configurations is then $(1 + \lceil b^2/\epsilon \rceil)^b =: D$, a constant; label these load configurations $1, 2, \ldots, D$. Let vector $(c_{1,1}, c_{1,2}, \ldots, c_{1,D}, c_{2,1}, c_{2,2}, \ldots, c_{2,D}, \ldots, c_{\delta,1}, c_{\delta,2}, \ldots, c_{\delta,D})$ be a *schedule configuration*, where $c_{v,\mu}$ is the number of machines with machine type v that have load configuration μ. There are m machines, so each $c_{v,\mu} \in \{0, 1, \ldots, m\}$ and because there are δD many elements in a schedule configuration, the total possible number of schedule configurations is $O(m^{\delta D})$, a polynomial function as δ and D are constant. A schedule configuration is *valid* if $\sum_{\upsilon=1}^{\delta} \sum_{\mu=1}^{D} c_{\upsilon,\mu} = m$, and $\sum_{\mu=1}^{D} c_{\upsilon,\mu} = N_t(\upsilon)$, for $\upsilon = 1, 2, \ldots, \delta$. For each valid schedule configuration there are exactly m load configurations, one for each machine. That is, for each $c_{\upsilon,\mu} > 0$, assign load configuration μ to $c_{\upsilon,\mu}$ machines of machine type υ. Then, the makespan of a valid schedule configuration is $\max_{1 \leq i \leq m} \left\{ \sum_{k=1}^{b} \ell_{i,k}(\frac{\tau\epsilon}{b^2}) \right\}$.

We compute all valid schedule configurations for makespan τ and choose one for which the jobs can be allocated to the machines according to that schedule configuration. If there is a feasible schedule, at least one such schedule configuration exists where for each $j \in J$ with $j \in B_k$, there is a machine M_i with $p_{i,j} \leq \ell_{i,k}(\frac{\tau\epsilon}{b^2}) \leq \lceil \frac{b^2}{\epsilon} \rceil (\frac{\tau\epsilon}{b^2})$ where $\frac{b^2}{\epsilon}(\frac{\tau\epsilon}{b^2}) = \tau \leq \lceil \frac{b^2}{\epsilon} \rceil (\frac{\tau\epsilon}{b^2})$. To find this schedule we proceed as follows. For each valid schedule configuration, assign to machine M_i a load configuration L_i as described above. Then, consider each bag B_k, $k = 1, 2, \ldots, b$, and build a bipartite graph $G_k = (B_k \cup M, E_k)$, where $E_k = \left\{ (j, M_i) \mid M_i \in M, \ j \in B_k, (\ell_{i,k} - 1)\left(\frac{\tau\epsilon}{b^2}\right) < p_{i,j} \leq \ell_{i,k}\left(\frac{\tau\epsilon}{b^2}\right) \right\}$. Compute a maximum matching of G_k, and for each arc (j, M_i) in the matching, schedule j on M_i. Discard the schedule if at least one job of bag B_k is not scheduled. Otherwise, for $k = 1, 2, \ldots, b$, a matching of size $|B_k|$ is computed for G_k and thus every job $j \in B_k$ is scheduled. This will assign at most one job from each bag B_k to each machine, so a feasible schedule is produced.

Let machine M_λ be a machine that finishes last in the schedule configuration with minimum makespan τ^* selected by the algorithm and let $L_\lambda^* = (\ell_{\lambda,1}^*, \ell_{\lambda,1}^*, \ldots, \ell_{\lambda,b}^*)$ be its load configuration. Note that for each job $j \in B_k$ on M_λ, $p_{\lambda,j} \leq \ell_{\lambda,k}^*(\tau^*\epsilon)/b^2$, but $p_{\lambda,j} > (\ell_{\lambda,k}^* - 1)(\tau^*\epsilon)/b^2$ as otherwise there would be another schedule configuration of lesser makespan where all the jobs can be allocated to the machines. Since $\tau^* \geq OPT$, then

$$\sum_{\substack{\text{job } j \text{ scheduled} \\ \text{on machine } M_\lambda}} p_{\lambda,j} \geq OPT > \sum_{k=1}^{b} \max \left\{ (\ell_{\lambda,k}^* - 1)\frac{\tau^*\epsilon}{b^2}, 0 \right\}.$$

Therefore,

$$\sum_{\substack{\text{job } j \text{ scheduled} \\ \text{on machine } M_\lambda}} p_{\lambda,j} \leq \sum_{k=1}^{b} \ell_{\lambda,k}^* \frac{\tau^*\epsilon}{b^2} \leq \sum_{k=1}^{b} \max \left\{ (\ell_{\lambda,k}^* - 1)\frac{\tau^*\epsilon}{b^2}, 0 \right\} + \sum_{k=1}^{b} \frac{\tau^*\epsilon}{b^2}$$

$$< OPT + \sum_{k=1}^{b} \frac{\tau^*\epsilon}{b^2},$$

and since $\rho/b \leq OPT \leq \tau^* \leq \rho$, the makespan is at most

$$OPT + \sum_{k=1}^{b} \frac{\tau^*\epsilon}{b^2} = OPT + \left(\frac{\tau^*}{b}\right)\epsilon \leq OPT + \left(\frac{\rho}{b}\right)\epsilon \leq (1+\epsilon)OPT.$$

Theorem 2. *There is a PTAS for $R|bag|C_{max}$ with machine types when both the number b of bags and the number δ of machine types are constant.*

Consider makespan minimization on uniform machines with bags ($Q|bag|C_{max}$). The processing time for a job on a machine depends on the speed of the machine. Therefore, the number of machine types is in fact the number of machine speeds.

Corollary 1. *There is a PTAS for $Q|bag|C_{max}$ when both the number of distinct machine speeds and the number of bags are constant.*

7   A $b/2$-Approximation Algorithm for the Graph Balancing Problem with $b \geq 2$ Bags

Recall that in the graph balancing problem with bags the jobs and machines can be represented as a weighted multigraph $G = (V, E)$, where the jobs are edges i.e. $E = \bigcup_{k=1}^{b} B_k$, each edge $e \in E$ has length $p_e \in \mathbb{Z}^+$, and the machines are the vertices. We continue to use m and n to be the number of machines and jobs, respectively. Let $G_{B_k} = (V_{B_k}, B_k)$ where vertex $v \in V_{B_k}$ if $v \in e \in B_k$. We call a maximally connected component of G_{B_k} a *bag component*. A *pseudoforest* is a collection of trees and graphs with at most one cycle called 1-*trees*.

Property 2. Consider the graph balancing problem with bags. If there is a feasible schedule S, then for every $B_k \in B$, G_{B_k} is a pseudoforest.

In the sequel we assume that the input multigraph G satisfies the conditions of Property 2. There are at most two possible orientations for a bag component that is a 1-tree: direct each edge to a unique vertex along the cycle of the 1-tree, and then direct all other edges away from the cycle. A tree $T = (V_T, E_T)$ however has at most $|V_T|$ possible orientations: select each vertex as the root of the tree and direct all edges away from it. We use these facts in our algorithm.

We use binary search and a $b/2$-relaxed decision procedure as described in Sect. 3 with $U = \sum_{e \in E} p_e$ to find the smallest value $d \in \mathbb{Z}^+$ for which there is a schedule with makespan at most $(b/2)d$ that satisfies the bag constraints. If the $b/2$-relaxed decision algorithm below returns FAIL, then there is no feasible schedule with makespan at most d for G; otherwise the algorithm computes a feasible schedule with makespan at most $(b/2)d$. Let $l_L(u)$ be the load contributed by the edge with the largest edge length directed toward vertex u in G; hence $l_L(u) = 0$ if no edge is directed toward u. We note that if $l_L(u) > d/2$ then no other edges with length larger than $d/2$ can be directed toward u without the makespan exceeding d; we call an edge e a *big* edge when its length $p_e > d/2$ and an edge is *small* if $p_e \leq d/2$.

The $b/2$-relaxed decision algorithm uses a set D to store the edges that have been assigned a direction. Initially $D = \varnothing$ and if a schedule exists, at the end D will contain all the edges in G. First, if any edge $e \in E$ has length larger than d return FAIL. While there is an edge in $E \setminus D$ do the following.

Compute a bag component C of $G \setminus D$ and perform an *expansion* of C by using the procedure described in Step (2) of the algorithm below. For each feasible orientation of the edges in C an expansion forces edges away from vertices in C if doing so does not violate the bag constraints and $l_L(u) + p_e \leq d$ for every $u \in C$ and edge e incident on u. The forcing of edges away "expands" C and this process will continue "expanding" C until no more edges need to be forced in a certain direction or infeasibility is determined. If an expansion is successfully computed, directions for a set C_E of edges is found and so we set $D = D \cup C_E$. The process is then repeated if there are any undirected edges left in $E \setminus D$. Otherwise, if no expansion was found another orientation for C is considered and another expansion is computed. The algorithm returns FAIL if there are no more orientations to try. We assume below that each $l_L(u)$ is updated as the direction of edges are changed. Now we formally describe the algorithm.

1. Set $D = \varnothing$. If any edge $e \in E$ has length $p_e > d$, return FAIL.
2. **While** $E \setminus D$ is not empty:
   (a) Compute a bag component C of $G \setminus D$.
   (b) Find a new orientation of the edges in C for which at most one edge from each bag is directed to the same vertex and any two edges e, e' directed to the same vertex satisfy $p_e + p_{e'} \leq d$. If there are no more new orientations to try for C **return** FAIL. Let C_v be the set of vertices u where an edge is directed toward u by this step.
   (c) **While** there is a vertex $u \in C_v$ and undirected edge $e = \{u, v\}$ in $E \setminus D$ where $l_L(u) + p_e > d$:
      i. Direct e from u to v; then direct all edges of the same bag as e that are reachable from v away from u. Add to C_v all vertices whose loads increased in this step.
      ii. **If** any vertex $w \in C_v$ has two edges from the same bag directed toward it or **if** there are two edges e and e' directed toward w so that $p_e + p_{e'} > d$ **then** reset all loads and edges directed by this iteration of Step (2) and go to Step (2b).
   (d) Let C_E be the set of edges that were assigned a direction in Steps (2b) and (2c). Set $D = D \cup C_E$.
3. **Return** schedule corresponding to the orientation of the edges.

The time complexity of this algorithm is $O(n^2 m + mn^2)$. Let $C_1, C_2, \ldots, C_h$ be the bag components selected by the algorithm in Step (2a) in the order they were chosen.

Lemma 1. *If the expansion of C_h is attempted by the algorithm and it returns FAIL, then there is no schedule with makespan at most d. Also, if the algorithm produces a schedule, the makespan of the schedule is at most $(b/2)d$.*

Theorem 3. *There is a $b/2$-approximation algorithm for the graph balancing problem with $b \geq 2$ bags.*

8   A Polynomial-Time Algorithm for $Q|bag, p_j = 1, \mathcal{M}_j|C_{max}$

In this section we consider the restricted assignment problem on uniform parallel machines with bags where every job has the same length. Without loss of generality we can assume that all the machines have speeds $s_1, \ldots, s_m \in \mathbb{Z}^+$. Let the least common multiple of the speeds $s_1, \ldots, s_m$ be c. The above problem is equivalent to when every job $j \in J$ has length $p_j = c$, so for convenience we assume below that every job has length $p_j = c$. Observe that since c is the least common multiple of the speeds, p_j/s_i is integral for all $j \in J$ and $i \in \{1, \ldots, m\}$. We employ binary search with upper bound $U = nc$ and a 1-relaxed decision procedure to compute the smallest value $d \in [1, nc]$ for which there is a schedule with makespan at most d. Note that in the special case when there is one bag for each job, our algorithm is exactly the algorithm of Lin and Li [17] for $Q|p_j = 1, \mathcal{M}_j|C_{max}$.

Let a *conflict machine set for bag* B_k be $\mathcal{C}(B_k) = \{M_i \in M \mid \exists j, j' \in B_k : M_i \in \mathcal{M}_j \cap \mathcal{M}_{j'}\}$, where $\mathcal{M}_j$ and $\mathcal{M}_{j'}$ are the sets of machines where for jobs j and j' can be scheduled, respectively. As a result of this definition, if there is a machine $M_i \notin \mathcal{C}(B_k)$, then at most one job in B_k can be scheduled on machine M_i. In our algorithm we first build a flow network N with a source s and sink t as follows. First, there will be a *job node* for each job $j \in J$; then for each $k = 1, \ldots, b$, create a *conflict machine node* for every machine $M_i' \in \mathcal{C}(B_k)$, and a *machine node* for each machine $M_i \in M$. To avoid ambiguity, we write M_i' whenever we refer to a conflict machine node of a machine M_i, and M_i when we refer to the machine node for machine M_i. We add arcs as follows: add arcs with capacity 1 from the source to each job node; if job $j \in B_k$ can be scheduled on machine M_i: (i) if $M_i \in \mathcal{C}(B_k)$ then add an arc from j to the machine conflict node M_i' of bag B_k with capacity 1, (ii) otherwise add an arc from j to machine node M_i with capacity 1. Add an arc with flow capacity 1 from each machine conflict node M_i' to its corresponding machine node M_i and include an arc from each machine node M_i to the sink with capacity $\lfloor (s_i d)/c \rfloor$.

Lemma 2. *There is an integral flow f that saturates all the arcs incident on s if and only if there is a feasible schedule with makespan at most d.*

The 1-relaxed decision algorithm is the following: build the flow network N described above and compute an integral maximum flow f; if f does not saturate at least one arc incident on the source, return FAIL; otherwise all the arcs incident on the source are saturated, and for each job $j \in J$, schedule job j on machine i if there is flow sent from job node j to machine node M_i.

Theorem 4. *There is a polynomial-time algorithm for $Q|bag, p_j = 1, \mathcal{M}_j|C_{max}$.*

9   Inapproximability and Complexity

9.1   Restricted Assignment Problem with $b = 2$ Bags

To begin, we prove that restricted assignment problem with $b = 2$ bags where the job lengths are either 1 or 2 has no approximation algorithm with approximation ratio less than $3/2$, unless $P = NP$ (Corollary 2). To do this we reduce from the 3-dimensional matching problem ([SP1] in [9]). In the 3-dimensional matching problem there are three disjoint sets $X = \{x_1, x_2, \ldots, x_{m'}\}, Y = \{y_1, \ldots, y_{m'}\}, Z = \{z_1, \ldots, z_{m'}\}$, and a set $T \subseteq X \times Y \times Z$ of triples, and the goal is to determine whether there is a set $T' \subseteq T$ containing m' triples, such that for any pair of triples $(x_k, y_k, z_k), (x_\ell, y_\ell, z_\ell) \in T'$, $x_k \neq x_\ell$, $y_k \neq y_\ell$, and $z_k \neq z_\ell$. We note that our reduction is similar to the one given by Lenstra et al. [16], but their reduction assumes there are $b = n$ bags.

Let us describe the reduction. For each triple $t \in T$ where element $z \in t$ and $z \in Z$, create a machine M_t of *type* z. Next, each element in $X \cup Y$ is a job j, where we place j in bag B_1 if $j \in X$ and in bag B_2 if $j \in Y$; j has $p_{t,j} = 1$ on machine M_t if $j \in t$, and $p_{t,j} = \infty$ otherwise. Let $deg(z)$ be the number of triples of T containing element $z \in Z$. Then for each element $z \in Z$, create $(deg(z) - 1)$ *dummy* jobs of *type* z, where each dummy job j takes 2 time units on machines of type z, otherwise $p_{t,j} = \infty$. Put all the dummy jobs in bag B_1.

Theorem 5. *It is* NP-*hard to decide whether there is a schedule with makespan at most 2 for the restricted assignment problem with $b = 2$ bags when the jobs have lengths either* 1 *or* 2.

Corollary 2. *There is no p-approximation algorithm with $p < 3/2$ for the restricted assignment problem with $b = 2$ bags where the job lengths are either* 1 *or* 2, *unless* $P = NP$.

9.2   Graph Balancing Problem with $b = 3$ Bags

In this section we show that when there are $b \geq 3$ bags, it is NP-hard to approximate the graph balancing problem with b bags with approximation ratio less than $3/2$. To do this we extend a reduction of Ebenlendr et al. [6], and reduce from a variant of 3-SAT called At-Most-3-SAT($2L$), which is known to be NP-complete [1]. More precisely, let there be a propositional logic formula ϕ in conjunctive normal form (CNF), where there are n' boolean variables $x_1, \ldots, x_{n'}$ and m' clauses $y_1, y_2, \ldots, y_{m'}$. There are at most three literals per clause, and each literal (a variable or its negation) occurs at most twice in ϕ. This problem asks if there is an assignment of values to the variables so that ϕ is satisfied.

Let us first briefly describe the original reduction. Create one vertex for each clause y_i, and two vertices, one for each literal of variable x_i, x_i and $\neg x_i$; let the former be called *clause vertices* and the latter be called *literal vertices*. For each variable x_i, add an edge $\{x_i, \neg x_i\}$ with length 2 called a *tautologous edge*; add a self-loop on clause vertex y_i with length $3 - |y_i|$ if $3 - |y_i| > 0$, where $|y_i|$ is the number of literals in clause y_i. Finally, for each clause y_i and literal l_j, add a *clause edge* $\{l_i, y_i\}$ if literal l_j is in clause y_i.

Now we describe our extension to this reduction that will assign each edge to a bag. Create a modified version of G called G', where, for each self loop incident on a clause vertex y_i in G, replace the self-loop with a new vertex y_i' and *self edge* $\{y_i, y_i'\}$ in G'; each self-edge in G' corresponds to a self-loop in G.

Lemma 3. *There is an edge colouring of G' that uses at most four colours η_1, η_2, η_3, η_4, this colouring can be computed in polynomial time.*

Proof. Assign colour η_4 to all tautologous edges, then consider the subgraph G'' of G' consisting of the same vertices but only the uncoloured edges. Observe that every edge in G'' either has a literal vertex and a clause vertex as its endpoints or is a self-edge with one endpoint that is a leaf, thus G'' is bipartite. Since G'' is bipartite and the maximum degree of any vertex in G'' is three, there is an edge colouring of G'' using three colours η_1, η_2, η_3, this edge colouring can be computed in polynomial time. □

Using Lemma 3 we can assign the edges in G to three bags: the edges coloured in G' using colours η_1, η_2, and η_3 are placed in bags B_1, B_2, and B_3, respectively. Finally, place the edges coloured with η_4 in any of the three bags.

Theorem 6. *There is no p-approximation algorithm with $p < 3/2$ for the graph balancing problem with $b \geq 3$ bags where job lengths are either 1 or 2, unless $P = NP$.*

9.3 $Q|bag|C_{max}$ with $b = 2$ Bags

For $P|bag|C_{max}$ with $b = 3$ bags and $Q|bag|C_{max}$ with $b = 2$ bags, we show that both are strongly NP-hard. We reduce from numerical 3-dimensional matching ([SP16] in [9]), which is known to be NP-complete in the strong sense. In the numerical 3-dimensional matching problem $3m'$ elements are contained in 3 disjoint sets $X = \{a_1, a_2, \ldots, a_{m'}\}, Y = \{a_{m'+1}, \ldots, a_{2m'}\}, Z = \{a_{2m'+1}, \ldots, a_{3m'}\}$, and every element $a_j \in X \cup Y \cup Z$ has a size $s(a_j) \in \mathbb{Z}^+$. Given a value $\beta \in \mathbb{Z}^+$ the goal is to determine whether there are disjoint triples $A_1, \ldots, A_{m'}$ where each triple A_i contains exactly one element of X, one element of Y, and one element of Z, such that $\sum_{a_j \in A_i} s(a_j) = \beta$.

Notice that if an instance of $P|bag|C_{max}$ with $b = 3$ bags has exactly $3m$ jobs, Property 1 implies that every machine in a feasible schedule processes 3 jobs. We obtain a straightforward reduction from numerical 3-dimensional matching to $P|bag|C_{max}$ with $b = 3$ bags: set $m = m'$, $n = 3m'$, every element $a_j \in X \cup Y \cup Z$ is a job $j \in J$ with length $p_j = s(a_j)$, and $B_1 = X$, $B_2 = Y$, and $B_3 = Z$. This reduction was independently presented by Dokka *et al.* [5].

Theorem 7 (Dokka et al. [5]). *$P|bag|C_{max}$ with $b = 3$ bags is strongly NP-hard.*

When the machines are uniform we can eliminate the third bag necessary in the above reduction. Instead $n = 2m'$, and associate each machine M_i with a unique element $z_i \in Z$ and set the speed of M_i to $s_i = (\beta - s(z_i))/\beta$.

Theorem 8. *$Q|bag|C_{max}$ with $b = 2$ bags is strongly NP-hard.*

References

1. Asahiro, Y., Jansson, J., Miyano, E., Ono, H., Zenmyo, K.: Approximation algorithms for the graph orientation minimizing the maximum weighted outdegree. J. Comb. Optim. **22**(1), 78–96 (2011)
2. Bodlaender, H., Jansen, K., Woeginger, G.: Scheduling with incompatible jobs. Discret. Appl. Math. **55**(3), 219–232 (1994)
3. Chakrabarty, D., Khanna, S., Li, S.: On $(1, \varepsilon)$-restricted assignment makespan minimization. In: 26th Annual ACM-SIAM Symposium on Discrete Algorithms, pp. 1087–1101 (2015)
4. Das, S., Wiese, A.: On minimizing the makespan when some jobs cannot be assigned on the same machine. In: 24th Annual European Symposium on Algorithms, LIPIcs, vol. 87, pp. 31:1–31:14 (2017)
5. Dokka, T., Kouvela, A., Spieksma, F.: Approximating the multi-level bottleneck assignment problem. Oper. Res. Lett. **40**, 282–286 (2012)
6. Ebenlendr, T., Krčál, M., Sgall, J.: Graph balancing: a special case of scheduling unrelated parallel machines. In: 19th Annual ACM-SIAM Symposium on Discrete Algorithms, pp. 483–490 (2008)
7. Eisenbrand, F.: Solving an avionics real-time scheduling problem by advanced IP-methods. In: de Berg, M., Meyer, U. (eds.) ESA 2010. LNCS, vol. 6346, pp. 11–22. Springer, Heidelberg (2010). https://doi.org/10.1007/978-3-642-15775-2_2
8. Gairing, M., Monien, B., Woclaw, A.: A faster combinatorial approximation algorithm for scheduling unrelated parallel machines. Theor. Comput. Sci. **380**(1), 87–99 (2007)
9. Garey, M., Johnson, D.: Computers and Intractability: A Guide to the Theory of NP-completeness (1979)
10. Gehrke, J., Jansen, K., Kraft, S., Schikowski, J.: A PTAS for scheduling unrelated machines of few different types. Int. J. Found. Comput. Sci. **29**, 591–621 (2018)
11. Graham, R., Lawler, E., Lenstra, J., Rinnooy, K.: Optimization and approximation in deterministic sequencing and scheduling: a survey. Ann. Discret. Math. **5**, 287–326 (1979)
12. Huang, C., Ott, S.: A combinatorial approximation algorithm for graph balancing with light hyper edges. In: 24th Annual European Symposium on Algorithms, LIPIcs, vol. 57, pp. 49:1–49:15 (2016)
13. Jansen, K., Maack, M.: An EPTAS for scheduling on unrelated machines of few different types. Algorithms and Data Structures. LNCS, vol. 10389, pp. 497–508. Springer, Cham (2017). https://doi.org/10.1007/978-3-319-62127-2_42
14. Jansen, K., Rohwedder, L.: A quasi-polynomial approximation for the restricted assignment problem. In: Eisenbrand, F., Koenemann, J. (eds.) IPCO 2017. LNCS, vol. 10328, pp. 305–316. Springer, Cham (2017). https://doi.org/10.1007/978-3-319-59250-3_25
15. Kolliopoulos, S., Moysoglou, Y.: The 2-valued case of makespan minimization with assignment constraints. Inf. Process. Lett. **113**(1), 39–43 (2013)
16. Lenstra, J., Shmoys, D., Tardos, E.: Approximation algorithms for scheduling unrelated parallel machines. Math. Program. **46**(1–3), 259–271 (1990)
17. Lin, Y., Li, W.: Parallel machine scheduling of machine-dependent jobs with unit-length. Eur. J. Oper. Res. **156**(1), 261–266 (2004)
18. Page, D., Solis-Oba, R.: A 3/2-approximation algorithm for the graph balancing problem with two weights. Algorithms **9**(2), 38 (2016)
19. Shchepin, E., Vakhania, N.: An optimal rounding gives a better approximation for scheduling unrelated machines. Oper. Res. Lett. **33**, 127–133 (2005)

Channel Assignment with r-Dynamic Coloring

Junlei Zhu[1,2](✉) [iD] and Yuehua Bu[2,3] [iD]

[1] College of Mathematics, Physics and Information Engineering, Jiaxing University, Jiaxing 314001, China
[2] College of Mathematics, Physics and Information Engineering, Zhejiang Normal University, Jinhua 321004, Zhejiang, China
zhujl-001@mail.zjxu.edu.cn
[3] Zhejiang Normal University Xingzhi College, Jinhua 321004, Zhejiang, China
yhbu@zjnu.edu.cn

Abstract. The channel assignment is an important problem with applications in optical networks. This problem was formulated to the $L(p, 1)$-labeling of graphs by Griggs and Yeh. The r-dynamic coloring is a generalization of the $L(1, 1)$-labeling. An r-dynamic k-coloring of a graph G is a proper k-coloring such that every vertex v is adjacent to at least $min\{d(v), r\}$ different colors. Denote $\chi_r(G) = min\{k \mid G$ has an r-dynamic k-coloring$\}$ and $ch_r(G) = min\{k \mid G$ has a list r-dynamic k-coloring$\}$. In this paper, we show upper bounds $ch_r(G) \leq r + 5$ for planar graphs G with $g(G) \geq 5$ and $r \geq 15$, $ch_r(G) \leq r + 10$ for graphs G with $mad(G) < \frac{10}{3}$.

Keywords: r-dynamic coloring · Planar graph
Maximum average degree · Girth

1 Introduction

The channel assignment problem is to assign channels to radio transmitters such that close transmitters do not interfere with each other. In 1991, Roberts [17] proposed to assign channels such that close transmitters receive different channels and very close transmitters receive channels that are at least two channels apart. Motivated by this problem, Griggs and Yeh [6] introduced the $L(2, 1)$-labeling problem where the transmitters are represented by vertices, the very close transmitters are adjacent vertices and the close transmitters are vertices at distance two. Later, the $L(2, 1)$-labeling problem was subsequently generalized to the $L(1, 1)$-labeling problem. In this paper we are going to study the r-dynamic coloring of graphs, which is an extension of $L(1, 1)$-labeling of graphs.

Let k, r be integers with $k > 0$, $r > 0$, $[k] = \{1, 2, \cdots, k\}$. Let c be a vertex coloring of G and then define $c(V') = \{c(v) | v \in V'\}$ for $V' \subseteq V(G)$.

Supported by NSFC 11771403.

A (k, r)-coloring of a graph G is a proper k-coloring c such that $|c(N_G(v))| \geq min\{d_G(v), r\}$ for any $v \in V(G)$. The condition $|c(N_G(v))| \geq min\{d_G(v), r\}$ for any $v \in V(G)$ is often referred to as the r-dynamic condition. Such coloring is also called as an r-dynamic k-coloring. $\chi_r(G) = min\{k \mid G$ has an r-dynamic k-coloring$\}$ is called the r-dynamic chromatic number of G. The concept of dynamic coloring was first introduced in [12, 16]. It is also studied under the name r-hued coloring [17–19].

The r-dynamic coloring is a generalization of the traditional vertex coloring for which $r = 1$. By definition, the square coloring of a graph, which is equivalent to the $L(1, 1)$-labeling, is the special case when $r = \Delta$. For integer $i > j > 0$, any (k, i)-coloring of G is also a (k, j)-coloring of G and so we have

$$\chi(G) = \chi_1(G) \leq \chi_2(G) \leq \cdots \leq \chi_r(G) \leq \cdots \leq \chi_\Delta(G) = \chi(G^2).$$

The r-dynamic chromatic numbers of some classed of graphs are known. For example, complete graphs, cycles, trees and complete bipartite graphs [11], K_4-minor free graphs [17], moon graphs [5, 14]. In [4], $\chi_2(G) \leq 5$ was shown for planar graphs G. In [7], Jahanbekam et al. proved that $\chi_r(G) \leq r\Delta(G) + 1$ for $r \geq 2$, also they studied bounds on $\chi_r(G)$ for k-regular graphs in terms of $\chi(G)$, the relationship between $\chi_r(G)$ and $\chi(G)$ when G has small diameter. Wegner [20] conjectured that if G is a planar graph, then $\chi_\Delta(G) \leq \Delta(G) + 5$ if $4 \leq \Delta(G) \leq 7$, $\chi_\Delta(G) \leq \lfloor 3\Delta(G)/2 \rfloor + 1$ if $\Delta(G) \geq 8$.

A conjecture similar to the above Wegner's conjecture is proposed in [17].

Conjecture 1. Let G be a planar graph, then $\chi_r(G) \leq r + 3$ if $1 \leq r \leq 2$, $\chi_r(G) \leq r + 5$ if $3 \leq r \leq 7$, $\chi_r(G) \leq \lfloor 3r/2 \rfloor +$ if $r \geq 8$.

Song et al. [18] proved the following theorem towards Conjecture 1, which is a generalization of the case $r = \Delta$ in [2].

Theorem 1. If $r \geq 3$ and G is a planar graph with $g(G) \geq 6$, then $\chi_r(G) \leq r + 5$.

A list assignment L for a graph G assigns to each vertex $v \in V(G)$ a set $L(v)$ of acceptable colors. An L-coloring c of G is a proper vertex coloring such that for every $v \in V(G)$, $c(v) \in L(v)$. G is L-colorable if G has an L-coloring.

A graph G is r-dynamically L-colorable if, for a list assignment L, G has an r-dynamic coloring c such that $c(v) \in L(v)$ for every $v \in V(G)$. G is r-dynamically k-choosable if G is r-dynamic L-colorable for any list assignment L satisfying $|L(v)| \geq k$ for all $v \in V(G)$. $ch_r(G) = min\{k | G$ is r-dynamic k-choosable$\}$. List r-dynamic coloring is an extension of r-dynamic coloring, where instead of having the same set of colors, every vertex is assigned some set of colors and has to be colored from it. In other words, r-dynamic coloring is a special case of list r-dynamic coloring, so for any graph G, we have $ch_r(G) \geq \chi_r(G)$.

A way to measure the sparseness of a graph G is through its girth $g(G)$, the length of a shortest cycle. Another way to measure the sparseness of a graph G is through its maximum average degree $mad(G)$, where $mad(G) = max\{\frac{2|E(H)|}{|V(H)|}|H \subseteq G\}$. A straightforward consequence of Euler's Formula is that every planar graph G satisfies $mad(G) < \frac{2g(G)}{g(G)-2}$.

The list 2-dynamic chromatic numbers of some classed of graphs have been studied. Kim et al. [8] proved that $ch_2(G) \leq 5$ for planar graphs G. Kim and

Park [9] proved that $ch_2(G) \leq 4$ for graphs G with $mad(G) < \frac{8}{3}$, for planar graphs G with $g(G) \geq 7$. Loeb et al. [15] proved that for toroidal graphs G, $ch_3(G) \leq 10$. More results on 3-dynamic coloring can be seen in [3,10,13,19].

In this paper, we prove the following two theorems.

Theorem 2. *If* $r \geq 15$ *and* G *is a planar graphs with* $g(G) \geq 5$, *then* $ch_r(G) \leq r + 5$.

Theorem 3. *If* G *is a graph with* $mad(G) < \frac{10}{3}$, *then* $ch_r(G) \leq r + 10$. *In particular, for every planar graph* G *with* $g(G) \geq 5$, *we have* $ch_r(G) \leq r + 10$.

2   Terminology and Notations

A vertex of degree k (resp. at least k, at most k) will be called a k-vertex (resp. k^+-vertex, k^--vertex). Let $n_i(v)$ $(n_{i+}(v))$ be the number of i-vertices (i^+-vertices) adjacent to v. For a 2-vertex v, the neighbors of v are called weak-adjacent. A $(d(v_1), d(v_2), \cdots, d(v_i))$-vertex is an i-vertex whose neighbors are degree of $d(v_1), d(v_2), \cdots, d(v_i)$ respectively. Let $k(i)$-vertex be a k-vertex adjacent to i 2-vertices. Undefined notations are referred to [1].

The key method in our proofs is discharging, which relies on reducible configurations. We call a configuration reducible if it cannot appear in a minimal counterexample. A graph is minimal for a property if it satisfied the property but none of its proper subgraph does.

Let $V' \subseteq V(G)$ and let c be a partial r-dynamic L-coloring of $G[V']$. V' is the support of c, denoted by $S(c)$. If c_1, c_2 are two partial dynamic L-coloring of G such that $S(c_1) \subseteq S(c_2)$ and $c_1(v) = c_2(v)$ for any $v \in S(c_1)$, then c_2 is an extension of c_1. Given a partial dynamic L-coloring c on V', we define $\{c(v)\} = \emptyset$ for any $v \in V - V'$ and $c(N_G(v)) = \bigcup\{c(z)|z \in N_G(v)\}$ for any vertex $v \in V$. Define $c[v] = \{c(v)\}$ if $|c(N_G(v))| \geq r$, $c[v] = \{c(v)\} \cup c(N_G(v))$ otherwise. Thus, it follows that $|c[v]| \leq r$ and we have the following claim [9].

Claim. Let c be a partial dynamic L-coloring of G with support $S(c)$. For any $u \notin S(c)$, and for any $v \in N_G(u)$, by the definition of $c[v]$, we have $|c[v]| \leq \min\{d(v), r\}$ and $c[v]$ represents the colors that cannot be used as $c(u)$ if one wants to extend the support of c to include u. In other words, the colors in $L(u) - \bigcup_{v \in N(u)} c[v]$ are available colors to define $c(u)$ in extending the support c from $S(c)$ to $S(c) \cup \{u\}$.

3   Proof of Theorem 2

We prove by contradiction. Let G be a minimal counterexample to Theorem 2. Then there exist a list assignment L such that $|L(v)| \geq r + 5 \geq 20$ for $\forall v \in V(G)$ and G is not r-dynamic L-colorable. By the minimality of G, any $H \subset G$ is r-dynamic L-colorable. We first prove that some configurations are reducible.

Let $d(v) = k$ and $N(v) = \{v_1, v_2, \cdots, v_k\}$, $d(v_1) \leq d(v_2) \leq \cdots \leq d(v_k)$. Let $|F(v)|$ denote the number of colors cannot be used on v. Let v be a 3^+-vertex, $v_i \in N(v)$ and $d(v_i) = 2$, then we denote $N(v_i) = \{v, v_i'\}$.

Lemma 1. *G has no 1^--vertex.*

Proof. Assume that G has a 1^--vertex v. By the minimality of G, $G - v$ has a dynamic L-coloring c. Since $|F(v)| \leq r < r + 5$, we can color v and thus we extend the support of c to $V(G)$, a contradiction.

Lemma 2. *G has no adjacent 2-vertices.*

Proof. Assume that G has two adjacent 2-vertex u and v. By the minimality of G, $G - \{u, v\}$ has a dynamic L-coloring. Since $|F(u)| \leq r + 1$, $|F(v)| \leq r + 1$, we can color them. Thus, we extend the coloring of $G - \{u, v\}$ to G, a contradiction.

Lemma 3

(1) *G has no 3-vertex v such that $d(v_1) = 2$ and $d(v_2) + d(v_3) \leq r + 3$, or $d(v_1) = 2$ and $d(v_i) \leq 3$, where $i = 2$ or 3.*

(2) *G has no 3-vertex v such that $d(v_1) = 3$ and $d(v_2) + d(v_3) \leq r + 1$, $d(v_1') + d(v_1'') \leq r + 2$, where $N(v_1) = \{v, v_1', v_1''\}$.*

Proof

(1) Assume that G has a 3-vertex v such that $d(v_1) = 2$ and $d(v_2) + d(v_3) \leq r + 3$ or $d(v_1) = 2$ and $d(v_i) \leq 3$, where $i = 2$ or 3. By the minimality of G, $G - vv_1$ has a dynamic L-coloring. We erase the colors on v and v_1. Since $|F(v)| \leq d(v_2) + d(v_3) + 1 \leq r + 3 + 1$, $|F(v_1)| \leq r + 2$, we can recolor v, v_1 in turn, a contradiction.

(2) Assume that G has such a 3-vertex v. By the minimality of G, $G - vv_1$ has a dynamic L-coloring. We erase the colors on v, v_1. Since $|F(v_1)| \leq r + 2 + 2 = r + 4$, $|F(v)| \leq r + 1 + 2 = r + 3$, we can recolor v_1, v in turn, a contradiction.

Lemma 4

(1) *G has no 4-vertex v such that $d(v_1) = d(v_2) = 2$ and v_3 is a 2-vertex or a 3(1)-vertex.*

(2) *G has no 4-vertex v such that $d(v_1) = d(v_2) = 2$ and $d(v_3) + d(v_4) \leq r + 2$.*

(3) *G has no 4-vertex v such that $d(v_1) = 2$ and $d(v_2) + d(v_3) + d(v_4) \leq r + 3$.*

(4) *G has no 4-vertex v such that v_1, v_2 are 3(1)-vertices and $d(v_3) + d(v_4) \leq r$.*

Proof

(1) Assume that G has such a 4-vertex. By the minimality of G, $G - vv_1$ has a dynamic L-coloring. Let v_3' be the 2-neighbor of v_3 if v_3 is a 3(1)-vertex. We erase the colors on vertices v, v_1, v_2 and v_3'. Since $|F(v)| \leq r + 2 + 2$, $|F(v_1)| \leq r + 2$, $|F(v_2)| \leq r + 2$, we can recolor v, v_1, v_2 in turn. After v, v_1, v_2 are recolored, $|F(v_3')| \leq r + 3$ and thus we can recolor it, a contradiction.

(2) Assume that G has a 4-vertex v such that $d(v_1) = d(v_2) = 2$ and $d(v_3) + d(v_4) \leq r + 2$. By the minimality of G, $G - vv_1$ has a dynamic L-coloring. We erase the colors on vertices v, v_1, v_2. Since $|F(v)| \leq r + 2 + 2$, $|F(v_1)| \leq r + 2$, $|F(v_2)| \leq r + 2$, we can recolor v, v_1, v_2 in turn, a contradiction.

(3) Assume that G has such a 4-vertex. By the minimality of G, $G - vv_1$ has a dynamic L-coloring. We erase the colors on vertices v and v_1. Since $|F(v)| \leq r + 3 + 1, |F(v_1)| \leq r + 3$, we can recolor v, v_1 in turn, a contradiction.

(4) Assume that G has such a 4-vertex. By the minimality of G, $G - vv_1$ has a dynamic L-coloring. Let v_i' be the 2-neighbor of v_i for $i = 1, 2$. We erase the colors on vertices v, v_1, v_2, v_1' and v_2'. Since $|F(v_1)| \leq r + 3, |F(v_2)| \leq r + 3$, $|F(v)| \leq r + 2$, we can recolor v_1, v_2, v in turn. After v_1, v_2, v are recolored, $|F(v_1')| \leq r + 3$, $|F(v_2')| \leq r + 3$ and thus we can recolor v_1' and v_2', a contradiction.

Lemma 5

(1) G has no 5-vertex v such that $d(v_1) = d(v_2) = \cdots = d(v_k) = 2$, $d(v_{k+1}) = d(v_{k+2}) = \cdots = d(v_4) = 3$ and $d(v_5) \leq k + 7$, where $1 \leq k \leq 4$.

(2) G has no 5-vertex v such that $d(v_1) = d(v_2) = d(v_3) = d(v_4) = 2$ and $d(v_i') \leq r - 1$ for some i, where $1 \leq i \leq 4$.

(3) G has no 5-vertex v such that $d(v_1) = d(v_2) = d(v_3) = 2$, $d(v_4) + d(v_5) \leq r + 1$ and $d(v_i') \leq r - 1$ for some i, where $1 \leq i \leq 3$.

Proof

(1) Assume that G has such a 5-vertex v. By the minimality of G, $G - vv_1$ has a dynamic L-coloring. We erase the colors on $v, v_1, v_2, \cdots, v_k$. Since $|F(v_i)| \leq r + 5 - k$ for $i = 1, 2, \cdots, k$, we can recolor $v_1, v_2, \cdots, v_k$. After $v_1, v_2, \cdots, v_k$ are recolored, $|F(v)| \leq k + 7 + 2k + 3(4 - k) = 19$ and thus we can recolor it, a contradiction.

(2) Without loss of generality, we assume that G has such a 5-vertex v such that $d(v_1) = d(v_2) = d(v_3) = d(v_4) = 2$ and $d(v_1') \leq r - 1$. By the minimality of G, $G - vv_1$ has a dynamic L-coloring. We erase the colors on v, v_1, v_2, v_3, v_4. Since $|F(v)| \leq r + 4, |F(v_1)| \leq r - 1 + 1, |F(v_2)| \leq r + 1, |F(v_3)| \leq r + 1, |F(v_4)| \leq r + 1$, we can recolor v, v_4, v_3, v_2, v_1 in turn, a contradiction.

(3) Without loss of generality, we assume that G has such a 5-vertex v such that $d(v_1) = d(v_2) = d(v_3) = 2$, $d(v_4) + d(v_5) \leq r + 1$ and $d(v_1') \leq r - 1$. By the minimality of G, $G - vv_1$ has a dynamic L-coloring. We erase the colors on v, v_1, v_2, v_3. Since $|F(v)| \leq r + 1 + 3, |F(v_1)| \leq r - 1 + 2, |F(v_2)| \leq r + 2, |F(v_3)| \leq r + 2$. Thus, we can recolor v, v_3, v_2, v_1 in turn, a contradiction.

Lemma 6

(1) G has no 6-vertex v such that $d(v_1) = d(v_2) = \cdots = d(v_k) = 2$, $d(v_{k+1}) = d(v_{k+2}) = \cdots = d(v_6) = 3$ and $d(v_i') \leq r - 1$ for some i, where $1 \leq i \leq k$, $3 \leq k \leq 6$.

(2) G has no 6-vertex v such that $d(v_1) = d(v_2) = \cdots = d(v_5) = 2$, $d(v_6) \leq r - 1$ and $d(v_i') \leq r - 1$, $d(v_j') \leq r - 2$ for some i, j, where $1 \leq i \neq j \leq 5$.

(3) G has no 6-vertex v such that $d(v_1) = d(v_2) = \cdots = d(v_4) = 2$, $d(v_5) + d(v_6) \leq 11$ and $d(v_i') \leq r - 1$ for some i, where $1 \leq i \leq 4$.

Proof

(1) Without loss of generality, we assume that G has such a 6-vertex v such that $d(v_1) = d(v_2) = \cdots = d(v_k) = 2$, $d(v_{k+1}) = d(v_{k+2}) = \cdots = d(v_6) = 3$ and $d(v_1') \leq r - 1$. By the minimality of G, $G - vv_1$ has a dynamic L-coloring. We erase the colors on v, v_1, v_2, v_3. Since $|F(v_1)| \leq r - 1 + 3, |F(v_2)| \leq r + 3, |F(v_3)| \leq r + 3$, we can recolor v_3, v_2, v_1 in turn. After v_1, v_2, v_3 are recolored, $|F(v)| \leq 15$. Thus, we can recolor it, a contraction.

(2) Without loss of generality, we assume that G has such a 6-vertex v such that $d(v_1) = d(v_2) = \cdots = d(v_5) = 2$, $d(v_6) \leq r - 1$ and $d(v_2') \leq r - 1$, $d(v_1') \leq r - 2$. By the minimality of G, $G - vv_1$ has a dynamic L-coloring. We erase the colors on $v, v_1, v_2, \cdots, v_5$. Since $|F(v)| \leq r - 1 + 5, |F(v_1)| \leq r - 2 + 1, |F(v_2)| \leq r - 1 + 1, |F(v_3)| \leq r + 1, |F(v_4)| \leq r + 1, |F(v_5)| \leq r + 1$, we can recolor $v, v_5, v_4, v_3, v_2, v_1$ in turn, a contraction.

(3) Without loss of generality, we assume that G has such a 6-vertex v such that $d(v_1) = d(v_2) = \cdots = d(v_4) = 2$, $d(v_5) + d(v_6) \leq 11$ and $d(v_1') \leq r - 1$. By the minimality of G, $G - vv_1$ has a dynamic L-coloring. We erase the colors on v, v_1, v_2, v_3, v_4. Since $|F(v)| \leq 15, |F(v_1)| \leq r - 1 + 2, |F(v_2)| \leq r + 2, |F(v_3)| \leq r + 2, |F(v_4)| \leq r + 2$, we can recolor v_4, v_3, v_2, v_1, v in turn, a contraction.

Lemma 7

(1) G *has no 7-vertex* v *such that* $d(v_1) = d(v_2) = \cdots = d(v_k) = 2$, $d(v_{k+1}) = d(v_{k+2}) = \cdots = d(v_7) = 3$ *and* $d(v_i') \leq r - 1$ $d(v_j') \leq r - 2$, *for some* i, j, *where* $1 \leq i \neq j \leq k$, $5 \leq k \leq 7$.

(2) G *has no 7-vertex* v *such that* $d(v_1) = d(v_2) = d(v_6) = 2$, $d(v_7) \leq r - 2$ *and* $d(v_i') \leq r - 1$, $d(v_j') \leq r - 2$, $d(v_k') \leq r - 3$ *for some* i, j, k, *where* $1 \leq i, j, k \leq 6$.

Proof

(1) Without loss of generality, we assume that G has such a 7-vertex v such that $d(v_1) = d(v_2) = \cdots = d(v_k) = 2$, $d(v_{k+1}) = d(v_{k+2}) = \cdots = d(v_7) = 3$ and $d(v_2') \leq r - 1$ $d(v_1') \leq r - 2$. By the minimality of G, $G - vv_1$ has a dynamic L-coloring. We erase the colors on $v, v_1, v_2, \cdots, v_5$. Since $|F(v_1)| \leq r - 2 + 2, |F(v_2)| \leq r - 1 + 2, |F(v_3)| \leq r + 2, |F(v_4)| \leq r + 2, |F(v_5)| \leq r + 2$, we can recolor v_5, v_4, v_3, v_2, v_1 in turn. After v_i are recolored, $1 \leq i \leq 5$, $|F(v)| \leq 16$ and thus we can recolor it, a contradiction.

(2) Without loss of generality, we assume that G has a 7-vertex v such that $d(v_1) = d(v_2) = \cdots = d(v_6) = 2$, $d(v_7) \leq r - 2$ and $d(v_3') \leq r - 1$, $d(v_2') \leq r - 2$, $d(v_1') \leq r - 3$. By the minimality of G, $G - vv_1$ has a dynamic L-coloring. We erase the colors on $v, v_1, v_2, \cdots, v_6$. Since $|F(v)| \leq r - 2 + 6, |F(v_1)| \leq r - 3 + 1, |F(v_2)| \leq r - 2 + 1, |F(v_3)| \leq r - 1 + 1, |F(v_4)| \leq r + 1, |F(v_5)| \leq r + 1, |F(v_6)| \leq r + 1$, we can recolor $v, v_6, v_5, \cdots, v_1$ in turn, a contradiction.

Lemma 8. G *has no 8-vertex* v *such that* $v_1, v_2, \cdots, v_7$ *are 2-vertices,* $d(v_8) \leq 5$ *and* $d(v_i') \leq r-3$, $d(v_j') \leq r-2$, $d(v_k') \leq r-1$ *for some* i, j, k, *where* $1 \leq i, j, k \leq 7$.

Proof. Without loss of generality, assume that G has a 8-vertex v such that $v_1, v_2, \cdots, v_7$ are 2-vertices, $d(v_8) \leq 5$ and $d(v_1') \leq r - 3$, $d(v_2') \leq r - 2$, $d(v_3') \leq$

$r - 1$. By the minimality of G, $G - vv_1$ has a dynamic L-coloring. We erase the colors on $v, v_1, v_2, \cdots, v_7$. Since $|F(v_7)| \leq r + 1, |F(v_6)| \leq r + 1, |F(v_5)| \leq r+1, |F(v_4)| \leq r+1, |F(v_3)| \leq r-1+1, |F(v_2)| \leq r-2+1, |F(v_1)| \leq r-3+1$, we can recolor $v_7, v_6, \cdots, v_1$ in turn. After $v_1, v_2, \cdots, v_7$ are recolored, $|F(v)| \leq 19$ and thus we can recolor it, a contradiction.

Lemma 9. *G has no 9-vertex v such that $v_1, v_2, \cdots, v_9$ are 2-vertices, and $d(v_i') \leq r - 4$, $d(v_j') \leq r - 3$, $d(v_k') \leq r - 2$, $d(v_l') \leq r - 1$ for some i, j, k, l, where $1 \leq i, j, k, l \leq 9$.*

Proof. Without loss of generality, assume that G has a 9-vertex v such that $v_1, v_2, \cdots, v_9$ are 2-vertices and $d(v_1') \leq r-4, d(v_2') \leq r-3, d(v_3') \leq r-2, d(v_4') \leq r - 1$. By the minimality of G, $G - vv_1$ has a dynamic L-coloring. We erase the colors on $v, v_1, v_2, \cdots, v_9$. Since $|F(v_9)| \leq r, |F(v_8)| \leq r, |F(v_7)| \leq r, |F(v_6)| \leq r, |F(v_5)| \leq r, |F(v_4)| \leq r-1, |F(v_3)| \leq r-2, |F(v_2)| \leq r-3, |F(v_1)| \leq r-4$, we can recolor $v_9, v_7, \cdots, v_1$ in turn. After $v_1, v_2, \cdots, v_9$ are recolored, $|F(v)| \leq 18$ and thus we can recolor it, a contradiction.

Proof. Since G is a minimal counterexample, G is connected. We define a weight function w by $w(v) = \frac{3}{2}d(v) - 5$ for $v \in V$ and $w(f) = d(f) - 5$ for $f \in F$. By Euler's formula $|V| - |E| + |F| = 2$ and formula $\sum_{v \in V} d(v) = 2|E| = \sum_{f \in F} d(f)$, we can derive $\sum_{x \in V \cup F} w(x) = -10$. We then design appropriate discharging rules and redistribute weights accordingly. Once the discharging is finished, a new weight function w' is produced. During the process, the total sum of weights is kept fixed. It follows that $\sum_{x \in V \cup F} w'(x) = \sum_{x \in V \cup F} w(x) = -10$. However, we will show that after the discharging is complete, the new weight function $w'(x) \geq 0$ for all $x \in V \cup F$. This leads to the following obvious contradiction

$$0 \leq \sum_{x \in V \cup F} w'(x) = \sum_{x \in V \cup F} w(x) = -10 < 0$$

In this section, let v be a 3-vertex and $N(v) = \{v_1, v_2, v_3\}$. If $d(v_1) = 3$ and $d(v_2) + d(v_3) \leq r + 1$, then v is called a weak 3-vertex. If $d(v_1) = 3$ and $d(v_2) + d(v_3) \geq r + 3$, then v is called a strong 3-vertex. A special face is a 5^+-face $[w_1 u_1 v u_2 w_2 \cdots]$ such that $d(u_1) = d(u_2) = 2$, $d(w_1) \geq 10$, $d(w_2) \geq 10$ and $5 \leq d(v) \leq 9$.

Discharging rules:

R1. Let $f = [w_1 u_1 v u_2 w_2 \cdots]$ be a special face. If $d(f) \geq 6$, then f gives $\frac{1}{2}$ to v. If $d(f) = 5$, then big vertices w_1 and w_2 gives $\frac{1}{4}$ to f along $w_1 w_2$, respectively and f gives $\frac{1}{2}$ to v.
R2. Every 2-vertex receives 1 from each of its adjacent 3^+-vertices.
R3. Every 3(1)-vertex receives $\frac{1}{2}$ from each of its adjacent k-vertices, $4 \leq k \leq 9$, 1 from each of its adjacent 10^+-vertices;
Every 3(0)-vertex receives $\frac{1}{6}$ from each of its adjacent k-vertices, $4 \leq k \leq 8$, $\frac{1}{2}$ from each of its adjacent k-vertices, $9 \leq k \leq r - 1$, 1 from each of its adjacent r^+-vertices;
Every weak 3-vertex receives $\frac{1}{6}$ from each of its ajacent strong 3-vertices.

R4. Every 4-vertex receives $\frac{1}{2}$ from each of its adjacent k-vertices, $k = 8, 9$, 1 from each of its adjacent k-vertices, $10 \leq k \leq r - 1$, $\frac{7}{6}$ from each of its adjacent r^+-vertices.

R5. Every 5-vertex receives $\frac{1}{2}$ from each of its adjacent 9-vertices, 1 from each of its adjacent 10^+-vertices.

R6. Every 6-vertex receives $\frac{1}{2}$ from each of its adjacent k-vertices, $9 \leq k \leq r-1$, 1 from each of its adjacent r^+-vertices, $\frac{1}{6}$ from each of its weak adjacent r^+-vertices.

R7. Every 7-vertex receives $\frac{1}{2}$ from each of its adjacent $(r - 1)^+$-vertices.

Checking $w'(f) \geq 0, f \in F$. If $d(f) = 5$, then $w'(f) \geq \min\{5 - 5, 5 - 5 + \frac{1}{4} \times 2 - \frac{1}{2}\} = 0$ by (R1). If $d(f) \geq 6$, then $w'(f) \geq \min\{d(f) - 5, d(f) - 5 - \frac{1}{2}\frac{d(f)}{4}\} > 0$ by R1.

Checking $w'(v) \geq 0, v \in V$. By Lemma 1, $\delta(G) \geq 2$.

Case $d(v) = 2$. By Lemma 2, $n_{3+}(v) = 2$ and then by R2, $w'(v) = 3 - 5 + 1 \times 2 = 0$.

Case $d(v) = 3$. By Lemma 3(1), 3-vertex is a 3(1)-vertex or a 3(0)-vertex. Furthermore, if v is a 3(1)-vertex, then $d(v_2) \geq 4, d(v_3) \geq 4$ and $d(v_2) + d(v_3) \geq r + 4 \geq 19$. Thus, v_3 is a 10^+-vertex. By R2 and R3, $w'(v) \geq -\frac{1}{2} - 1 + \frac{1}{2} + 1 = 0$. If v is a 3(0)-vertex and $d(v_1) = 3, d(v_2) + d(v_3) \leq r + 1$, by Lemma 3(2), v is a weak 3-vertex, v_1 is a strong 3-vertex, v_i $(i = 2, 3)$ are strong 3-vertices or 4^+-vertices. Thus, $w'(v) \geq -\frac{1}{2} + \frac{1}{6} \times 3 = 0$ by R3. If v is a 3(0)-vertex and $d(v_1) = 3, d(v_2) + d(v_3) = r + 2 \geq 17$, then v is neither a weak 3-vertex nor strong 3-vertex and v_3 is a 9^+-vertex. By R3, $w'(v) \geq -\frac{1}{2} + \frac{1}{2} = 0$. If v is a 3(0)-vertex and $d(v_1) = 3, d(v_2) + d(v_3) \geq r + 3 \geq 18$, then v is a strong 3-vertex and v_3 is a 9^+-vertex. If $d(v_1) = d(v_2) = 3$, then $d(v_3) \geq r$ and $w'(v) \geq -\frac{1}{2} - \frac{1}{6} \times 2 + 1 > 0$ by R3. Otherwise, $w'(v) \geq -\frac{1}{2} - \frac{1}{6} + \frac{1}{2} + \frac{1}{6} = 0$ by R3. If $d(v_i) \geq 4$ for $i = 1, 2, 3$, then $w'(v) \geq -\frac{1}{2} + \frac{1}{6} \times 3 = 0$ by R3.

Case $d(v) = 4$. By Lemma 4(1), $n_2(v) \leq 2$. Furthermore, if $d(v_1) = d(v_2) = 2$, then v_3, v_4 cannot be 3(1)-vertices by Lemma 4(1) and $d(v_3) + d(v_4) \geq r + 3 \geq 18$ by Lemma 4(2). Thus, if v_3 is a 3(0)-vertex, then v_4 is a r^+-vertex. Moreover, v_3 and v_4 are two 9^+-vertices or v_4 is a 10^+-vertex. By R2-R4, $w'(v) \geq \min\{1 - 1 \times 2 - \frac{1}{6} + \frac{7}{6}, 1 - 1 \times 2 + \frac{1}{2} \times 2, 1 - 1 \times 2 + 1\} = 0$. If $n_2(v) = 1$, then by Lemma 4(3), $d(v_2) + d(v_3) + d(v_4) \geq r + 4 \geq 19$. If $d(v_2) = d(v_3) = 3$, then $d(v_4) \geq 13$ and thus $w'(v) \geq 1 - 1 - \frac{1}{2} \times 2 + 1 = 0$ by R2-R4. If $d(v_2) = 3, d(v_3) \geq 4$ and $d(v_4) \geq 4$, then $d(v_4) \geq 8$ and thus $w'(v) \geq 1 - 1 - \frac{1}{2} + \frac{1}{2} = 0$ by R2-R4. If $d(v_2) \geq 4$ for $i = 2, 3, 4$, then $w'(v) \geq 1 - 1 = 0$ by R2. If $n_2(v) = 0$ and there are at least two 3(1)-vertices, then by Lemma 4(4), $d(v_3) + d(v_4) \geq r + 1 \geq 16$. If $d(v_3) = 3$, then $d(v_4) \geq 13$ and thus $w'(v) \geq 1 - \frac{1}{2} \times 3 + 1 > 0$ by R3 and R4. If $d(v_3) \geq 4, d(v_4) \geq 4$, then $w'(v) \geq 1 - \frac{1}{2} \times 2 = 0$ by (R3). If $n_2(v) = 0$ and there is at most one 3(1)-vertex, then $w'(v) \geq 1 - \frac{1}{2} - \frac{1}{6} \times 3 = 0$ by R3.

Case $d(v) = 5$. By Lemma 5(1), $n_2(v) \leq 4$. If $n_2(v) = 4$, then by Lemma 5(1)(2), $d(v_5) \geq 12$ and $d(v_i') \geq r$ for $i = 1, 2, 3, 4$. Thus, there are at least three special faces and $w'(v) \geq \frac{5}{2} - 1 \times 4 + 1 + \frac{1}{2} \times 3 > 0$ by R1, R2 and R5. If

$n_2(v) = 3$ and $d(v_4) = 3$, then by Lemma 5(1), $d(v_5) \geq 11$ and thus $w'(v) \geq \frac{5}{2} - 1 \times 3 - \frac{1}{2} + 1 = 0$ by R2, R3 and R5. If $n_2(v) = 3$ and $d(v_4) \geq 4$, then by Lemma 5(3), $d(v_4) + d(v_5) \geq r + 2$ or $d(v_4) + d(v_5) \leq r + 1$ and $d(v_i') \geq r$ for $i = 1, 2, 3$. If $d(v_4) + d(v_5) \geq r + 2 \geq 17$, then $d(v_5) \geq 9$. Thus, $w'(v) \geq \frac{5}{2} - 1 \times 3 + \frac{1}{2} = 0$ by R2 and R5. If $d(v_4) + d(v_5) \leq r + 1$ and $d(v_i') \geq r$ for $i = 1, 2, 3$, then v is incident to at least one special face and $w'(v) \geq \frac{5}{2} - 1 \times 3 + \frac{1}{2} = 0$ by R1 and R2. If $n_2(v) = 2$ and $d(v_3) = d(v_4) = 3$, then by Lemma 5(1), $d(v_5) \geq 10$ and thus $w'(v) \geq \frac{5}{2} - 1 \times 2 - \frac{1}{2} \times 2 + 1 > 0$ by (R2), (R3) and (R5). If $n_2(v) = 2$ and $n_3(v) \leq 1$, then $w'(v) \geq \frac{5}{2} - 1 \times 2 - \frac{1}{2} = 0$ by R2 and R3. If $n_2(v) = 1$ and $d(v_2) = d(v_3) = d(v_4) = 3$, then by Lemma 5(1), $d(v_5) \geq 9$ and thus $w'(v) \geq \frac{5}{2} - 1 - \frac{1}{2} \times 3 + \frac{1}{2} > 0$ by R2, R3 and R5. If $n_2(v) = 1$ and $n_3(v) \leq 2$, then $w'(v) \geq \frac{5}{2} - 1 - \frac{1}{2} \times 2 > 0$ by R2 and R3. If $n_2(v) = 0$, then $w'(v) \geq \frac{5}{2} - \frac{1}{2} \times 5 = 0$ by R3.

Case $d(v) = 6$. If $n_2(v) = 6$, then by Lemma 6(1), $d(v_i') \geq r$ for $i = 1, 2, \cdots, 6$. Thus, v is incident to six special faces and $w'(v) \geq 4 - 1 \times 6 + \frac{1}{2} \times 6 > 0$ by R1 and R2. If $n_2(v) = 5$ and $d(v_6) \geq r$, then $w'(v) \geq 4 - 1 \times 5 + 1 = 0$ by (R2) and (R6). If $n_2(v) = 5$ and $d(v_6) \leq r - 1$, then by Lemma 6(2), v is weak adjacent to at least four $(r-1)^+$-vertices and thus v is incident to at least two special faces. Therefore, if $4 \leq d(v_6) \leq r - 1$, then $w'(v) \geq 4 - 1 \times 5 + \frac{1}{2} \times 2 = 0$ by R1 and R2. If $d(v_6) = 3$, then by Lemma 6(1), $d(v_i) \geq r$ for $i = 1, 2, \cdots, 5$. Thus, v is incident to at least four special faces and $w'(v) \geq 4 - 1 \times 5 - \frac{1}{2} + \frac{1}{2} \times 4 > 0$ by R1-R3. If $n_2(v) = 4$ and $d(v_5) + d(v_6) \geq 12$, then $d(v_5) = 3$ and $d(v_6) \geq 9$ or $d(v_5) \geq 4$ and $d(v_6) \geq 4$. Thus, $w'(v) \geq min\{4 - 1 \times 4 - \frac{1}{2} + \frac{1}{2}, 4 - 1 \times 4\} = 0$ by R2, (R3) and (R6). Let $n_2(v) = 4$ and $d(v_5) + d(v_6) \leq 11$. By Lemma 6(3), $d(v_i') \geq r$ for $i = 1, 2, 3, 4$ and thus v is incident to at least two special faces. Therefore, $w'(v) \geq 4 - 1 \times 4 - \frac{1}{2} \times 2 + \frac{1}{2} \times 2 = 0$ by R1-R3. If $n_2(v) = 3$ and $n_3(v) = 3$, then by Lemma 6(1), $d(v_i') \geq r$ for $i = 1, 2, 3$ and thus $w'(v) \geq 4 - 1 \times 3 - \frac{1}{2} \times 3 + \frac{1}{6} \times 3 = 0$ by R2, R3 and R6. If $n_2(v) = 3$ and $n_3(v) \leq 2$, then $w'(v) \geq 4 - 1 \times 3 - \frac{1}{2} \times 2 = 0$ by R2 and R3. If $n_2(v) \leq 2$, then $w'(v) \geq 4 - 1 \times 2 - \frac{1}{2} \times 4 = 0$ by R2 and R3.

Case $d(v) = 7$. If $n_2(v) = 7$, then by Lemma 7(1), v is weak adjacent to at least six $(r-1)^+$-vertices and thus incident to at least five special faces. By R1 and R2, $w'(v) \geq \frac{11}{2} - 1 \times 7 + \frac{1}{2} \times 5 > 0$. If $n_2(v) = 6$ and $d(v_7) = 3$, then by Lemma 7(1), v is weak adjacent to at least five $(r-1)^+$-vertices and thus incident to at least three special faces. By R1-R3, $w'(v) \geq \frac{11}{2} - 1 \times 6 - \frac{1}{2} + \frac{1}{2} \times 3 > 0$. If $n_2(v) = 6$ and $4 \leq d(v_7) \leq r - 2$, then by Lemma 7(2), v is weak adjacent to at least four $(r-2)^+$-vertices and thus incident to at least one special face. By R1 and R2, $w'(v) \geq \frac{11}{2} - 1 \times 6 + \frac{1}{2} = 0$. If $n_2(v) = 6$ and $d(v_7) \geq r - 1$, then by R2 and R7, $w'(v) \geq \frac{11}{2} - 1 \times 6 + \frac{1}{2} = 0$. If $n_2(v) = 5$ and $n_3(v) = 2$, then by Lemma 7(1), v is weak adjacent to at least four $(r-1)^+$-vertices and thus incident to at least one special face. By R1-R3, $w'(v) \geq \frac{11}{2} - 1 \times 5 - \frac{1}{2} \times 2 + \frac{1}{2} = 0$. If $n_2(v) = 5$ and $n_3(v) \leq 1$, then by R2 and R3, $w'(v) \geq \frac{11}{2} - 1 \times 5 - \frac{1}{2} = 0$. If $n_2(v) \leq 4$, then $w'(v) \geq \frac{11}{2} - 1 \times 4 - \frac{1}{2} \times 3 = 0$ by R2 and R3.

Case $d(v) = 8$. If $n_2(v) = 8$, then by Lemma 8, v is adjacent to at least five $(r-2)^+$-vertices and thus incident to at least two special faces. By R1 and R2, $w'(v) \geq 7 - 1 \times 8 + \frac{1}{2} \times 2 = 0$. If $n_2(v) = 7$, then by Lemma 8, v is adjacent to at least five $(r-2)^+$-vertices and thus incident to at least two special faces. If $3 \leq d(v_8) \leq 5$, then by R1-R3, $w'(v) \geq 7 - 1 \times 7 - \frac{1}{2} + \frac{1}{2} \times 2 > 0$. Otherwise, by R1-R3, $w'(v) \geq 7 - 1 \times 7 = 0$. If $n_2(v) \leq 6$, then by R2 and R3, $w'(v) \geq 7 - 1 \times 6 - \frac{1}{2} \times 2 = 0$.

Case $d(v) = 9$. If $n_2(v) = 9$, then by Lemma 9, v is weak adjacent to at least six $(r-3)^+$-vertices and thus incident to at least three special faces. By R1 and R2, $w'(v) \geq \frac{17}{2} - 1 \times 9 + \frac{1}{2} \times 3 > 0$. If $n_2(v) \leq 8$, then by R2 and R3 $w'(v) \geq \frac{17}{2} - 1 \times 8 - \frac{1}{2} = 0$.

Case $10 \leq d(v) \leq r - 1$. By R2-R7, v sends at most 1 to each of its neighbors and thus $w'(v) \geq \frac{3}{2}d(v) - 5 - d(v) \geq 0$.

Case $d(v) \geq r$. By R2-R7, v sends at most 1 to each of its neighbors and at most $\frac{1}{6}$ to each of its weak-adjacent neighbors. Hence, $w'(v) \geq \frac{3}{2}d(v) - 5 - \frac{7}{6}d(v) \geq 0$.

Hence, $w'(x) \geq 0$ for all $x \in V \cup F$ after application the discharging rules. Together with (1), this concludes the proof of Theorem 2.

4   Proof of Theorem 3

We prove by contradiction. Let G be a minimal counterexample to Theorem 3. Then there exist a list assignment L such that $|L(v)| \geq r + 10$ for every vertex v of G and G is not r-dynamic L-colorable. By the minimality of G, any proper subgraph H of G is r-dynamic L-colorable. We first prove that some configurations are reducible.

Let v be a 3-vertex and $N(v) = \{v_1, v_2, v_3\}$. In this section, we say 3-vertex v good if $d(v_1) = 3$ and $min\{d(v_2), d(v_3)\} \geq 7$. We say 3-vertex v is bad if $d(v_1) = 3$ and $min\{d(v_2), d(v_3)\} \leq 6$.

Lemma 10

(1) G has no 1^--vertex v;

(2) G has no two adjacent 2-vertices u and v;

(3) G has no 3-vertex v adjacent to a 2-vertex v_1 and a 8^--vertex v_2;

(4) G has no 3-vertex v adjacent to a weak 3-vertex v_1 and a 7^--vertex v_2;

(5) G has no k-vertex v adjacent to $k-2$ 2-vertices $v_1, v_2, \cdots, v_{k-2}$ and a $(11-k)^-$-vertex v_{k-1}, where $4 \leq k \leq 8$;

(6) G has no 4-vertex v adjacent to a 2-vertex v_1, 3-vertex v_2 and a 5^--vertex v_3;

(7) G has no 9-vertex v adjacent to eight 2-vertices $v_1, v_2, \cdots, v_8$.

Proof

(1) Assume that G has a 1^--vertex v. By the minimality of G, $G - v$ has a dynamic L-coloring c. Since $|F(v)| \leq r < r + 10$, we can color v and thus we extend the support of c to $V(G)$, a contradiction.

(2) Assume that G has two adjacent 2-vertices u and v. By the minimality of G, $G - \{u, v\}$ has a dynamic L-coloring. Since $|F(u)| \leq r+1$, $|F(v)| \leq r+1$, we can color them. Thus, we extend the coloring of $G - \{u, v\}$ to G, a contradiction.

(3) Assume that G has a 3-vertex v adjacent to a 2-vertex v_1 and a 8^--vertex v_2. By the minimality of G, $G - vv_1$ has a dynamic L-coloring. We erase the colors on v and v_1. Since $|F(v)| \leq r+8+1 = r+9$, $|F(v_1)| \leq r+2$, we can color v and v_1 in turn, a contradiction.

(4) Assume that G has a 3-vertex v adjacent to a weak 3-vertex v_1 and a 7^--vertex v_2. By the minimality of G, $G - vv_1$ has a dynamic L-coloring. We erase the colors on v and v_1. Since $|F(v)| \leq r+7+2 = r+9$, $|F(v_1)| \leq r+6+2 = r+8$, we can color v and v_1 in turn, a contradiction.

(5) Assume that G has a k-vertex v adjacent to $k-2$ 2-vertices $v_1, v_2, \cdots, v_{k-2}$ and a $(11-k)^-$-vertex v_{k-1}, where $4 \leq k \leq 8$. By the minimality of G, $G - vv_1$ has a dynamic L-coloring. We erase the colors on v and $v_1, v_2, \cdots, v_{k-2}$. Since $|F(v)| \leq r+(11-k)+(k-2) = r+9$, we can color it. Then $|F(v_i)| \leq r+3$ for each $i = 1, 2, \cdots k-2$, so we can color them, a contradiction.

(6) Assume that G has a 4-vertex v adjacent to a 2-vertex v_1, 3-vertex v_2 and a 5^--vertex v_3. By the minimality of G, $G - vv_1$ has a dynamic L-coloring. We erase the colors on v and v_1. Since $|F(v)| \leq r+5+3+1 = r+9$, $|F(v_1)| \leq r+3$ forbidden colors, so we can color v and v_1 in turn, a contradiction.

(7) Assume that G has a 9-vertex v adjacent to eight 2-vertices $v_1, v_2, \cdots, v_8$. By the minimality of G, $G - \{v, v_1, v_2, \cdots, v_8\}$ has a dynamic L-coloring. Since $|F(v)| \leq r+8$, we can color it. Then $|F(v_i)| \leq r+2$ for each $i = 1, 2, \cdots, 8$, so we can color them, a contradiction.

Proof. For $v \in V(G)$, we define its initial weight w by $w(v) = d(v)$. Let $R1 - R8$ be eight discharging rules. We will use them and Lemma 4.1 to show that for every vertex v, its final weight $w'(v) \geq \frac{10}{3}$ after the discharging finished. This will leads to a contradiction.

Discharging Rules

R1. Every 3^+-vertex sends $\frac{2}{3}$ to each adjacent 2-vertex.
R2. Every good 3-vertex sends $\frac{1}{3}$ to each adjacent bad 3-vertex.
R3. Every 4^+-vertex sends $\frac{1}{9}$ to each adjacent 3-vertex.
R4. Every 6^+-vertex sends $\frac{1}{18}$ to each adjacent 4(1)-vertex.
R5. Every 7-vertex sends $\frac{1}{6}$ to each adjacent good 3-vertex.
R6. Every 7^+-vertex sends $\frac{1}{6}$ to each adjacent 5(3)-vertex.
R7. Every 8^+-vertex sends $\frac{1}{3}$ to each adjacent good 3-vertex and 4(2)-vertex.
R8. Every 9^+-vertex sends $\frac{1}{2}$ to each adjacent 3(1)-vertex.

By Lemma 10(1), we have $\delta(G) \geq 2$.

Case $d(v) = 2$. By Lemma 10(2), $n_{3+}(v) = 2$ and then by R1, $w'(v) = 2 + \frac{2}{3} \times 2 = \frac{10}{3}$.

Case $d(v) = 3$. By Lemma 10(3), $n_2(v) \leq 1$. Moreover, if $n_2(v) = 1$, then $n_{9+}(v) = 2$. By R2 and R8, $w'(v) = 3 - \frac{2}{3} + \frac{1}{2} \times 2 = \frac{10}{3}$. If v is a $(4^+, 4^+, 4^+)$-vertex, then by R3, $w'(v) \geq 3 + \frac{1}{9} \times 3 = \frac{10}{3}$. If v is a $(3, 4^+, 4^+)$-vertex and its

3-neighbor v_1 has a 6^--neighbor other than v, then according to Lemma 10(4), v is a $(3, 8^+, 8^+)$-vertex, which implies v is a good 3-vertex. By R2 and R7, $w'(v) \geq 3 - \frac{1}{3} + \frac{1}{3} \times 2 = \frac{10}{3}$. Now we can assume v is a $(3, 4^+, 4^+)$-vertex and its 3-neighbor v_1 have two 7^+-neighbors, i.e. v_1 is a $(3, 7^+, 7^+)$-vertex. If v has a 6^--vertex other than v_1, i.e. v is a bad 3-vertex, then by R2, $w'(v) \geq 3 + \frac{1}{3} = \frac{10}{3}$. Otherwise, v is a good 3-vertex, by R5 and R7, $w'(v) \geq 3 + \frac{1}{6} \times 2 = \frac{10}{3}$.

Case $d(v) = 4$. By Lemma 10(5), $n_2(v) \leq 2$. Moreover, if $n_2(v) = 2$, then $n_{8+}(v) = 2$. By R1 and R7, $w'(v) = 4 - \frac{2}{3} \times 2 + \frac{1}{3} \times 2 = \frac{10}{3}$. If $n_2(v) = 1$ and $n_3(v) \geq 1$, then according to Lemma 10(6), $n_{6+}(v) = 2$. By R1, R3 and R4, $w'(v) \geq 4 - \frac{2}{3} - \frac{1}{9} + \frac{1}{18} \times 2 = \frac{10}{3}$. If $n_2(v) = 1$ and $n_3(v) = 0$, then by R1, $w'(v) \geq 4 - \frac{2}{3} = \frac{10}{3}$. If $n_2(v) = 0$, then by R3, $w'(v) \geq 4 - \frac{1}{9} \times 4 > \frac{10}{3}$.

Case $d(v) = 5$. By Lemma 10(5), $n_2(v) \leq 3$. Moreover, if $n_2(v) = 3$, then $n_{7+}(v) = 2$. By R1 and R6, $w'(v) = 5 - \frac{2}{3} \times 3 + \frac{1}{6} \times 2 = \frac{10}{3}$. If $n_2(v) \leq 2$, then by R1 and R3, $w'(v) \geq 5 - \frac{2}{3} \times 2 - \frac{1}{9} \times 3 = \frac{10}{3}$.

Case $d(v) = 6$. By Lemma 10(5), $n_2(v) \leq 4$. Moreover, if $n_2(v) = 4$, then $n_{6+}(v) = 2$. By R1, $w'(v) = 6 - \frac{2}{3} \times 4 = \frac{10}{3}$. If $n_2(v) \leq 3$, then by R1, R3 and R4, $w'(v) \geq 6 - \frac{2}{3} \times 3 - \frac{1}{9} \times 3 > \frac{10}{3}$.

Case $d(v) = 7$. By Lemma 10(5), $n_2(v) \leq 5$. Moreover, if $n_2(v) = 5$, then $n_{5+}(v) = 2$. By R1 and R6, $w'(v) \geq 7 - \frac{2}{3} \times 5 - \frac{1}{6} \times 2 = \frac{10}{3}$. If $n_2(v) \leq 4$, then by R1 and R3-R6, $w'(v) \geq 7 - \frac{2}{3} \times 4 - \frac{1}{6} \times 3 > \frac{10}{3}$.

Case $d(v) = 8$. By Lemma 10(5), $n_2(v) \leq 6$. Moreover, if $n_2(v) = 6$, then $n_{4+}(v) = 2$. By R1, R4, R6 and R7, $w'(v) \geq 8 - \frac{2}{3} \times 6 - \frac{1}{3} \times 2 = \frac{10}{3}$. If $n_2(v) \leq 5$, then by R1 and R3, $w'(v) \geq 8 - \frac{2}{3} \times 5 - \frac{1}{3} \times 3 > \frac{10}{3}$.

Case $d(v) = 9$. By Lemma 10(7), $n_2(v) \leq 7$. By R1-R8, v sends $\frac{2}{3}$ to each of its 2-neighbor and at most $\frac{1}{2}$ to each of its 3^+-neighbors, $w'(v) \geq 9 - \frac{2}{3} \times 7 - \frac{1}{2} \times 2 = \frac{10}{3}$.

Case $d(v) \geq 10$. By R1-R8, v sends at most $\frac{2}{3}$ to each of its neighbors and then $w'(v) \geq d(v) - \frac{2}{3} \times d(v) \geq \frac{10}{3}$.

Hence, $w'(v) \geq \frac{10}{3}$ for all $v \in V$ after application the discharging rules. As a result, $\frac{10}{3}|V(G)| \leq \Sigma_{v \in V} w'(v) = \Sigma_{v \in V} w(v) = \Sigma_{v \in V} d(v) \leq |V(G)| mad(G) < \frac{10}{3}|V(G)|$. Consequently, we have $ch_r(G) \leq r + 10$.

For every planar graph G, we have $mad(G) < \frac{2g(G)}{g(G)-2}$. Thus, every planar graph G with girth $g(G) \geq 5$ satisfies $mad(G) \geq \frac{10}{3}$. This concludes the proof of Theorem 3.

References

1. Bondy, J.A., Murty, U.S.R.: Graph Theory. Springer, New York (2008)
2. Bu, Y.H., Zhu, X.B.: An optimal square coloring of planar graphs. J. Comb. Optim. **24**, 580–592 (2012)
3. Cranston, D.W., Kim, S.J.: List-coloring the square of a subcubic graph. J. Graph Theory **57**, 65–87 (2008)

4. Chen, Y., Fan, S.H., Lai, H.J., Song, H.M., Sun, L.: On dynamic coloring for planar graphs and graphs of higher genus. Discrete Appl. Math. **160**, 1064–1071 (2012)
5. Ding, C., Fan, S.H., Lai, H.J.: Upper bound on conditional chromatic number of graphs. Jinan Univ. **29**, 7–14 (2008)
6. Griggs, J.R., Yeh, R.K.: Labelling graphs with a condition at distance 2. SIAM J. Discrete Math. **5**, 586–595 (1992)
7. Jahanbekam, S., Kim, J., Suil, O., West, D.B.: On r-dynamic colorings of graphs. Discrete Appl. Math. **206**, 65–72 (2016)
8. Kim, S.J., Lee, S.J., Park, W.J.: Dynamic coloring and list dynamic coloring of planar graphs. Discrete Appl. Math. **161**, 2207–2212 (2013)
9. Kim, S.-J., Park, W.-J.: List dynamic coloring of sparse graphs. In: Wang, W., Zhu, X., Du, D.-Z. (eds.) COCOA 2011. LNCS, vol. 6831, pp. 156–162. Springer, Heidelberg (2011). https://doi.org/10.1007/978-3-642-22616-8_13
10. Kim, S.J., Park, B.: List 3-dynamic coloring of graphs with small maximum average degree. Discrete Math. **341**, 1406–1418 (2018)
11. Lai, H.J., Lin, J., Montgomery, B., Tao, Z., Fan, S.H.: Conditional colorings of graphs. Discrete Math. **306**, 1997–2004 (2006)
12. Lai, H.J., Montgomery, B., Poon, H.: Upper bounds of dynamic coloring number. Ars Combin. **68**, 193–201 (2003)
13. Li, H., Lai, H.J.: 3-dynamic coloring and list 3-dynamic coloring of K1,3-free graphs. Discrete Appl. Math. **222**, 166–171 (2017)
14. Lin,Y.: Upper bounds of conditional chromatics number. Master Thesis, Jinan University (2008)
15. Loeb, S., Mahoney, T., Reiniger, B., Wise, J.: Dynamic coloring parameters for grphs with given genus. Discrete Appl. Math. **235**, 129–141 (2018)
16. Montgomery, B.: (PhD Dissertation). West Virginia University (2001)
17. Roberts, F.S.: T-colorings of graphs: recent results and open problems. Discrete Math. **93**, 229–245 (1991)
18. Song, H.M., Fan, S.H., Chen, Y., Sun, L., Lai, H.J.: On r-hued coloring of K4-minor free graphs. Discrete Math. **315–316**, 47–52 (2014)
19. Song, H.M., Lai, H.J., Wu, J.L.: On r-hued coloring of planar graphs with girth at least 6. Discrete Appl. Math. **198**, 251–263 (2016)
20. Wegner, G.: Graphs with given diameter and a coloring problem. Technical report. University of Dortmund (1997)

Profit Maximization Problem
with Coupons in Social Networks

Bin Liu[1], Xiao Li[1], Huijuan Wang[2(✉)], Qizhi Fang[1], Junyu Dong[3],
and Weili Wu[4]

[1] School of Mathematical Sciences, Ocean University of China, Qingdao, China
[2] School of Mathematics and Statistics, Qingdao University, Qingdao, China
sduwhj@163.com
[3] College of Information Science and Engineering, Ocean University of China,
Qingdao, China
[4] Department of Computer Science, University of Texas at Dallas, Richardson, USA

Abstract. Viral marketing has become one of the most effective marketing strategies. In the process of real commercialization, in order to let some seed individuals know the products, companies can provide free samples to them. However, for some companies, especially famous ones, they are more willing to offer coupons than give samples. In this paper, we consider the Profit Maximization problem with Coupons (PM-C) in our new diffusion model named the Independent Cascade Model with Coupons and Valuations (IC-CV). To solve this problem, we propose the PMCA algorithm which can return a $\left(\frac{1}{3} - \varepsilon\right)$-approximate solution with at least $1 - 2n^{-l}$ probability, and runs in $O(\log(np) \cdot mn^3 \log n(l \log n + n \log 2)/\varepsilon^3)$ expected time. Further more, during the analysis we provide a method to estimate the non-monotone submodular function.

Keywords: Profit Maximization · Social network
Approximation algorithm

1 Introduction

Social network has become a hot topic nowadays with the opening of the new industry of online networks. The diffusion of information such as the diffusion of news, viewpoints, rumors, etc, has always been addressed theoretically by researchers. The Influence Maximization (IM) problem is one of the fundamental issue during the propagation process. In [1], Kempe et al. describe this optimization problem: use a graph $G = (V, E)$ to represent a social network where nodes in V represent individuals in this network and edges represent the relationships between individuals. Also give a positive integer k, the problem is to find k initially influenced nodes such that the expected number of influenced nodes after the propagation is maximized under a certain diffusion model. Two basic diffusion models we usually use are the *Linear threshold* (LT) and the *Independent Cascade* (IC).

© Springer Nature Switzerland AG 2018
S. Tang et al. (Eds.): AAIM 2018, LNCS 11343, pp. 49–61, 2018.
https://doi.org/10.1007/978-3-030-04618-7_5

There are several different approximation algorithms have been showed to solve the influence maximization problem. The greedy approach [1] uses a monte carlo method to estimate the expected influence, but the time complexity is too huge to use in practical application. Then in [6], Borgs et al. proposed a breakthrough method named the Reverse Influence Sampling (*RIS*) method. Their algorithm can return a $(1 - 1/e - \varepsilon)$-approximate solution in a practical efficiency way. Borrowing ideas from *RIS*, Tang et al. present the *TIM* and the *IMM* method [2,3] which can return a $(1 - 1/e - \varepsilon)$-approximate solution while cutting down the computation costs.

In the marketing strategies, when it comes to selling products for companies, price plays an important role in people's decisions of adopting the product or not, thus influence and adopting (then it brings profit to the company) are two different problems. So under the real marketing scenario, the models are always more complexity, and the IC or LT model can be extended to more new propagation models. For instance, Zhu et al. [10] considered the relationship between influence and profit, then showed a Balanced Influence and Profit (*BIP*) under two price related models named the PR-I and the PR-L. To distinguish whether an individual will actually adopt the product or just be influenced by others, Lu et al. [5] formulate a valuation of the product for each person. It proposed the problem of the Profit Maximization over social networks by incorporating both prices and valuations, under a diffusion model called the Linear Threshold Model with User Valuations (LT-V).

For the existing studies, no matter what they are aiming at (e.g. maximizing the influence, maximizing the profit, minimizing the cost, etc.), almost all the models assume that there is a set of individuals at the very beginning, usually called the *seed set*, who adopt the information or product and can influence their neighbors with some probabilities. So considering the initial motivation of the diffusion, in the real world, a widely used strategy is providing free samples [5,10,11,13]. But for companies, especially for famous companies, they would prefer to issue coupons rather than give free samples [12,14].

Though the model of the Profit Maximization is enriched and perfected continuously, there are still not many constant algorithms been provided. Recently, Zhang et al. [15] give a *PMCE* algorithm for the Profit Maximization with Multiple Adoptions (PM^{2}A) problem, which can get a $\frac{1}{2}(1 - e^2)$-approximation solution. They build the model under the situation that companies sale various products given a budget and items' profits, but the nature of PM^{2}A is still an Influence Maximization problem as they didn't take the cost of initial activation into account when estimating the profit. In the real market, giving free samples or coupons, will lead to some loss of profit called seeding expense, so the profit function is always not a monotone function.

In this paper, we consider the Profit Maximization problem with Coupons (PM-C) in our new diffusion model named the Independent Cascade Model with Coupons and Valuations (IC-CV). Our main contribution is:

- We present PM-C problem a PMCA algorithm runs in $O(\log np \cdot mn^3 \log n$
  $(l \log n + n \log 2)/\varepsilon^3)$ expected time and can return a $(\frac{1}{3} - \varepsilon)$-approximate
  solution with a high probability.
- We propose M-RR set which is more suitable under the LT-CV diffusion
  model, by modifying the traditional RR set.
- We provide a method to estimate the profit function instead of calling the
  value oracle while utilize the Local Search algorithm [4].

2 Preliminary

In this section, we will introduce a more realistic diffusion model in marketing
strategy, named the *Independent Cascade model with Coupons and Valuations*
(IC-CV), and give the definition of the PM-C problem.

When considering the marketing strategies in managing science, we always
describe the *social network* as a set of directed graph $G = (V, E)$, where V is the
set of nodes representing users and E is the set of directed edges representing
relationships between users. Each edge $e \in E$ is associated with an influence
probability $p(e)$ defined by function $p : E \rightarrow [0, 1]$, if $[i, j] \notin E$, then define
$p([i, j]) = 0$. Each node $i \in V$ is associated with a valuation $v_i \geq 0$, and the
distribution function of v_i is $F(\cdot)$ with domain $[0, b]$. Also, we assume that the
price of a product is $p > 0$ and the value of each coupon is $c \in [0, p]$.

There are following assumptions in this social network:

1. We focus on the selling of one item in the network, and the price p of it will
   not change in the propagation period. For convenience, we also assume that
   the production cost will not change, so p can denote *the profit of selling one
   product* if the coupons aren't used.
2. The company select the initial individuals as potential customers just by
   offering coupons. And furthermore, each of the coupon is of the same discount.
   Since the price of each product will not be changed, we will use a constant c
   to represent the value of the coupon in this paper. It is different from most of
   the existing work in which they select the initial individuals by offering free
   samples. Note that, if one who receive the coupon does not finally adopt the
   product, then there is no actual expenditure for the company.
3. Each individual i has his own valuation v_i of the product. The valuation for
   not adopting is defined to be zero. One will purchase the product if the price
   is not exceeding his own valuation, and will reject it otherwise. Following
   the literature [5,16,17], we make the *independent private value* assumption,
   under which the valuation of each user is drawn independently at random
   from a certain distribution.
4. Only the person who have bought the product can propagate it.

2.1 The Diffusion Model

In this special social network, we will describe a diffusion model (named IC-CV)
which is an extension of the IC model that incorporate coupons and valuations.

A diffusion under the IC-CV model proceeds in discrete time steps. Each node in the graph can be in one of four states: *inactive*, *active*, *influenced* and *adopting*. Initially, all nodes are inactive, and the propagation process as follows:

1. At timestamp 1, a seed set S is targeted (i.e., each node in S is given a coupon with value c), and all nodes in S become active. Then the actual price of the product for each seed node becomes $p - c$. For a node $i \in S$, its state immediately transforms to *adopting* if $v_i \geq p - c$, otherwise, transforms to *influenced*.
2. If node i's *adopting* state first appears at timestamp t, then for each directed edge e that points from i to an inactivate node j, i can activate j at timestamp $t+1$ with $p(e)$ probability. After timestamp $t+1$, i *cannot* activate any other nodes.
3. Once a node i becomes activated, it immediately becomes *adopting* if $v_i \geq p$, or it becomes *influenced* otherwise. The *adopting* (or *influenced*) nodes will remain as adopters (or influenced).
4. The diffusion ends if no more nodes can change its state.

In this model, active nodes includes both influenced and adopting nodes. For each node $i \in V$, let $P_{ac}(i|S)$ and $P_{ad}(i|S)$ respectively be the probabilities of i being active and adopting at the end of the propagation process with S being the seed set. Then, for each $i \in S$, we have $P_{ac}(i|S) = 1$ and $P_{ad}(i|S) = 1 - F(p-c)$. And for each $i \in V \setminus S$, $P_{ad}(i|S) = P_{ac}(i|S) \cdot [1 - F(p)]$.

2.2 Problem Definition

Under the strategy that the company provides coupons to customs, the Profit Maximization with Coupons (PM-C) problem can be defined as follows:

Definition 1. *(Profit Maximization with Coupons) Given a social network $G = (V, E)$, a distributed function $F(\cdot)$, an edge weight $p(e) \in [0, 1]$ for each edge $e \in E$, and real numbers c, p with $0 \leq c \leq p$. The problem is to find a seed set $S \in V$, such that the expected total profit, denoted by $\pi(S)$, is maximized.*

Consider that at the end of the process, the profit of an adopting seed node is $p - c$, and is p otherwise. Therefore, we have the following equations.

$$\pi(S) = \sum_{i \in S} (p - c) \cdot P_{ad}(i|S) + \sum_{i \in V - S} p \cdot P_{ad}(i|S)$$

$$= (p - c) \cdot [1 - F(p - c)] \cdot |S| + p \cdot [1 - F(p)] \cdot \sum_{i \in V \setminus S} P_{ac}(i|S)$$

For convenience, we denote $\sum_{i \in V - S} P_{ad}(i|S)$ and $\sum_{i \in V \setminus S} P_{ac}(i|S)$ by $P_{ad}(S)$ and $P_{ac}(S)$, respectively. It's clear that the IM problem under the IC model is a special case of the PM-C problem with $c = 0$, $p = 1$ and $v_i = 1$ for each $i \in V$, which has been proved to be NP-hard. So we have the following hardness result.

Claim 1 *The PM-C problem under the IC-CV model is NP-hard.*

2.3   Properties of $\pi(S)$

It is shown in [1] that a greedy algorithm can achieve $(1 - 1/e)$-approximation by exploiting the monotonicity and submodularity properties of the influence function. Submodularity and monotonicity are two key theoretical properties for optimization problem and are defined as follows.

Given a ground set V, a set function $f : 2^V \to \mathbb{R}$ is *monotone* if $f(S_1) \leq f(S_2)$ for all subsets $S_1 \subseteq S_2 \subseteq V$. Also, the function is *submodular* if $f(S_1 \cup \{x\}) - f(S_1) \geq f(S_2 \cup \{x\}) - f(S_2)$ for all subsets $S_1 \subseteq S_2 \subseteq V$ and all $x \in V \setminus S_2$. Intuitively, a function is submodular if it satisfies the diminishing return property. This property states that the marginal gain from adding an element to a set S is at least as high as the marginal gain from adding that element to the superset T.

In order to make better analyses for profit function $\pi(S)$, we will focus on the submodularity and monotonicity of it.

Non-monotonicity. Unlike the conventional influence maximization problem, the objective function $\pi(\cdot)$ is *non-monotone* under the IC-CV model. For a simple example, let $G = (V, E)$ be a directed graph with vertex set $V = \{1, 2, 3\}$ and edge set $E = \{(1, 2), (1, 3)\}$. The price of each product is $p = 8$ and the value of each coupon is $c = 2$. Let $v_1 = 7$, $v_2 = v_3 = 9$, $p(1, 2) = 0.9$ and $p(1, 3) = 0.8$. If we choose $S_1 = \{1\}$ as the seed set, then the expected total profit is $\pi(S_1) = (8 - 2) + 8 \times 0.9 + 8 \times 0.8 = 19.6$. If we choose $S_2 = \{1, 2\}$ as the seed set, then the expected total profit is $\pi(S_2) = (8 - 2) \times 2 + 8 \times 0.8 = 18.4$. Note that $S_1 \subset S_2$ and $\pi(S_1) > \pi(S_2)$, which indicates that $\pi(\cdot)$ non-monotone.

Submodularity. Consider the situation in the cascade process when a node i has just changes its state to *adopting*, and starts to activate a neighbor j with succeeding probability $p([i, j])$. If j is activated successfully, it will change its state to *adopting* or *influenced* according to its valuation v_j and price p. This kind of random event can be viewed as being determined by flipping a coin of bias $p([i, j])$ and allocating value v_j according to a distribution function $F(\cdot)$ randomly. Looking from the whole process, it obviously does not matter whether we flip the coin at the moment i become *adopting* or at the beginning of the whole process, also does not matter whether we allocate the valuation of j at the moment j is activated successfully or at the beginning.

With all coins flipped and valuation allocated, we can view the process as follows. Firstly, for each edge $e \in E$, we flip a coin with bias $p(e)$, denote the edge "live" with probability $p(e)$, and "blocked" with probability $1 - p(e)$. Then for each node $i \in V$, we allocate value v_i according to the distribute function $F(\cdot)$. Therefore we can get a certain resulting graph and denote it as g.

When it comes to the random event whether a node i will adopt the product or not after being active, it is just related to the numerical relationship between v_i and $p - c$ if i is chosen as a seed node, or that of v_i and p otherwise. Firstly, if $v_i < p - c$, then the probability of this event is $F(p - c)$. In this case, i will not adopt the product after being active regardless of the actual price is p or $p - c$, and we call i a *non-potential-adopter*. Secondly, if $p - c \leq v_i < p$, then the probability is $F(p) - F(p - c)$, in which case i will not adopt the product if it is not a seed node, but will become adopting otherwise. Then we call i a *semi-potential-adopter* in this case. Finally, if $v_i \geq p$, then the probability is $1 - F(p)$. In this case, i is certain to adopt the product after being active, and we call i a *potential-adopter*.

After all the coins are flipped, the types of all the edges and nodes are deterministic, so we can get a certain graph g. To describe which kind of nodes can adopt the product under a fixed g, we will introduce the following definition and claim.

Definition 2. *(Live path) For a path under a certain graph g, if the initial node is semi-potential-adopter or potential-adopter, all the remaining nodes are potential-adopters, and all the edges are live, then we call the path a live path. And for nodes $i, j \in V$, we say that i can "reach" j if there is a live path start from i to j.*

Claim 2. *Under a certain graph g, a node i will adopt the product, if it is reachable from a seed node in S.*

Claim 3. *The profit function $\pi(S)$ is non-monotone and submodular.*

3 Algorithm and Its Analysis

3.1 Framework

Now our problem is to maximize a nonnegative, non-monotone submodular function. Feige [4] give a deterministic local-search algorithm called LS which guarantees $(\frac{1}{3} - \frac{\varepsilon}{n})$-approximation while using value oracle access, that means, the function value of a set S can be finded in an oracle. Though LS algorithm can help to find the approximate solution, the computation of $\pi(S)$ is #P-hard, because the computation of $P_{ac}(i|S)$ is #P-hard [1].

So we need to use a function $\hat{\pi}(S)$ which can be computed in polynomial time to estimate $\pi(S)$. In general, our ideas of solving the PM-C problem are showed as the following steps.

- $\boldsymbol{\pi(S)}$ **Estimation.** Using a method which is similar to the *IMM* method, we consider a maximum coverage problem [8] and obtain a result $\hat{\pi}(S)$. Then use $\hat{\pi}(S)$ to estimate $\pi(S)$, such that for any $S \in 2^V$, $|\hat{\pi}(S) - \pi(S)| \leq \frac{\varepsilon}{2} \cdot OPT$, OPT denotes the optimal solution of PM-C problem.
- **Solve the PM-C problem.** As $\hat{\pi}(S)$ is non-monotone and submodular, use the Local Search algorithm to solve $\max\limits_{S \in 2^V} \hat{\pi}(S)$. And we can show the algorithm can return a $(\frac{1}{3} - \varepsilon)$-approximate solution with a high probability.

3.2   Estimation of $\pi(S)$

To estimate the profit function, we borrow ideas from the *IMM* technique [3] and make some change according to the IC-CV model. *IMM* is an influence maximization algorithm which has a practical high efficiency and returns a $(1 - 1/e - \varepsilon)$-approximate solution with a high probability. We will first introduce the following two concepts.

Definition 3. *(Modified-Reverse Reachable Set) Let g be a resulting graph obtained by removing each edge e in G with $1 - p(e)$ probability and allocating v_i to each $i \in V$ according to $F(\cdot)$. Let i be a node in G, the Modified-Reverse Reachable (M-RR) Set for i is the set of nodes that can reach i in g.*

Definition 4. *(Random M-RR set) Let $\mathcal{G}$ be the distribution of resulting graphs, denote g as an instance sampled from $\mathcal{G}$ randomly. A random M-RR set is a M-RR set generated on g (for a node selected uniformly at random from g).*

Algorithm 1. M-RR Sets Generation (Generation)

Input: Graph G, $\mathcal{R}$, and a positive integer θ.
Output: A set $\mathcal{R}$ with at least θ M-RR sets.
 1: **while do**$|\mathcal{R}| < \theta$.
 2:     Generate a random M-RR set and insert it into $\mathcal{R}$.
 3: return $\mathcal{R}$

Let $\mathcal{R}$ be the set of all random M-RR sets generated in Algorithm 1, i.e., $\mathcal{R} = \{R_1, R_2, \cdots, R_\theta\}$. For any node set S, let x_i ($i \in [1, \theta]$) be a random variable, its value is 1 if $S \cap R_i \neq \emptyset$, and 0 otherwise. Let $F_\mathcal{R}(S)$ denote the fraction of M-RR sets in $\mathcal{R}$ covered by S, that is

$$F_\mathcal{R}(S) = \frac{\{R_i \in \mathcal{R} \mid S \cap R_i \neq \emptyset\}}{\theta} = \frac{1}{\theta} \cdot \sum_{i=1}^{\theta} x_i$$

To show the relationship between M-RR sets and the probability for a node of becoming adopting, we have the following lemma.

Lemma 1. *For any node set $S \subseteq V$,*

$$\mathbb{E}[F_\mathcal{R}(S)] = (\mathrm{P_{ad}}(S) + [1 - F(p - c)]|S|)/n. \tag{1}$$

Denote $\hat{\pi}(S) = pnF_\mathcal{R}(S) - c[1 - F(p - c)]|S|$, then according to the former lemma, it's obvious that $\hat{\pi}(S)$ is an unbiased estimat of $\pi(S)$.

Corollary 1. *For any node set $S \subseteq V$, $\mathbb{E}[\hat{\pi}(S)] = \pi(S)$.*

Lemma 2. *$\hat{\pi}(S)$ is a submodular function.*

Due to the submodularity of $\hat{\pi}(S)$, we can easily test that whether $\hat{\pi}(S) < 0$ will appear under a fixed $\mathcal{R}$. If $\hat{\pi}(S) < 0$ holds for all single node set S, then according to the property of submodular function that the marginal benefit decrease, no matter how many nodes we add into S, $\hat{\pi}(S) < 0$ always holds. In this case we predicate the $\mathcal{R}$ is not a good sample and can reject it. So in all of the proofs that follow, we always regard the $\mathcal{R}$ we use is "good".

A Martingale View of Dependent M-RR Sets. Tang [3] shows that, for $\mathcal{R} = \{R_1, R_2, \ldots, R_\theta\}$ and node set S, the random variable x_i can establish a connection with martingales. For any $i \in [1, \theta]$, $\mathbb{E}[x_i | x_1, x_2, \ldots, x_{i-1}] = \mathbb{E}[x_i]$. Let $\rho = \mathbb{E}[x_i]$, and $M_i = \sum_{j-1}^{i}(x_j - \rho)$. As $M_1, M_2, \ldots, M_\theta$ is a martingale [3], then by the property of martingale we have the following frequently used lemma which is similar with Chernoff bounds.

Lemma 3. ([3,7,9]) For any $\varepsilon > 0$,

$$\Pr\left[\sum_{i=1}^{\theta} x_i - \theta\rho \geq \varepsilon \cdot \theta\rho\right] \leq \exp\left(-\frac{\varepsilon^2}{2 + \frac{2}{3}\varepsilon} \cdot \theta\rho\right),$$

$$\Pr\left[\sum_{i=1}^{\theta} x_i - \theta\rho \leq -\varepsilon \cdot \theta\rho\right] \leq \exp\left(-\frac{\varepsilon^2}{2} \cdot \theta\rho\right) \tag{2}$$

Approximation Guarantees of Sampling Phase. We need to prove the result we get from Algorithm 2 will not deviate significantly from $\pi(S)$ when θ is enough large. Using Lemma 2, we will show that if θ is sizable, the result we obtain from Algorithm 2 for any $S \subseteq V$ is an accurate estimate of $\pi(S)$.

Algorithm 2. Profit Estimation (P-E)

Input: A node set S, $\mathcal{R} = \{R_1, R_2, \ldots R_\theta\}$, $0 < \varepsilon < 1$.
Output: $\hat{\pi}(S)$.
  1: Initialize a set $F_{\mathcal{R}}(S) = \emptyset$.
  2: **for** $i = 1$ to θ **do**
  3:     $F_{\mathcal{R}}(S) = F_{\mathcal{R}}(S) + \frac{\min\{|S^{(i)} \cap R_k|, 1\}}{\theta}$
  4: Return $\hat{\pi}(S) = npF_{\mathcal{R}}(S) - c[1 - F(p - c)]|S|$

Lemma 4. Let $\varepsilon \geq 0$, suppose that θ satisfies

$$\theta \geq \left[\frac{8np + \frac{4}{3}np\varepsilon}{\varepsilon^2 \cdot OPT} + \frac{8n^2pc[1 - F(p - c)]}{\varepsilon^2 \cdot OPT^2}\right] \cdot (l\ln n + n\ln 2 + \ln 2) \tag{3}$$

then, for any node set $S \subseteq V$, $|\hat{\pi}(S) - \pi(S)| < \frac{\varepsilon}{2} \cdot OPT$ holds with at least $1 - n^{-l}/2^n$ probability.

Now we have prove that $\hat{\pi}(S)$ is a good estimation of $\pi(S)$, and can be computed in polynomial time by Algorithm 2. So while ensuring that θ satisfies Eq. (3), we will turn to solve the problem:

$$\max \hat{\pi}(S) = pnF_{\mathcal{R}}(S) - c[1 - F(p - c)|S|. \tag{4}$$

3.3 Use Local Search Algorithm to Solve Problem max $\hat{\pi}(S)$

Algorithm 3. Local Search (LS)

Input: Graph G, $\mathcal{R} = \{R_1, R_2, \ldots R_\theta\}$, $0 < \varepsilon < 1$,
Output: A node set $S \subseteq V$ that is a $(\frac{1}{3} - \varepsilon)$-approximate solution for the problem of
    maximizing $\hat{\pi}(S)$.
1: **if** For any node $v \in V$, max P-E($\{v\}, \mathcal{R}, \varepsilon$)¡0 **then**
2:      Set $S = \emptyset$
3: **else**
4:      **repeat**
5:          **repeat**
6:              $S = S \cup \{u\}$
7:          **until** There is no node $u \in V$, s.t. P-E($S \cup \{u\}, \mathcal{R}, \varepsilon$) $> (1 + \frac{\varepsilon}{n^2})$P-E($S, \mathcal{R}, \varepsilon$)
8:          $S = S \setminus \{u\}$
9:      **until** There is no node $u \in S$, s.t. P-E($S \setminus \{u\}, \mathcal{R}, \varepsilon$) $> (1 + \frac{\varepsilon}{n^2})$P-E($S, \mathcal{R}, \varepsilon$)
10:     Let $S = \arg\max\{$P-E($S, \mathcal{R}, \varepsilon$), P-E($V \setminus S, \mathcal{R}, \varepsilon$)$\}$
11: **return** S

In this section, we will provide a Local Search algorithm which can achieve a $(\frac{1}{3} - \frac{\varepsilon}{n})$-approximation solution for problem in Eq. (4). And then we will prove that Algorithm 3 can return a $(\frac{1}{3} - \varepsilon)$-approximate solution for the original PM-C problem.

Theorem 1. *Given a graph G, $0 < \varepsilon < 1$, $l > 0$ and a set of M-RR set $\mathcal{R} = \{R_1, R_2, \ldots R_\theta\}$ where θ satisfies Eq. (3), Algorithm 3 returns a $(\frac{1}{3} - \varepsilon)$-approximate solution of the PM-C problem with at least $1 - n^{-l}$ probability.*

3.4 Estimation of θ

Notice that in the estimation of $\pi(S)$ section, the number θ of random M-RR set is required to satisfy Eq. (3) in order to guarantee the approximation. For convenient we simplify Eq. (3) as $\theta \geq \frac{\lambda_1}{OPT} + \frac{\lambda_2}{OPT^2}$, where

$$\lambda_1 = \frac{8np + \frac{4}{3}np\varepsilon}{\varepsilon^2} \cdot (l\ln n + n\ln 2 + \ln 2)$$

$$\lambda_2 = \frac{8n^2pc[1 - F(p-c)]}{\varepsilon^2} \cdot (l\ln n + n\ln 2 + \ln 2) \tag{5}$$

But it is difficult to give θ directly according to Eq. (3) since OPT is unknown ahead of time. So we aim to find a lower bound LB of OPT and then let $\theta = \frac{\lambda_1}{LB} + \frac{\lambda_2}{LB^2} \geq \frac{\lambda_1}{OPT} + \frac{\lambda_2}{OPT^2}$. To save the computation time of generating random M-RR sets, LB should be as proximate to OPT as possible. To solve this challenge, we will design a statistical treatment $T(x)$, and make sure that $T(x) = false$ when $OPT < x$ with high probability.

Consider the value range of OPT, the worst case is we choose one seed node and it can't active any other nodes in V, then the expected profit is $(p - c)[1 - F(p - c)]$, and denote this value as A. And the best situation is that we choose one seed node and it active all other nodes in V, the expected profit is $(p - c)[1 - F(p - c)] + p(n - 1)[1 - F(p)]$, and denote this value as B. Obviously $OPT \in [A, B]$, so we will find the lower bound of OPT by testing $T(x)$ on an enough large number of values of x in interval $[A, B]$. In Algorithm 4, the method of bisection will be used. Algorithm 4 will keep calling Algorithm 3 to provide the method to implement $T(x)$. So we have the following lemma.

Lemma 5. *Let $x \in [A, B]$, and ε', $\delta_1 \in [0, 1]$. Suppose that we run Algorithm 3 inputting a set $\mathcal{R}$ with θ' random M-RR set, where*

$$\theta' \geq \left(\frac{4np + \frac{4}{3}np\varepsilon'}{\varepsilon'^2 x} + \frac{2n^2pc[1 - F(p - c)]}{\varepsilon'^2 x^2} \right) \left(\log(\frac{1}{\delta_1}) + n\log 2 \right). \tag{6}$$

Let S_A be the set outputted by Algorithm 3. If $OPT \leq x$, then with at least $1 - \delta_1$ probability we have $\hat{\pi}(S_A) < (1 + \varepsilon')x$.

Algorithm 4. OPT Estimation (O-E)

Input: Graph G, ε, l.
Output: $\hat{\pi}(S)$.
1: Initialize $\mathcal{R} = \emptyset$ and $LB = A$
2: Let $\varepsilon' = \sqrt{2}\varepsilon$, $t = \lceil \log_2(B - A) \rceil$.
3: **for** $i = 1$ to t **do**
4:     Let $x = \frac{B - A}{2^i} + A$
5:     Let $\theta_i = \frac{\lambda'_1}{x} + \frac{\lambda'_2}{x^2}$, where λ'_1 and λ'_2 are as defined in Eq. (7)
6:     **while** $|\mathcal{R}| < \theta_i$ **do**
7:         Generate a random M-RR set and insert it into $\mathcal{R}$.
8:     Let $S_i =$LS$(G, \mathcal{R}, \varepsilon)$
9:     **if** $\hat{\pi}(S_i) \geq (1 + \varepsilon')x$ **then**
10:         $LB = \frac{\hat{\pi}(S_i)}{1 + \varepsilon'}$
11: **return** LB and $\mathcal{R}$

In Algorithm 4, for given G, ε, p, c and l, we first set the initial set $\mathcal{R} = \emptyset$ and the initial lower bound $LB = A$, then we keep dichotomise the value range of OPT, that means we start a for loop with no more than $t = \lceil \log_2(B - A) \rceil$ times.

In the i-th iteration, algorithm set $x = \frac{B - A}{2^i} + A$ and get $\theta_i = \frac{\lambda'_1}{x} + \frac{\lambda'_2}{x^2}$, where

$$\begin{aligned}
\lambda'_1 &= \frac{4np + \frac{4}{3}np\varepsilon'}{\varepsilon'^2} \cdot (l\log n + \log t + n\log 2) \\
\lambda'_2 &= \frac{2n^2pc[1 - F(p - c)]}{\varepsilon'^2} \cdot (l\log n + \log t + n\log 2)
\end{aligned} \tag{7}$$

Here θ_i is the smallest θ' that satisfies Lemma 5 when $\delta_1 = n^{-l}/t$. Then the algorithm generate random M-RR sets and add them into $\mathcal{R}$ until $|\mathcal{R}| = \theta_i$. After invoking Algorithm 3 and get a node set S_i, it use Algorithm 2 to compute $\hat{\pi}(S_i)$. By Lemma 5, if $\hat{\pi}(S_i) \geq (1 + \varepsilon')x$, we will have $OPT \geq x$ with at least $(1 - \frac{n^{-l}}{t})$-probability. So once $\hat{\pi}(S_i) \geq (1 + \varepsilon')x$ holds, we can stop the loop and regard x as a lower bound of OPT. But in Algorithm 4 we set $\frac{\hat{\pi}(S_i)}{1+\varepsilon'}$ as the lower bound rather than x, next we will show that our choice is a tighter lower bound than x.

Lemma 6. *Let x, ε', $\delta_1 \in [0,1]$, $\mathcal{R}$, and S_A be all as defined in Lemma 5. If $x \leq OPT < 2x$, then with at least $1 - \delta_1$ probability we have $OPT \geq \frac{\hat{\pi}(S_A)}{1+\varepsilon'}$.*

The previous discussion assume that $\hat{\pi}(S_i) \geq (1 + \varepsilon')x$ holds in one iteration in Algorithm 4. But if it doesn't hold in all the t iterations, we set $LB = A$. Once the lower bound LB is determined, the algorithm get $\theta = \frac{\lambda_1}{LB} + \frac{\lambda_2}{LB^2}$ where λ_1 and λ_2 are as defined in Eq. (5), then it generate more random M-RR sets and add them into $\mathcal{R}$ until $|\mathcal{R}| = \theta$. In the end it return a collection of M-RR sets $\mathcal{R}$. Therefore we have the following theorem.

Theorem 2. *With at least $1 - n^{-l}$ probability, Algorithm 4 returns a set of random M-RR sets $\mathcal{R}$ that satisfies $|\mathcal{R}| \geq \frac{\lambda_1}{OPT} + \frac{\lambda_2}{OPT^2}$, here λ_1 and λ_2 are as defined in Eq. (5).*

Furthermore, we will prove that LB is close to OPT with a relatively high probability. Firstly we show that Algorithm 4 will terminate the for loop after the j-th iteration with a high probability in the following lemma.

Lemma 7. *Let x, ε', $\delta_1 \in [0,1]$, $\mathcal{R}$, and S_A be all as defined in Lemma 5, ε is as defined in Lemma 3, set $d \geq 1$. Then if $OPT \geq \frac{(1+\varepsilon')^2}{\frac{1}{3}-\varepsilon}x$ holds, then the probability that $\hat{\pi}(S_A) \leq (1 + \varepsilon')x$ holds will not proceed δ_1.*

In addition, based on Lemma 7, we have the following result.

Lemma 8. *Let LB be as defined in Algorithm 4, then with at least $1 - n^{-l}$ probability we have $LB \geq \frac{\frac{1}{3}-\varepsilon}{(1+\varepsilon')^2} \cdot OPT$ (when $\delta_1 = \frac{n^{-l}}{t}$). Furthermore, when $n^{-l} \leq \frac{1}{2}$, have*

$$\mathbb{E}[|(R)|] = O\left(\frac{\max\{\lambda_1, \lambda_1'\}(1 + \varepsilon')^2}{OPT} + \frac{\max\{\lambda_2, \lambda_2'\}(1 + \varepsilon')^4}{(OPT)^2}\right)$$

As we can see in Lemma 8, we should try to minimize $\max\{\lambda_1, \lambda_1'\}(1 + \varepsilon')^2$ and $\max\{\lambda_2, \lambda_2'\}(1 + \varepsilon')^4$ in order to reduce the running time of generating M-RR sets. But it is difficult due to the complex expressions of $\lambda_1, \lambda_1', \lambda_2, \lambda_2'$. So we roughly approximate with a simple function and set $\varepsilon' = \sqrt{2}\varepsilon$ as the optimal choice of the function.

Algorithm 5. PMCA

Input: Graph G, ε, l.
Output: A set S that is a $(\frac{1}{3} - \varepsilon)$-approximate solution for PM-C problem with at
    least $1 - 2n^{-l}$ probability.
1: Let $(LB,\mathcal{R})$=O-E(G,ε,l)
2: Let $\theta = \frac{\lambda_1}{LB} + \frac{\lambda_2}{LB^2}$, where λ_1 and λ_2 are as defined in Eq. (5)
3: $\mathcal{R}$=Generation$(G,\mathcal{R},\theta)$
4: $S_{\mathcal{A}}$ =LS$(G, \mathcal{R}, \varepsilon)$
5: return $S_{\mathcal{A}}$

3.5 Time Complexity

In this section, we will show the time complexity of the PMCA algorithm (Algorithm 5). In summary, the PMCA algorithm use Algorithm 4 to generate a set of M-RR set $\mathcal{R}$, and then use Algorithm 3 (the Local Search method) to find a node set $S_{\mathcal{A}}$ which can maximize $\hat{\pi}(S)$. According to Theorems 1 and 2, and the union bound, PMCA algorithm can return a $(\frac{1}{3} - \varepsilon)$-approximate solution to the PM-C problem with at least $1 - 2n^{-l}$ probability.

Theorem 3. *The PMCA algorithm returns a $(\frac{1}{3} - \varepsilon)$-approximate solution for PM-C problem with at least $1 - 2n^{-l}$ probability, and runs in $O(\log np \cdot mn^3 \log n(l \log n + n \log 2)/\varepsilon^3)$ expected time.*

Acknowledgement. This work was supported in part by National Natural Science Foundation of China (11501316, 11871442), China Postdoctoral Science Foundation (2016M600556), Qingdao Postdoctoral Application Research Project (2016156), and Natural Science Foundation of Shandong Province of China (ZR2017QA010).

References

1. Kempe, D., Kleinberg, J.M., Tardos, É.: Maximizing the spread of influence through a social network. In: KDD, pp. 137–146 (2003)
2. Tang, Y., Xiao, X., Shi, Y.: Influence maximization: near-optimal time complexity meets practical efficiency. In: SIGMOD, pp. 75–86 (2014)
3. Tang, Y., Shi, Y., Xiao, X.: Influence maximization in near-linear time: a martingale approach. In: SIGMOD, pp. 1539–1554 (2015)
4. Feige, U., Mirrokni, V.S., Vondrak, J.: Maximizing non-monotone submodular functions. SIAM J. Comput. **40**(4), 2053–2078 (2011)
5. Lu, W., Lakshmanan, L.V.S.: Profit maximization over social networks. In: ICDM, pp. 479–488 (2012)
6. Borgs, C., Brautbar, M., Chayes, J., Lucier, B.: Maximizing Social Influence in nearly optimal time. In: SODA, pp. 946–957 (2014)
7. Chung, F., Lu, L.: Concentration inequalities and martingale inequalities: a survey. Internet Math. **3**(1), 79–127 (2006)
8. Vazirani, V.V.: Approximation Algorithms. Springer, Heidelberg (2003). https://doi.org/10.1007/978-3-662-04565-7

9. Williams, D.: Probability with Martingales. Cambridge University Press, Cambridge (1991)
10. Zhu, Y., Lu, Z., Bi, Y., Wu, W., Jiang, Y., Li, D.: Influence and profit: two sides of the coin. In: ICDM, pp. 1301–1306 (2013)
11. Hartline, J., Mirrokni, V., Sundararajan, M.: Optimal marketing strategies over social networks. In: WWW, pp. 189–198 (2008)
12. Arthur, D., Motwani, R., Sharma, A., Xu, Y.: Pricing strategies for viral marketing on social networks. In: Leonardi, S. (ed.) WINE 2009. LNCS, vol. 5929, pp. 101–112. Springer, Heidelberg (2009). https://doi.org/10.1007/978-3-642-10841-9_11. abs/0902.3485
13. Leskovec, J., Singh, A., Kleinberg, J.: Patterns of influence in a recommendation network. In: Ng, W.-K., Kitsuregawa, M., Li, J., Chang, K. (eds.) PAKDD 2006. LNCS (LNAI), vol. 3918, pp. 380–389. Springer, Heidelberg (2006). https://doi.org/10.1007/11731139_44
14. Tang, J., Tang, X., Yuan, J.: Profit maximization for viral marketing in online social networks. In: ICNP, pp. 1–10 (2016)
15. Zhang, H., Zhang, H., Kuhnle, A., Thai, M.T.: Profit maximization for multiple products in online social networks. In: INFOCOM, pp. 1–9 (2016)
16. Shoham, Y., Leyton-Brown, K.: Multiagent Systems - Algorithmic, Game-Theoretic, and Logical Foundations. Cambridge University Press, Cambridge (2009)
17. Kleinberg, R., Leighton, T.: The value of knowing a demand curve: bounds on regret for online posted-price auctions. In: FOCS, pp. 594–605 (2003)

A Bicriteria Approximation Algorithm for Minimum Submodular Cost Partial Multi-Cover Problem

Yishuo Shi[1], Zhao Zhang[2]($\boxtimes$) (iD), and Ding-Zhu Du[3] (iD)

[1] College of Mathematics and System Sciences, Xinjiang University, Urumqi 830046, Xinjiang, China
[2] College of Mathematics and Computer Science, Zhejiang Normal University, Jinhua 321004, Zhejiang, China
hxhzz@sina.com
[3] Department of Computer Science, University of Texas at Dallas, Richardson, TX 75080, USA

Abstract. This paper presents a bicriteria approximation algorithm for the *minimum submodular cost partial multi-cover problem* (SCPMC), the goal of which is to find a minimum cost sub-collection of sets to fully cover q percentage of total profit of all elements, where the cost on sub-collections is a submodular function, and an element e with covering requirement r_e is fully covered if it belongs to at least r_e picked sets. Such a problem occurs naturally in a social network influence problem.

Assuming that the maximum covering requirement $r_{\max} = \max_e r_e$ is a constant and the cost function is nonnegative, monotone nondecreasing, and submodular, we give the first $(O(b/q\varepsilon), (1-\varepsilon))$-bicriteria algorithm for SCPMC, the output of which fully covers at least $(1-\varepsilon)q$-percentage of the total profit of all elements and the performance ratio is $O(b/q\varepsilon)$, where $b = \max_e \binom{f_e}{r_e}$ and f_e is the number of sets containing element e. In the case $r \equiv 1$, an $(O(f/q\varepsilon), 1 - \varepsilon)$-bicriteria solution can be achieved even when monotonicity requirement is dropped off from the cost function, where f is the maximum number of sets containing a common element.

Keywords: Partial cover · Multi-cover · Submodular cover
Bicriteria algorithm

1 Introduction

This paper studies the *minimum submodular cost partial multi-cover* problem (SCPMC), which is a variant of the set cover problem. The *minimum set cover problem* (SC) is one of the most important combinatorial optimization problems in both the theoretical field and the application field, the goal of which is to

Supported by NSFC (11771013, 11531011, 61751303) and Major projects of Zhejiang Science Foundation (D19A010003).

S. Tang et al. (Eds.): AAIM 2018, LNCS 11343, pp. 62–73, 2018.
https://doi.org/10.1007/978-3-030-04618-7_6

find a sub-collection of sets with the minimum cost to cover all elements. There are a lot of variants of the set cover problem. The minimum *partial set cover* problem (PSC) is to find a minimum cost sub-collection of sets to cover at least q-percentage of all elements. One motivation of PSC comes from the phenomenon that in a real world, "satisfying all requirements" would be too costly or even impossible, because of resource limitation or political policy. Another variant is the minimum *multi-cover* problem (MC), which comes from the requirement of fault tolerance in practice. In MC, each element e has a covering requirement r_e, and the goal is to find a minimum cost sub-collection $\mathcal{S}'$ to fully cover all elements, where element e is fully covered by $\mathcal{S}'$ if e belongs to at least r_e sets of $\mathcal{S}'$. Another generalization of set cover is *submodular cost set cover* (SCSC), in which the cost function on sub-collection of sets is submodular and the goal is to find a set cover with the minimum cost. The SCPMC problem is a combination of the above three problems, in which each element has a profit as well as a covering requirement, the goal is to find a minimum submodular cost sub-collection of sets such that the profit of fully covered elements is at least a fixed percentage of the total profit.

The SCPMC problem has a background in influence problems of a social network. In [35], Wang *et al.* proposed a *positive dominating set problem* (PDS) under the following consideration. Suppose an opinion is to be injected into a social network. A person will adopt the opinion if at least half of his friends hold this opinion. The problem is to select the minimum number of individuals to inject the opinion such that all people in the network will adopt the opinion under the above influence mechanism. This problem is extremely hard [7] . But if we only consider *one-step* influence, that is, we aim at the adoption of the opinion in just one step of influence, then it can be viewed as an MC problem. In fact, in the language of set cover, every individual is an element to be covered, and every individual also corresponds to a set which covers all his friends (a person is viewed as a friend of himself). The problem is to select the minimum number of sets such that every individual v can be covered by at least $\lceil d(v)/2 \rceil$ sets, where $d(v)$ is the number of friends of v. This is exactly the MC problem. If we relax the requirement such that only a fraction of individuals are to be influenced, then it is the partial multi-cover problem [29], which is a special SCPMC problem.

1.1 Related Work

For SC, Hochbaum [13] gave an f–approximation algorithm based on LP rounding where f is the maximum number of sets containing a common element. Khot and Regev [21] showed that SC cannot be approximated within $f - \varepsilon$ for any constant $\varepsilon > 0$ assuming unique games conjecture. Another classic result on SC is that greedy strategy yields a $\ln \Delta$-approximation [9,17,26], where Δ is the maximum cardinality of a set. Dinur and Steurer [5] showed that SC cannot be approximated to $(1 - o(1)) \ln n$ unless $P = NP$, where n is the number of elements.

Dobson [10] studied a generalization of MC, namely the *minimum multi-set multi-cover problem* (MSMC), and gave an H_K-approximation, where K is the maximum size of a multi-set (recall that $H_K \approx \ln K$). Rajagopalan and Vazirani [28] gave a greedy algorithm for MSMC achieving the same performance ratio, using dual fitting analysis. For the *minimum set k-cover problem* in which the covering requirement of every element is k, Berman *et al.* [2] gave a randomized algorithm achieving expected performance ratio at most $\ln(\frac{\Delta}{k})$. In fact, MSMC is a special case of the *covering integer program* problem (CIP), which can be modeled as $\min\{c^T x \colon Ax \geq b, 0 \leq x \leq d, x \in \mathbb{Z}\}$, where A is a non-negative matrix, and c, b, d are positive vectors. There are large quantities of studies on CIP [6,23,32]. In particular, Chekuri and Quanrud [4] obtained currently best known approximation for CIP which depends logarithmically on the Δ_0-sparsity (the maximum number of nonzero entries in a column) and the Δ_1-sparsity (the maximum sum of entries in a column).

For PSC, Kearns [20] gave the first greedy algorithm achieving performance ratio $(2H_n + 3)$. Slavík [31] improved the ratio to $H_{\min\{\lceil qn \rceil, \Delta\}}$, where q is the desired covering ratio. Using primal dual method, Gandhi *et al.* [12] obtained an f-approximation. Bar-Yehuda [1] studied a generalized version of the partial cover problem in which each element has a profit. Using local ratio method, he also obtained an f-approximation. Proposing an Lagrangian relaxation framework, Konemann *et al.* [22] gave a $(\frac{4}{3} + \varepsilon)H_\Delta$-approximation for the generalized partial cover problem. A mixed partial cover problem (where the number of covered elements is an integer but every element only needs to be fractionally covered up to its requirement) is studied by Dinitz and Gupta [11].

From the above related work, it can be seen that both PSC and MC admit performance ratios which match those best ratios for the classic set cover problem. However, combining partial cover with multi-cover seems to enormously increase the difficulty of studies. Ran *et al.* [29] were the first to study approximation algorithm for the *minimum partial multi-cover problem* (PMC). Using greedy strategy and dual fitting analysis, they gave the first approximation algorithm with a theoretically guaranteed performance ratio. However, this ratio is meaningful only when the covering percentage q is very close to 1. In [30], Ran *et al.* presented a simple greedy algorithm achieving performance ratio Δ. Recall that in terms of Δ, greedy algorithm for Set Cover achieves performance ratio $\ln \Delta$. So, ratio Δ for PMC is exponentially larger than the one for SC. In the same paper, they also presented a local ratio algorithm which reveals an interesting phenomenon which is called "shock wave phenomenon" in their paper: the performance ratio is f for both PSC (that is, when $r \equiv 1$ which is the partial *single* cover problem) and MC (that is, when $q = 1$ which is the *full* multi-cover problem); however, when q is smaller than 1 by an arbitrarily small constant, the ratio jumps abruptly to $O(n)$.

It should be noticed that the *Minimum k-Union problem* (MkU) studied by Chlamtáč *et al.* [8] is a special case of PMC. In an MkU problem, given a hypergraph, the goal is to choose k hyperedges to minimize the number of vertices in their union. It is equivalent to choosing the minimum number of vertices such that the number of hyperedges which are completely contained in the chosen vertex set is at least p. This is a PMC problem in which $r_e = f_e$ holds for each element e, where f_e is the number of sets (or hyperedges) containing element e. It is highly believed that MkU does not admit a better than polynomial approximation ratio [8].

The submodular cost set cover problem was first proposed by Iwata and Nagano [14]. They gave an f-approximation algorithm for nonnegative submodular functions. In paper [24], Koufogiannakis and Young generalized set cover constraint to arbitrary covering constraints and gave an f-approximation algorithm for monotone non-decreasing nonnegative submodular functions. For other submodular minimization problems with various constraints, refer to [15,16,19]. In particular, in paper [18], Kamiyama studied nonnegative submodular function minimization problem with covering type linear constraints and obtained an approximation ratio depending on Γ_1 (the maximum number of nonzero entries in a row) and Γ_2 (the second maximum number of nonzero entries in a row). Notice that in [36], Wolsey studied the problem of minimizing a linear function with the constraint that the cost (which is submodular) of the chosen set reaches the maximum possible value (namely the cost of the whole element set). This goal is different from the above problems whose goal is to minimize a submodular cost.

In this paper we study approximation algorithms for the set cover problem combining the submodular cost function with partial multi-cover constraint.

1.2 Our Contribution

We study the SCPMC problem with a profit on each element and the goal is to find a minimum cost sub-collection of sets such that the profit of fully covered elements is at least q-percentage of total profit. A randomized $(O(\frac{b}{q\varepsilon}), 1 - \varepsilon)$-bicriteria algorithm is given, that is, the algorithm produces a solution covering at least $(1 - \varepsilon)q$-percentage of the total covering requirement, and achieves performance ratio $O(\frac{b}{q\varepsilon})$ with a high probability, where $b = \max_e \binom{f_e}{r_e}$, and f_e is the number of sets containing element e.

It should be noticed that SCPMC is not a submodular cost submodular cover problem (SCSC): the profit of elements which are fully covered is not a submodular function. Hence previous methods for SCSC can not be used.

Before presenting this algorithm, we show that a natural integer program for SCPMC does not work since its integrality gap is arbitrarily large. Hence, to obtain a good approximation, we propose a novel integer program. The relaxation of the integer program uses Lovász extension [25]. Our algorithm consists of two stages of rounding. The first stage is a deterministic rounding. The second stage is a random rounding, the analysis of which is based on an equivalent expression of Lovász extension in view of expectation [3].

We show that for the special case when the covering requirement $r \equiv 1$ (the special case is abbreviated as SCPSC), our method can be adapted to yield an $(O(f/q\varepsilon), 1 - \varepsilon)$-bicriteria algorithm, where $f = \max\{f_e : e \in E\}$, *even when monotonicity is dropped off* from the requirement of the cost function.

This paper is organized as follows. In Sect. 2, we introduce formal definition of the SCPMC problem, as well as some technical results. The bicriteria algorithm for SCPMC is presented and analyzed in Sect. 3. At the end of Sect. 3, we show how to adapt the algorithm to deal with SCPSC without monotonicity. The last section concludes the paper and discusses some future work. Because of limited space, detailed proofs are omitted which can be found in [33].

2   Preliminaries

Definition 1 (Submodular Cost Partial Multi-Cover (SCPMC)). Suppose E is an element set and $\mathcal{S} \subseteq 2^E$ is a collection of subsets of E with $\bigcup_{S \in \mathcal{S}} S = E$; each element $e \in E$ has a positive covering requirement r_e and a positive profit p_e; cost function $\rho_0 : 2^{\mathcal{S}} \mapsto \mathbb{R}$ is defined on sub-collections of $\mathcal{S}$, which is nonnegative, monotone non-decreasing, and submodular. Given a constant $q \in (0, 1]$ called *covering ratio*, the SCPMC problem is to find a minimum cost sub-collection $\mathcal{S}'$ such that $\sum_{e \sim \mathcal{S}'} p_e \geq qP$, where $P = \sum_{e \in E} p_e$ is the total profit, $e \sim \mathcal{S}'$ means that e is *fully covered* by $\mathcal{S}'$, that is, $|\{S \in \mathcal{S}' : e \in S\}| \geq r_e$. An instance of SCPMC is denoted as $(E, \mathcal{S}, r, p, q, \rho_0)$.

When $r_{\max} = 1$, we call the problem a *submodular cost partial set cover problem* (SCPSC). When the cost function is linear, that is, every set $S \in \mathcal{S}$ has a cost c_S and the cost of a sub-collection $\mathcal{S}'$ is $\rho_0(\mathcal{S}') = \sum_{S \in \mathcal{S}'} c_S$, the problem is exactly the minimum *partial multi-cover problem* (PMC).

An algorithm is a (σ, ζ)-*bicriteria algorithm* for SCPMC if the profit of fully covered elements is at least ζqP and the cost of the sub-collection is at most σ times that of the optimal cost.

Definition 2 (submodular function). Given a ground set E, a set function $\rho : 2^E \mapsto \mathbb{R}$ is *submodular* if for any $E'' \subseteq E' \subseteq E$ and $E_0 \subseteq E \setminus E'$, we have

$$\rho(E' \cup E_0) - \rho(E') \leq \rho(E'' \cup E_0) - \rho(E''). \tag{1}$$

A set $S \subseteq E$ can be indicated by its characteristic vector $x_S = (x_1, \ldots, x_n)$, where $n = |E|$, $E = \{e_1, \ldots, e_n\}$, and $x_i = 1$ if $e_i \in S$ and $x_i = 0$ if $e_i \notin S$. So, in the following, we shall use notation $\{0, 1\}^n \mapsto \mathbb{R}$ to refer to a set function. An important relaxation of a submodular function is the Lovász extension.

Definition 3 (Lovász extension [25]). For a set function $\rho : \{0, 1\}^n \mapsto \mathbb{R}$, the Lovász extension $\hat{\rho} : \mathbb{R}^n \to \mathbb{R}$ is defined as follows. For any vector $x \in \mathbb{R}^n$, order elements as $e_{j_1}, \ldots, e_{j_n}$ such that $x_{j_1} \geq x_{j_2} \geq \ldots \geq x_{j_n}$, where x_{j_i} is the coordinate of x indexed by e_{j_i}. Let $E_i = \{e_{j_1}, e_{j_2}, \ldots, e_{j_i}\}$. The value of $\hat{\rho}$ at x is

$$\hat{\rho}(x) = \sum_{i=1}^{n-1} (x_{j_i} - x_{j_{i+1}}) \rho(E_i) + x_{j_n} \rho(E_n). \tag{2}$$

The following is a relation between submodularity and convexity.

Theorem 1. *A set function ρ is submodular if and only if its Lovász extension $\hat{\rho}$ is a convex function.*

The following is an equivalent expression of Lovász extension in range $[0,1]^n$.

Theorem 2 ([3]). Let ρ be a set function $\{0,1\}^n \mapsto \mathbb{R}$. The Lovász extension $\hat{\rho}$ of ρ in range $[0,1]^n$ can be equivalently expressed as

$$\hat{\rho}(x) = \mathop{\mathbb{E}}_{\theta \in [0,1]} [\rho(x^\theta)] = \int_0^1 \rho(x^\theta) d\theta, \tag{3}$$

where $\mathbb{E}$ represent the expectation, $x_i^\theta = 1$ if $x_i \geq \theta$ and $x_i^\theta = 0$ otherwise.

We study SCPMC under the following assumptions.

(**Assumption 1**) The maximum covering requirement $r_{\max} = \max\{r_e \colon e \in E\}$ has a constant upper bound.

(**Assumption 2**) Since submodular cost (full) multi-cover problem is already studied in [14,24], we only consider the partial version, assuming that $q < 1$.

3 Approximation Algorithm for SCPMC

In this section, we study SCPMC. A natural idea to model the SCPMC problem is to use the following integer programm:

$$\min \rho_0(x)$$

$$s.t. \quad \sum_{e \colon e \in E} p_e y_e \geq qP,$$

$$\sum_{S \colon e \in S} x_S \geq r_e y_e, \text{ for any } e \in E \tag{4}$$

$$x_S \in \{0,1\} \text{ for } S \in \mathcal{S}$$

$$y_e \in \{0,1\} \text{ for } e \in E$$

Here x_S indicates whether set S is selected and y_e indicates whether element e is fully covered. The second constraint says that if $y_e = 1$ then at least r_e sets containing e must be selected and thus e is fully covered. However, the following example shows that the integrality gap of the above program is arbitrarily large.

Example 1. Let $E = \{e_1, e_2\}$, $\mathcal{S} = \{S_1, S_2, S_3\}$ with $S_1 = \{e_1\}$, $S_2 = \{e_2\}$, $S_3 = \{e_1, e_2\}$, $c(S_1) = c(S_2) = 1$, $c(S_3) = M$ where M is a large positive number, $r(e_1) = r(e_2) = 2$, $p(e_1) = p(e_2) = 1$, $q = 1/2$, and the cost function $\rho_0(x) = \sum_{S \in \mathcal{S}} c(S) x_S$. Then $x_{S_1} = x_{S_2} = 1$, $x_{S_3} = 0$, $y_{e_1} = y_{e_2} = 1/2$ form a feasible solution to the relaxation of (4) with objective value 2, while any integral feasible solution to (4) has cost at least $M + 1$.

Hence, to obtain a good approximation, we need to find another program.

3.1 Integer Program and Convex Relaxation

For an element e, an r_e-*cover* is a sub-collection $\mathcal{A} \subseteq \mathcal{S}$ with $|\mathcal{A}| = r_e$ such that $e \in S$ for every $S \in \mathcal{A}$. Denote by Ω_e the family of all r_e-covers and $\Omega = \bigcup_{e \in E} \Omega_e$. The following example illustrates these concepts.

Example 2. Let $E = \{e_1, e_2, e_3\}$. $\mathcal{S} = \{S_1, S_2, S_3\}$ with $S_1 = \{e_1, e_2\}$, $S_2 = \{e_1, e_2, e_3\}$, $S_3 = \{e_2, e_3\}$, $S_4 = \{e_1, e_3\}$, and $r(e_1) = 2$, $r(e_2) = r(e_3) = 1$. For this example, $\Omega_{e_1} = \{\{S_1, S_2\}, \{S_1, S_4\}, \{S_2, S_4\}\}$, $\Omega_{e_2} = \{\{S_1\}, \{S_2\}, \{S_3\}\}$, $\Omega_{e_3} = \{\{S_2\}, \{S_3\}, \{S_4\}\}$, $\Omega = \{\{S_1\}, \{S_2\}, \{S_3\}, \{S_4\}, \{S_1, S_2\}, \{S_1, S_4\}, \{S_2, S_4\}\}$.

Let $\rho \colon 2^{\Omega} \to \mathbb{R}$ be the function on sub-families of Ω defined by

$$\rho(\Omega') = \rho_0 \Big(\bigcup_{\mathcal{A} \in \Omega'} \mathcal{A} \Big) \text{ for } \Omega' \subseteq \Omega. \tag{5}$$

For example, $\rho(\{\{S_1\}, \{S_1, S_2\}\}) = \rho_0(\{S_1, S_2\})$. The SCPMC problem can be modeled as the following integer program:

$$\min \ \rho(x)$$

$$s.t. \ \sum_{e : e \in E} p_e y_e \geq qP,$$

$$\sum_{\mathcal{A} : \mathcal{A} \in \Omega_e} x_{\mathcal{A}} \geq y_e, \text{ for any } e \in E \tag{6}$$

$$x_{\mathcal{A}} \in \{0, 1\} \text{ for } \mathcal{A} \in \Omega$$

$$y_e \in \{0, 1\} \text{ for } e \in E$$

Here, $x_{\mathcal{A}}$ indicates whether cover $\mathcal{A}$ is selected and y_e indicates whether element e is fully covered. The second constraint says that if $y_e = 1$, then at least one r_e-cover must be selected and thus e is fully covered.

Example 3. For the example in Example 2, suppose $p_{e_i} \equiv 1$ for $i = 1, 2, 3$ and $q = 2/3$. Consider a feasible solution to (6): $x_{\mathcal{A}_1} = x_{\mathcal{A}_2} = 1$ for $\mathcal{A}_1 = \{S_1, S_2\}$, $\mathcal{A}_2 = \{S_2\}$, and $x_{\mathcal{A}} = 0$ for all other $\mathcal{A} \in \Omega \setminus \{\mathcal{A}_1, \mathcal{A}_2\}$, we have $y_{e_1} = y_{e_2} = 1$ and $y_{e_3} = 0$. This feasible solution to (6) has objective value $\rho(\{\mathcal{A}_1, \mathcal{A}_2\}) = \rho_0(S_1, S_2)$, which corresponds to a feasible solution $\{S_1, S_2\}$ to SCPMC with the same cost. Conversely, for the feasible solution $\{S_1, S_2\}$ to SCPMC, it is natural to set $x_{\mathcal{A}_1} = 1$ and all other $x_{\mathcal{A}}$ to be zeros. However, this is not a feasible solution to (6). Nevertheless, one can construct a feasible solution to (6) having the same cost by setting $x_{\mathcal{A}_1} = x_{\mathcal{A}_2} = 1$ and all other $x_{\mathcal{A}}$ to be zeros.

In general, for a feasible solution $\mathcal{S}'$ to SCPMC, one can construct a feasible solution to (6) as follows: for each element e which is fully covered by $\mathcal{S}'$, let $y_e = 1$ and let $x_{\mathcal{A}_e} = 1$ for exactly one r_e-cover $\mathcal{A}_e$ which contains r_e subsets of $\mathcal{S}'$ (such $\mathcal{A}_e$ exists since e is fully covered by $\mathcal{S}'$); all other variables are set to be zeros. Such a construction clearly results in a feasible solution to (6) whose objective value is at most $\rho_0(\mathcal{S}')$ (by the monotonicity of ρ_0). So, (6) is indeed a characterization of the SCPMC problem.

We can prove the following property of ρ.

Lemma 3. *If cost function ρ_0 is nonnegative, monotone non-decreasing, and submodular, then the function ρ defined in (5) is also nonnegative, monotone non-decreasing, and submodular.*

Remark 1. If ρ_0 is nonnegative and submodular but is not monotone non-decreasing, then ρ is not necessarily submodular.

Let $\hat{\rho}$ be the Lovász extension of ρ. By Theorem 1, $\hat{\rho}$ is convex. Relaxing (6), we have the following convex program:

$$\min \hat{\rho}(x)$$

$$s.t. \sum_{e:\ e \in E} p_e y_e \geq qP,$$

$$\sum_{A:\ A \in \Omega_e} x_A \geq y_e, \text{ for any } e \in E \tag{7}$$

$$x_A \geq 0 \text{ for } A \in \Omega$$

$$1 \geq y_e \geq 0 \text{ for } e \in E$$

It can be shown that this convex program is polynomial-time solvable. In fact, using the fact that the Lovász extension of a submodular function coincides with its convex closure, we can rewrite (7) as a linear program with exponential number of variables. Writing out its dual program and constructing a separation oracle (using an efficient algorithm for submodular function minimization), the linear program can be solved in polynomial-time.

Since (7) is a relaxation of (6), we have $opt_{cp} \leq opt$, where opt_{cp} is the optimal (fractional) value of (7) and opt is the optimal (integer) value of (6) (which is also the optimal value of SCPMC).

3.2 Rounding Algorithm

For a sub-collection $S' \subseteq S$, denote by $C(S')$ the set of elements fully covered by S'. The algorithm is presented in Algorithm 1. Two parameters s, t are needed which are chosen in Theorem 7 to guarantee the desired ratio with high probability. The rounding algorithm consists of two phases. In the first phase, a deterministic rounding is executed to form a sub-collection S_1. In the second phase, a randomized rounding is executed to form a sub-collection S_2. The output is the union of S_1 and S_2.

3.3 Approximation Analysis

The following three lemmas show the performance of collections S_1 and S_2.

Lemma 4. *For the collection of sets S_1 computed by Algorithm 1, $\rho_0(S_1) \leq bs \cdot opt_{cp}$. Furthermore, all elements with $y_e^* \geq \frac{1}{s}$ are fully covered by S_1.*

Lemma 5. *For the collection of sets S_2 computed by Algorithm 1, the expected cost of S_2 satisfies $\mathbb{E}[\rho_0(S_2)] \leq bs \ln(\frac{s}{s-t}) opt_{cp}$.*

Algorithm 1. Algorithm for SCPMC

Input: An SCPMC instance $(E, \mathcal{S}, r, p, q, \rho_0)$, two parameters s, t satisfying $1 < t < s \le 1/q$, and a real positive number $\varepsilon < 1$.
Output: A sub-collection $\mathcal{S}'$ which has total covering profit at least $(1 - \varepsilon)qP$.
1: Find an optimal solution (x^*, y^*) to (7).
2: $\mathcal{S}_1 \leftarrow \emptyset, \mathcal{S}_2 \leftarrow \emptyset$.
3: **for** all e with $y_e^* \ge \frac{1}{s}$ **do**
4:　　For each $\mathcal{A} \in \Omega_e$ with $x_{\mathcal{A}}^* \ge \frac{1}{bs}$, let $\hat{x}_{\mathcal{A}} \leftarrow 1$.
5: **end for**
6: For all $x_{\mathcal{A}}^*$ which is not rounded up to 1, set $\hat{x}_{\mathcal{A}} \leftarrow 0$.
7: $\mathcal{S}_1 \leftarrow \{S : S \in \mathcal{A} \text{ with } \hat{x}_{\mathcal{A}} = 1\}$.
8: If $\mathcal{S}_1$ has total covering profit at least $(1 - \varepsilon)qP$ then output $\mathcal{S}' \leftarrow \mathcal{S}_1$ and stop.
9: $E' \leftarrow E - \mathcal{C}(\mathcal{S}_1), q' \leftarrow (qP - p(\mathcal{C}(\mathcal{S}_1)))/P$.
10: **for** $i = 1$ to $bs \ln(\frac{s}{s-t})$ **do**
11:　　Pick $\theta \in [0, 1]$ randomly uniformly.
12:　　For each remaining $\mathcal{A}$ with $x_{\mathcal{A}}^* \ge \theta$, set $\hat{x}_{\mathcal{A}} \leftarrow 1$ and $\mathcal{S}_2 \leftarrow \mathcal{S}_2 \cup \{S : S \in \mathcal{A}\}$.
13: **end for**
14: Output $\mathcal{S}' = \mathcal{S}_2 \cup \mathcal{S}_2$.

Lemma 6. *For the collection of sets $\mathcal{S}_2$ computed by Algorithm 1, the expected profit of $\mathcal{S}_2$ satisfies $E[p(\mathcal{S}_2)] \ge tq'P$, where q' is the ratio defined in line 9 of Algorithm 1.*

It should be noticed that if we only care about an expected result, then we may obtain a randomized algorithm producing a sub-collection $\mathcal{S}'$ with $\mathbb{E}[\rho_0(\mathcal{S}')] \le bs(1 + \ln \frac{s}{s-t})opt$ and $\mathbb{E}[p(\mathcal{S}')] \ge qP$. This can be achieved by modifying $(1 - \varepsilon)qP$ in Line 8 of Algorithm 1 into qP. However, to obtain a randomized algorithm with guaranteed performance with high probability is more complicated. We can show that by choosing suitable parameters s and t, Algorithm 1 produces a bicriteria solution with high probability.

Theorem 7. *Setting $s = 1/q$ and $t = 1/\sqrt{q}$, then with a high probability, Algorithm 1 produces a $(O(\frac{b}{q\varepsilon}), 1 - \varepsilon)$-bicriteria solution, where $b = \max_e \left(\frac{f}{r_e}\right)$.*

3.4　Approximation Algorithm for SCPSC

As a corollary of Theorem 7, the minimum submodular cost partial set cover problem (SCPSC for short, in which the covering requirement for each element is one) admits a bicriteria randomized $(O(\frac{f}{q\varepsilon}), 1 - \varepsilon)$-bicriteria approximation, where $f = \max\{f_e : e \in E\}$. In the following, we show that an adaptation of our method can yield the same approximation for SCPSC even if the submodular function ρ_0 is non-monotone. The idea behind the adaptation is that in this case, a natural constraint is sufficient (we do not need to use the more complicated r_e-covers), and thus a technique similar to that in [14] dealing with non-monotone submodular function can be used.

Theorem 8. *For any nonnegative submodular function (not necessarily mono-tone non-decreasing), the SCPSC problem has a randomized $(O(\frac{f}{q\varepsilon}), 1 - \varepsilon)$-bicriteria algorithm with high probability.*

4   Conclusion

This paper gives a bicriteria approximation algorithm for the minimum submodular cost partial multi-cover problem (SCPMC), which is based on a novel convex program and turns out to be an $(O(\frac{b}{q\varepsilon}), 1 - \varepsilon)$-bicriteria approximation, where $b = \max_e \left(\frac{f_e}{r_e}\right)$. Notice that for the M$k$U problem, $b = 1$. Hence our algorithm gives an $O(\frac{1}{q\varepsilon}, 1 - \varepsilon)$-bicriteria approximation for MkU.

It should be noticed that the algorithm can be derandomized, using an observation that the number of distinct sub-collections for $\mathcal{S}_2$ is at most $|\Omega|$, which is polynomial under the assumption that $r_{\max}$ is a constant. However, a tricky problem is that good expectations of two properties (namely feasibility and guraranteed performance ratio) do not ensure the existence of a sub-collection satisfying both good properties. Currently, we only obtained a bicriteria algorithm for SCPMC. Is it possible to find an algorithm without violation?

Furthermore, our algorithm depends on the assumption that $r_{\max}$ is upper bounded by a constant. How to deal with the problem without such an assumption? These are problems deserving to be further explored.

References

1. Bar-Yehuda, R.: Using homogeneous weights for approximating the partial cover problem. J. Algorithms **39**(2), 137–144 (2001)
2. Berman, P., DasGupta, B., Sontag, E.: Randomized approximation algorithms for set multicover problems with applications to reverse engineering of protein and gene networks. Discret. Appl. Math. **155**(6–7), 733–749 (2007)
3. Chekuri, C., Ene, A.: Submodular cost allocation problem and applications. In: Aceto, L., Henzinger, M., Sgall, J. (eds.) ICALP 2011. LNCS, vol. 6755, pp. 354–366. Springer, Heidelberg (2011). https://doi.org/10.1007/978-3-642-22006-7_30
4. Chekuri, C., Quanrud, K.: On approximating (sparse) covering integer programs. ArXiv:1807.11538 [cs.DS]
5. Dinur, I., Steurer, D.: Analytical approach to parallel repetition. In: STOC 2014, pp. 624–633. ACM, New York (2014)
6. Chen, A., Harris, D.G., Srinivasan, A.: Partial resampling to approximate covering integer programs. In: Proceedings of 27th ACM-SIAM SODA, pp. 1984–2003 (2016)
7. Chen, N.: On the approximability of influence in social networks. SIAM J. Discret. Math. **23**(3), 1400–1415 (2009)
8. Chlamtáč, E., Dinitz, M., Makarychev, Y.: Minimizing the union: tight approximations for small set bipartite vertex expansion. In: SODA 2017, pp. 881–899. SIAM, Philadelphia (2017)
9. Chvatal, V.: A greedy heuristic for the set covering problem. Math. Oper. Res. **4**(3), 233–235 (1979)

10. Dobson, G.: Worst-case analysis of greedy heuristics for integer program with non-negative data. Math. Oper. Res. **7**(4), 515–531 (1982)
11. Dinitz, M., Gupta, A.: Packing interdiction and partial covering problems. In: Goemans, M., Correa, J. (eds.) Integer Programming and Combinatorial Optimization IPCO 2013. LNCS, vol. 7801. Springer, Heidelberg (2013). https://doi.org/10.1007/978-3-642-36694-9_14
12. Gandhi, R., Khuller, S., Srinivasan, A.: Approximation algorithms for partial covering problems. J. Algorithms **53**(1), 55–84 (2004)
13. Hochbaum, D.S.: Approximation algorithms for the set covering and vertex cover problems. SIAM J. Comput. **11**(3), 555–556 (1982)
14. Iwata, S., Nagano, K.: Submodular function minimization under covering constraints. In: FOCS 2009, pp. 671–680. IEEE Computer Society (2009)
15. Iyer, R.K., Bilmes, J.A.: Submodular optimization with submodular cover and submodular knapsack constraints. Adv. Neural Inf. Process. Syst. **26**, 2436–2444 (2013)
16. Iyer, R.K., Jegelka, S., Bilmes, J.A.: Monotone closure of relaxed constraints in submodular optimization: connections between minimization and maximization. In: Proceedings of the 30th Conference on Uncertainty in Artificial Intelligence, pp. 360–369 (2014)
17. Johnson, D.: Approximation algorithms for combinatorial problems. J. Comput. Syst. Sci. **9**(3), 256–278 (1974)
18. Kamiyama, N.: A note on submodular function minimization with covering type linear constraints. Algorithmica **80**(10), 2957–2971 (2018)
19. Kamiyama, N.: Submodular function minimization under a submodular set covering constraint. In: Ogihara, M., Tarui, J. (eds.) TAMC 2011. LNCS, vol. 6648. Springer, Heidelberg (2011). https://doi.org/10.1007/978-3-642-20877-5_14
20. Kearns, M.: The Computational Complexity of Machine Learning. MIT Press, Cambridge (1990)
21. Khot, S., Regev, O.: Vertex cover might be hard to approximate to within $2 - \varepsilon$. J. Comput. Syst. Sci. **74**(3), 335–349 (2008)
22. Konemann, J., Parekh, O., Segev, D.: A uinifed approach to approximating partial covering problems. Algorithmica **59**(4), 489–509 (2011)
23. Kolliopoulos, S.G., Young, N.E.: Approximation algorithms for covering/packing integer programs. J. Comput. Syst. Sci. **71**(4), 495–505 (2005). Preliminary version in FOCS (2001)
24. Koufogiannakis, C., Young, N.: Greedy Δ-approximation algorithm for covering with arbitrary constraints and submodular cost. Algorithmica **66**(1), 113–152 (2013)
25. Lovász, L.: Submodular functions and convexity. In: Bachem, A., Korte, B., Grötschel, M. (eds.) Mathematical Programming the State of the Art. Springer, Heidelberg (1983). https://doi.org/10.1007/978-3-642-68874-4_10
26. Lovász, L.: On the ratio of the optimal integral and fractional covers. Discret. Math. **13**(4), 383–390 (1975)
27. Mitzenmacher, M., Upfal, E.: Probability and Computing: Randomization and Probabilistic Techniques in Algorithms and Data Analysis. The Press Syndicate of the University of Cambridge, Cambridge (2005)
28. Rajagopalan, S., Vazirani, V.: Primal-dual RNC approximation algorithms for set cover and covering integer programs. SIAM J. Comput. **28**(2), 525–540 (1998)
29. Ran, Y., Zhang, Z., Du, H., Zhu, Y.: Approximation algorithm for partial positive influence problem in social network. J. Comb. Optim. **33**(2), 791–802 (2017)

30. Ran, Y., Shi, Y., Zhang, Z.: Local ratio method on partial set multi-cover. J. Comb. Optim. **34**(1), 302–313 (2017)
31. Slavík, P.: Improved performance of the greedy algorithm for partial cover. Inf. Process. Lett. **64**(5), 251–254 (1997)
32. Srinivasan, A.: An extension of the Lovsz local lemma, and its applications to integer programming. SIAM J. Comput. **36**(3), 609–634 (2006)
33. Shi, Y., Zhang, Z., Du, D.-Z.: Randomized bicriteria approximation algorithm for minimum submodular cost partial multi-cover problem. https://arxiv.org/abs/1701.05339
34. Shor, N.Z.: Cut-off method with space extension in convex programming problems. Cybern. Syst. Anal. **13**(1), 94–96 (1977)
35. Wang, F., et al.: On positive influence dominating sets in social networks. Theor. Comput. Sci. **412**, 265–269 (2011)
36. Wolsey, L.A.: An analysis of the greedy algorithm for the submodular set covering problem. Combinatorica **2**(4), 385–393 (1982)

A Novel Approach to Verifying Context Free Properties of Programs

Nan Zhang[1], Zhenhua Duan[1(✉)], Cong Tian[1], and Hongwei Du[2]

[1] Institute of Computing Theory and Technology, and ISN Laboratory,
Xidian University, Xi'an 710071, China
zhhduan@mail.xidian.edu.cn
[2] Department of Computer Science and Technology, Harbin Institute of Technology
Shenzhen Graduate School, Shenzhen 518055, China

Abstract. This paper proposes an approach to verifying programs against context free properties. To this end, the system to be verified is modeled by a program m in Modeling, Simulation and Verification Language (MSVL), and the desired property is also specified by an MSVL program m'. Then program m and formula $\neg m'$ are interpreted by means of executing programs m and m'. If an acceptable execution path is generated, a counterexample is found, otherwise the property is valid. To show how the proposed approach works, an example is given.

Keywords: Runtime verification · Model checking
Temporal logic · Automata

1 Introduction

To improve the correctness and reliability of software, model checking has been widely used in both academia and industry. With the traditional automata-theoretic model checking [5], a system S is modeled by an automaton A_1 and a desired property is specified by a temporal logic formula P; then $\neg P$ is converted to an automaton A_2. Further, the product automaton A of A_1 and A_2 is produced. If the language accepted by A is an empty set, the system S satisfying the property is valid, otherwise a counterexample is found. With this approach, LTL [10] and CTL [5] are widely used temporal logics for specifying desired properties. However, both LTL and CTL are not full regular, so some full regular properties such as interval related properties and periodically repeated properties cannot be verified. To solve this problem, some full regular temporal logics such as Propositional Projection Temporal Logic (PPTL) [7], Property Specification Language (PSL) [1] and μ-Calculus [11] are used to specify desired properties. As we can see, all the temporal logics we used cannot describe context free properties. However, these properties are required to be verified in some

The research is supported by National Natural Science Foundation of China under Grant Nos. 61420106004, 61572386, 61732013 and 61751207.

S. Tang et al. (Eds.): AAIM 2018, LNCS 11343, pp. 74–87, 2018.
https://doi.org/10.1007/978-3-030-04618-7_7

circumstances. Accordingly, in this paper, we are motivated to employ Modeling, Simulation and Verification Language (MSVL) [15] to specify desired properties as well as the systems to be verified since MSVL is capable of expressing context free properties.

To do so, a system to be verified is modeled by an MSVL program m, and a desired property is also specified by an MSVL program m'. Further, program m and formula $\neg m'$ are interpreted by means of executing programs m and m'. Actually, if m and m' can be reduced to their normal forms $m_e \wedge \varepsilon \vee \bigvee_i m_{ci} \wedge \bigcirc m_{fi}$ and $m'_e \wedge \varepsilon \vee \bigvee_j m'_{cj} \wedge \bigcirc m'_{fj}$, respectively, then according to the evaluation of m_e, m_{ci}, m'_e and m'_{cj} an execution state s of $m \wedge \neg m'$ can be generated since we assume each variable in m and m' is well evaluated at the current state. This process can repeatedly be conducted so as to produce at least one state sequence $\langle s_0, s_1, \ldots \rangle$ with true or terminate with false. For the latter case, m satisfying m' is valid while for the former case, the state sequence is a counterexample. In this way, a context free property can be verified by executing MSVL programs m and m' at code level.

The proposed approach is of three advantages: (1) it allows us to specify desired context free properties in MSVL so as to verify a program w.r.t. the property; (2) since both a property m' and a system model m are expressed in the same language MSVL, a new verification approach based on the Unified Model Checking [8] can be employed; (3) the verification proceeds by executing programs m and property m' at code level in a synchronized and more efficient way.

The contribution of the paper is as follows: (1) a novel approach to verifying programs against context free properties is proposed; (2) an example is given to illustrate how the proposed approach works; (3) a tool has been developed to support the proposed approach.

The paper is organized as follows: in the next section, MSVL and traceable automata are briefly introduced. In Sect. 3, the proposed approach to verifying programs against context free properties is elaborated in detail. In Sect. 4, an example is given to illustrate the verification process. Finally, conclusions are drawn in Sect. 5.

2 Preliminaries

2.1 Modeling, Simulation and Verification Language

Modeling, Simulation and Verification Language (MSVL) is an executable subset of Projection Temporal Logic (PTL) [7]. The following is a snapshot of the simple kernel of MSVL. For more details, please refer to [7]. With MSVL, expressions can be treated as terms and statements can be treated as formulas in PTL. The arithmetic and boolean expressions of MSVL can be inductively defined as follows:

$$e ::= n \mid x \mid \bigcirc e \mid \ominus e \mid f(e_1, \ldots, e_n)$$
$$b ::= \text{true} \mid \neg b \mid b_0 \wedge b_1 \mid e_0 = e_1 \mid e_0 < e_1$$

where $n \in \mathbb{R}$, set of real numbers, and $x \in \mathbb{V}$, set of variables. The $f()$ is a state function. The usual arithmetic operations such as $+$, $-$, $*$ and $\%$ can be viewed as two-arity functions. One may refer to the value of a variable at the previous state or the next one.

Some key statements of MSVL are inductively defined in Table 1. MSVL supports structured programming and covers some basic control flow statements such as sequential statement, conditional statement, while-loop statement and so on. Further, MSVL also supports non-determinism and concurrent programming since it includes selection, conjunction and parallel statements. Moreover, a framing technique is introduced to improve the efficiency of programs and synchronize communication for parallel processes. In addition, MSVL has been extended in a variety of ways recently. For instance, multi-types including integer, float, string, char, pointer, array and struct, have been recently formalized and implemented. Hence, typed variables and typed functions over the extended data domain can be defined [13]. To interpret MSVL programs, the normal form of MSVL programs has been defined in [7] and a compiler for MSVL has been developed [14].

Table 1. MSVL statements

1. Termination: empty $\stackrel{\text{def}}{=} \varepsilon$

2. Assignment: $x := e \stackrel{\text{def}}{=} \bigcirc x = e \wedge \mathsf{len}(1)$

3. Positive Immediate Assignment: $x <== e \stackrel{\text{def}}{=} x = e \wedge p_x$

4. State Frame: $\mathsf{lbf}(x) \stackrel{\text{def}}{=} \neg \mathsf{af}(x) \to \exists b : (\ominus x = b \wedge x = b)$

5. Interval Frame: $\mathsf{frame}(x) \stackrel{\text{def}}{=} \square(\mathsf{more} \to \bigcirc \mathsf{lbf}(x))$

6. Next: $\mathsf{next}\ \phi \stackrel{\text{def}}{=} \bigcirc \phi$

7. Always: $\mathsf{always}\ \phi \stackrel{\text{def}}{=} \square \phi$

8. Conditional: $\mathsf{if}\ b\ \mathsf{then}\ \phi_0\ \mathsf{else}\ \phi_1 \stackrel{\text{def}}{=} (b \to \phi_0) \wedge (\neg b \to \phi_1)$

9. Existential Quantification: $\mathsf{exist}\ x : \phi(x) \stackrel{\text{def}}{=} \exists x : \phi(x)$

10. Sequential: $\phi_0\ ;\ \phi_1 \stackrel{\text{def}}{=} \phi_0 ; \phi_1$

11. Conjunction: $\phi_0\ \mathsf{and}\ \phi_1 \stackrel{\text{def}}{=} \phi_0 \wedge \phi_1$

12. While: $\mathsf{while}\ b\ \{\ \phi\ \} \stackrel{\text{def}}{=} (b \wedge \phi)^* \wedge \square(\varepsilon \to \neg b)$

13. Selection: $\phi_0\ \mathsf{or}\ \phi_1 \stackrel{\text{def}}{=} \phi_0 \vee \phi_1$

14. Parallel: $\phi_0 \parallel \phi_1 \stackrel{\text{def}}{=} \phi_0 \wedge (\phi_1; \mathsf{true}) \vee (\phi_0; \mathsf{true}) \wedge \phi_1$

15. Projection: $(\phi_1, \ldots, \phi_m)\ \mathsf{prj}\ \phi \stackrel{\text{def}}{=} (\phi_1, \ldots, \phi_m)\ prj\ \phi$

16. Synchronous Communication: $\mathsf{await}(c) \stackrel{\text{def}}{=} \mathsf{frame}(x_1, x_2, \ldots, x_n) \wedge \square(\varepsilon \leftrightarrow c)$

Definition 1 (Normal Form of MSVL Programs). *An MSVL program φ is in normal form if*

$$\varphi \stackrel{\text{def}}{=} \left(\bigvee_{i=1}^{k} \varphi_{ei} \wedge \text{empty}\right) \vee \left(\bigvee_{j=1}^{h} \varphi_{cj} \wedge \bigcirc \varphi_{fj}\right)$$

where $k + h \geq 1$ and the following hold:

- *φ_{fj} is an interval program, that is, one in which variables may refer to the previous states but not beyond the first state of the current interval over which the program is executed.*
- *each φ_{cj} and φ_{ei} is either* true *or a state formula of the form $p_1 \wedge \ldots \wedge p_m$ $(m \geq 1)$ such that each p_l $(1 \leq l \leq m)$ is either $(x = e)$ with $e \in \mathbb{D}$, $x \in \mathbb{V}$, or p_x, or $\neg p_x$.*

2.2 Deterministic Traceable Automata

Deterministic Traceable Automata (DTA) [9] extends Deterministic Finite Automata (DFA) with an unbounded state stack used to store partial history. A DTA determines its next state transition based on its current state, current input symbol and current symbol at the top of the stack. It has been proved that the expressiveness of DTA falls in between DFA and Deterministic Pushdown Automata (DPA). For more details, please refer to [9].

Definition 2 (Formal Definition of Deterministic Traceable Automaton). A Deterministic Traceable Automaton A is a 5-tuple, $(Q, \Sigma, \delta, q_0, F)$, consisting of a non-empty finite set of states $Q = \{q_0, q_1, \ldots, q_n\}$, a finite set of input symbols called the alphabet Σ, a transition function $\delta : Q \times \Sigma \rightarrow Q \cup \{trace\}$, an initial or start state $q_0 \in Q$ and a non-empty set of final states $F \subseteq Q$.

Similar to non-deterministic finite automata, if transition function δ is modified as: $Q \times \Sigma \rightarrow 2^{Q \cup \{trace\}}$, such kind of traceable automata is called Non-deterministic Traceable Automata (NTA).

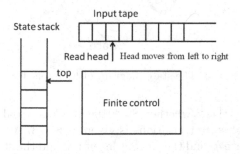

Fig. 1. A diagram of a traceable automaton

A diagram of a traceable automaton is given in Fig. 1. A traceable automaton consists of an input tape, a state stack and a finite control. It reads a given input string from left to right. Initially, the read head points to the leftmost symbol in the input tape, the state stack is empty and the finite control is in initial state q_0. In each step, the finite control chooses an action s.t. current state q and input symbol a:

1. If $\delta(q,a) = q'$, push q into the stack, change the current state of the finite control to state q', and the read head moves to the right cell.
2. If $\delta(q,a) = trace$ and the stack is not empty, change the current state of the finite control to the state p on the top of the stack, pop the stack, and the head moves to the right cell. If $\delta(q,a) = trace$ and the stack is empty, halt and reject the input string.
3. If a is a blank symbol, halt and accept the input string if current state q is a final state in F, otherwise the input string is rejected.
4. If $\delta(q,a)$ is not defined, halt and reject the input string.

The language $L(A)$ accepted by DTA A is the set of all the string accepted by A. A DTA can also be displayed intuitively by a state diagram. Nodes in a state diagram denote states or trace in the DTA. If $\delta(q,a) = p$, there is an edge labeled by the symbol a from node q to node p; if $\delta(q,a) = trace$, there is an edge labeled by the symbol a from node q to node $trace$. There may be several trace nodes in a state diagram, indicating that the related next state needs to be determined by the top state of the stack.

Example 1. *A deterministic traceable automaton* $A = (Q, \Sigma, \delta, q_0, F)$, *where* $Q = \{q_0, q_1\}$, $\Sigma = \{0, 1\}$, $F = \{q_0\}$ *and* δ *is defined as follows:* $\delta(q_0, 0) = q_1, \delta(q_1, 0) = q_1$ *and* $\delta(q_1, 1) = trace$.

The state diagram is given in Fig. 2. This DTA recognizes all the strings consisting of balanced parentheses, where 0 denotes left parenthesis while 1 denotes right parenthesis. Such language can be generated by the following Context Free Grammar G[B]: B $\rightarrow$ 0B1 | BB | ϵ.

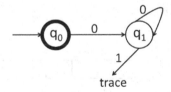

trace

Fig. 2. The state diagram of Example 1

Now we use A to check whether S="01001011" is a balanced parentheses string. The process of state transitions is given in Fig. 3. At the beginning, the DTA is in its start state q_0 and the current input is 0 and the stack is empty. Since $\delta(q_0, 0) = q_1$, the finite control pushes q_0 into the stack and the current state is changed to q_1. After that, the current symbol comes 1, since $\delta(q_1, 1) = trace$, the current state is changed to the top state q_0 of the stack, and q_0 is popped. When the last symbol has been scanned, the DTA is in state q_0 which is a final state, so string S is accepted.

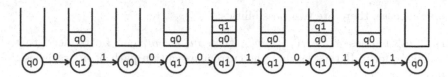

Fig. 3. The process of state transition of "01001011"

3   A Novel Approach to Verifying Context Free Properties

In this section, we propose a new approach to verifying context free properties. Given a system S and a desired property P, we need check whether S satisfying P is valid, denoted by $S \models P$. If we write an MSVL program m to implement system S and another MSVL program m' to characterize property P, the satisfaction problem is converted to the problem of checking whether there is a common model for program m and formula $\neg m'$. In order to execute program m, the input values of variables in m are required. Therefore, dynamic symbolic execution tool Cloud9 [3] and constraint solver Z3 [6] as well as static analysis tools based on CEGAR [4] and CPAChecker [2] are employed to generate these values.

To illustrate the proposed approach, some notations are needed. Let $Var(m)$ and $Var(m')$ denote the sets of all variables in m and m', respectively. An acceptable execution path (model) of an MSVL program is a non-empty finite or infinite sequence $\langle s_0, s_1, \ldots \rangle$ of states. Each state s_i is a map, $s_i : Var(m) \cup Var(m') \to \mathbb{D}$, where $\mathbb{D}$ is the data domain including integers, lists, sets etc. The following is a skeleton of the checking algorithm.

Step 1: System S is implemented by an MSVL program m_0, or translated from a C program into m_0 using tool C2M in MSV toolkit [12].

Step 2: The desired property P is also specified by an MSVL program m'_0.

Step 3: Transform m_i and m'_i into their normal forms as follows:

$$m_i = m_e^i \wedge \varepsilon \vee \bigvee_{j=1}^{h} m_{cj}^i \wedge \bigcirc m_{fj}^i$$

$$m'_i = m_e'^i \wedge \varepsilon \vee \bigvee_{j=1}^{l} m_{cj}'^i \wedge \bigcirc m_{fj}'^i$$

In the normal form, each disjunction represents a possible branch of program execution. We choose one disjunction to generate current state s_i and program m_{i+1} to be executed at the next state in the execution path. We make the following assumption: truth values of all state formulas, $m_e'^i$, $m_{cj}'^i$, m_e^i and m_{cj}^i, in the normal forms can well be evaluated.

To reduce programs m_i and m'_i, the following cases are taken into account: Case 1: If m_e^i is true, then a new state s_i is generated and the branch $m_e^i \wedge \varepsilon$ is chosen. It means that *system program m_0 terminates at current state s_i, which is defined as follows:*

$$s_i : Var(m) \to \mathbb{D} \text{ and } s_i(v) = d \text{ iff } m_e^i \to (v = d) \text{ for each } v \in Var(m).$$

In this case, there are two possibilities:

(1.1) If m_e^i implies $m_e'^i \wedge \bigwedge_{j=1}^l \neg m_{cj}'^i$, then $m_i \wedge \neg m_i'$ can be reduced to the following form. It tells us that the current execution path of m_0 satisfies the desired property.

$$m_i \wedge \neg m_i' \Rightarrow \varepsilon \wedge \neg \varepsilon \equiv false$$

(1.2) If m_e^i implies $\neg m_e'^i \wedge m_{cj_0}'^i \wedge \bigwedge_{j \neq j_0} \neg m_{cj}'^i$, then $m_i \wedge \neg m_i'$ can be reduced to "empty". It shows that the current execution path of m_0 doesn't satisfy the desired property and a counterexample is found.

$$m_i \wedge \neg m_i' \Rightarrow \varepsilon \wedge \neg \bigcirc m_{fj_0}'^i \equiv \varepsilon$$

Case 2: If m_{cj}^i is true, then a new state s_i is generated and branch $m_{cj}^i \wedge \bigcirc m_{fj}^i$ is chosen. It means that s_i *is not the last state of program* m_i, *in other word, system program* m_0 *doesn't terminate at state* s_i defined as follows:

$$s_i : Var(m) \to \mathbb{D} \text{ and } s_i(v) = d \text{ iff } m_{cj}^i \to (v = d) \text{ for each } v \in Var(m).$$

In this case, there are also two possibilities:

(2.1) If $m_{cj}^i \wedge m_e'^i \wedge \bigwedge_{j=1}^l \neg m_{cj}'^i$ is consistent, state s_i is extended as follows:

$$s_i : Var(m) \cup Var(m') \to \mathbb{D} \text{ and}$$
$$s_i(v) = d \text{ iff } m_{cj}^i \to (v = d) \text{ for each } v \in Var(m) \text{ and}$$
$$s_i(v') = d \text{ iff } m_e'^i \to (v' = d) \text{ for each } v' \in Var(m') \setminus Var(m).$$

Thus, $m_i \wedge \neg m_i'$ is reduced to the following form. Whether m_0 satisfies the desired property depends on the future execution after current state s_i.

$$m_i \wedge \neg m_i' \Rightarrow \bigcirc m_{fj}^i \wedge \neg \varepsilon \equiv \bigcirc m_{fj}^i \wedge \bigcirc \neg false \equiv \bigcirc (m_{fj}^i \wedge \neg false)$$
$$\equiv \bigcirc (m_{i+1} \wedge \neg m_{i+1}')$$

Let m_{i+1} be m_{fj}^i, m_{i+1}' be $false$ and $i := i + 1$, then go to Step 3.

(2.2) If $m_{cj}^i \wedge \neg m_e'^i \wedge m_{cj_0}'^i \wedge \bigwedge_{j \neq j_0} \neg m_{cj}'^i$ is consistent, state s_i is extended as follows:

$$s_i : Var(m) \cup Var(m') \to \mathbb{D} \text{ and}$$
$$s_i(v) = d \text{ iff } m_{cj}^i \to (v = d) \text{ for each } v \in Var(m) \text{ and}$$
$$s_i(v') = d \text{ iff } m_{cj_0}'^i \to (v' = d) \text{ for each } v' \in Var(m') \setminus Var(m).$$

Thus, $m_i \wedge \neg m_i'$ is reduced to the following form. Whether m_0 satisfies the desired property is up to the future execution after current state s_i.

$$m_i \wedge \neg m_i' \Rightarrow \bigcirc m_{fj}^i \wedge \neg \bigcirc m_{fj_0}'^i \equiv \bigcirc m_{fj}^i \wedge \bigcirc \neg m_{fj_0}'^i \equiv \bigcirc (m_{fj}^i \wedge \neg m_{fj_0}'^i)$$
$$\equiv \bigcirc (m_{i+1} \wedge \neg m_{i+1}')$$

Let m_{i+1} be m_{fj}^i, m_{i+1}' be $m_{fj_0}'^i$ and $i := i + 1$, then go to Step 3.

The runtime verification approach given above allows us to generate a sequence of states by applying normal form transformation on MSVL programs repeatedly. If an acceptable execution path can be found, it is actually a model common to m and $\neg m'$. Then we can conclude that the system doesn't satisfy the property. In this case, the common model is actually a counterexample. If for all possible inputs, m and $\neg m'$ have no any common model, the property is valid.

4 Verification

In this section, to illustrate how to use the new method to verify whether a program satisfies a desired property, Example 1 is presented to check whether a string consists of balanced parentheses as shown in Sect. 2.2.

System Implementation. A program M checking whether a given string consists of balanced parentheses ('(' denoted by 0 and ')' denoted by 1) is presented in this subsection. First, data structures and functions required by M are defined as follows:

As shown in Table 2, in the *struct* StackNode, two data fields are defined: (1) data: an item stored in the stack; (2) next: a pointer pointing to its previous node in the linked list. In the *struct* LinkStack, two data fields are defined: (1) top: a pointer pointing to the top element in the stack; (2) count: the current number of elements in the stack. Moreover, two operations over stack are defined: (1) push(LinkStack *s, char e): push the character e into stack s; (2) pop(LinkStack *s, char RValue): remove the top element from stack s and store it in variable RValue.

In program M, some variables are also required: (1) str: used to store the input string; (2) c: the next character from the input string; (3) stk1: used to store states of a traceable automaton; (4) t0: denotes the number of character '0's which have been scanned; (5) t1: denotes the number of character '1's which have been scanned; (6) result1: used to indicate whether the input string is well balanced. If it is 1, the string is balanced; otherwise it is not balanced with result1=0 or uncertain with result1=2; (7) current_st: used to record the current state of the traceable automaton; (8) p: used to point to the next character in the input string; (9) g: used to jump out of the loop when g='n'.

Armed with the above data structures, operations and variables, program M is implemented as shown in Table 3. Initially, t0 and t1 are set to 0; result1 is assigned 2; stk1 points to a stack LinkStack with stk1.top being null and stk1.count being 0; p points to the starting address of array str; g is assigned 'y'; and current_st is set to 0 denoting starting state q0 of the traceable automaton. Program M reads a string from the input stream and stores it into char array str[]. Then it reads one character from str[] each time using pointer p. If the character is 0, the current state is pushed into stack stk1 and t0 is increased by 1; if the character is 1 and the stack is not empty, the top state is popped and t1 is increased by 1; whereas if the character is 1 and the stack is empty, result1 is assigned 0; if the character is neither 0 nor 1, result1 is assigned 0. If the stack is

Table 2. Data structures and operations

```
1   struct StackNode{
2       char data and
3       struct StackNode *next
4   };
5   struct LinkStack{
6       StackNode *top and
7       int count
8   };
9   function push(LinkStack *s, char e){
10      frame( p ) and (
11          StackNode *p <== (StackNode *)malloc(sizeof(struct StackNode)) and
12          p->data := e and
13          p->next := s->top and
14          s->top := p and
15          s->count := s->count + 1
16          )
17  };
18  function pop(LinkStack *s, char RValue){
19      frame( e, p ) and (
20          char e <== s->top->data and StackNode *p <== s->top and
21          s->top := s->top->next;
22          free(p) and
23          s->count := s->count - 1 and
24          RValue := e )
25  };
```

Table 3. Implementation

```
1   #include<string.h>
2   #include<stdio.h>
3   #include<stdlib.h>
4   frame( c, stk1, t0, t1, result1, current_st, str, p, g ) and (
5       char c and LinkStack stk1 and int t0 <== 0 and int t1 <== 0
6       and int result1 <== 2 and stk1.top <== NULL and stk1.count <== 0 and
7       char current_st <== '0' and char str[128] and char *p <== str and
8       char g <== 'y' and ext gets( str ) and skip;
9       //gets() receives the string until '\n' is reached
10      c := *p and p := p + 1;
11      while( ( c = '0' or c = '1' ) and g = 'y' ){
12          if( c = '0' )
13          then{
14                  t0 := t0 + 1 and
15                  push( &stk1, current_st ) and
16                  if ( current_st = '0' )
17                  then { current_st := '1' }
18                  else { skip } and
19                  c := *p and p := p + 1
20          }
21          else{
22                  t1 := t1 + 1;
23                  if ( stk1.top = NULL )
24                  then { g := 'n'; }
25                  else {
26                          current_st := pop( &stk1, RValue ) and
27                          c := *p and p := p + 1
28                  }
29          }
30      };
31      if( stk1.top = NULL and c = '\0' and g = 'y' )
32      then { result1 := 1 and output( "Yes\n" ) }
33      else { result1 := 0 and output( "No\n" ) }
34  )
```

empty when the whole input string has been scanned, result1 is assigned 1 and the string is accepted.

Property Characterization. In this subsection, a context free property is specified.

Property 1: *A strict embedded balanced parentheses string can be accepted by M.*

In fact, a strict embedded balanced parentheses language can be generated by a CFG G[B]: B → 0B1 | ε. Property 1 is described by M'' in MSVL as shown in Table 4. The *ebp-check()* function implemented in a recursive manner (see Table 4) checks whether a string represented by a char array str[*begin..end*] consists of strict embedded balanced parentheses. As we can see, during the execution of the function *ebp-check()*, whenever *begin>end* holds, RValue is assigned 1 indicating the input string is well balanced while *begin<end* and str[*begin*] = 0 and str[*end*] = 1 holds, *ebp-check*(str, *begin*+1, *end*-1, RValue) is called recursively. In other cases, RValue is assigned 0 to indicate an error occurs. The *ebp-check()* function is called in M''.

Table 4. Program M'' for Property 1

```
1   function ebp-check( char str[], int begin, int end, int RValue ){
2       if ( begin > end )
3       then { RValue := 1 }
4       else {
5               if ( begin = end )
6               then { RValue := 0 }
7               else {
8                       if ( str[begin] = '0' and str[end] = '1' )
9                       then { RValue := ebp-check( str, begin+1, end-1, RValue )}
10                      else { RValue := 0 }
11              }
12      }
13  };
14  alw(
15          if ( result1 = 0 )
16          then {
17                  if ( ebp-check( str, 0, strlen(str)-1, RValue ) = 1 )
18                  then { false }
19                  else { true }
20          }
21          else { true }
22  )
```

Program M can be viewed as the following form: $M \overset{\text{def}}{=} F$ and $(I; W; R)$. F is defined as statement frame(c, stk1, t0, t1, result1, current_st, str, p, g); I is defined as the statement in M from line 5 to 10; W is defined as the while statement in M from line 11 to 30; R is defined as the statement in M from line 31 to 33.

Verification of Property 1. Let M'' denote the program describing Property 1. Suppose that program M is executed with input string "0011". Then the verification proceeds as follows.

[**step 1**] Transform M into its normal form.

$$M \stackrel{\text{def}}{=} M_0 \equiv M_{0c} \wedge \bigcirc M_{0f}$$
$$M_{0c} \equiv t_0 = 0 \wedge t_1 = 0 \wedge result1 = 2 \wedge stk1 = (< null >, 0) \wedge current\_st = '0'$$
$$\wedge p = \&str \wedge g = 'y'$$
$$M_{0f} \equiv F \wedge (t_0 = 0 \wedge t_1 = 0 \wedge result1 = 2 \wedge stk1 = (< null >, 0) \wedge current\_st = '0'$$
$$\wedge p = \&str \wedge g = 'y' \wedge str = \text{``0011''} \wedge c := *p \wedge p := p + 1; W; R)$$

[**step 2**] Transform M'' into its normal form.

$$M'' \stackrel{\text{def}}{=} M_0'' \equiv M_{0c}'' \wedge \odot M_{0f}''$$
$$M_{0c}'' \equiv (result1 \neq 0 \vee \text{ebp-check}(str, 0, strlen(str) - 1, RValue) \neq 1)$$
$$M_{0f}'' \equiv \square(result1 \neq 0 \vee \text{ebp-check}(str, 0, strlen(str) - 1, RValue) \neq 1)$$

[**step 3**] Choose one branch in the normal form of M to generate state s_0 and then reduce the normal form of M'' according to the assignments in the chosen branch. Since M has only one branch, from M_{0c} state s_0 is defined by a set of pairs as follows:

$$s_0 = \{(t_0, 0), (t_1, 0), (result1, 2), (stk1, (< null >, 0)), (current\_st, '0'), (p, \&str), (g, 'y')\}$$

Thus, program M_{0f} denoted by M_1 is executed at the next state.

$$M_1 \equiv F \wedge (t_0 = 0 \wedge t_1 = 0 \wedge result1 = 2 \wedge stk1 = (< null >, 0) \wedge current\_st = '0'$$
$$\wedge p = \&str \wedge g = 'y' \wedge str = \text{``0011''} \wedge c := *p \wedge p := p + 1; W; R)$$

With the assignments in state s_0, M_{0c}'' is reduced to $true$, while $\neg M''$ is reduced to $\neg \odot M_{0f}'' \equiv \bigcirc \neg M_{0f}''$. Further, M_{0f}'' denoted by M_1'' is executed at the next state.

As we can see, since M''(denoted by M_0'') is

$$\square(result1 \neq 0 \vee \text{ebp-check}(str, 0, strlen(str) - 1, RValue) \neq 1),$$

to reduce each M_i'' (defined as $M_{(i-1)f}''$) $(i = 1, \ldots, 7)$ into its normal form, we always obtain the same form:

$$M_i'' \equiv M_{ic}'' \wedge \odot M_{if}'' \tag{mi-1}$$
$$M_{ic}'' \equiv result1 \neq 0 \vee \text{ebp-check}(str, 0, strlen(str) - 1, RValue) \neq 1 \tag{mi-2}$$
$$M_{if}'' \equiv \square(result1 \neq 0 \vee \text{ebp-check}(str, 0, strlen(str) - 1, RValue) \neq 1) \tag{mi-3}$$

[**step 4**] Transform M_1 into its normal form, we obtain the following:

$$M_1 \stackrel{\text{def}}{=} M_{1c} \wedge \bigcirc M_{1f}$$
$$M_{1c} \equiv t_0 = 0 \wedge t_1 = 0 \wedge result1 = 2 \wedge stk1 = (< null >, 0) \wedge current\_st = '0'$$
$$\wedge p = \&str \wedge g = 'y' \wedge str = \text{``0011''}$$
$$M_{1f} \equiv F \wedge (t_0 = 0 \wedge t_1 = 0 \wedge result1 = 2 \wedge stk1 = (< null >, 0) \wedge current\_st = '0'$$
$$\wedge p = \&str[1] \wedge g = 'y' \wedge str = \text{``0011''} \wedge c = '0' \wedge W; R)$$

[**step 5**] M_1'' is reduced into its normal form as shown in formulas (mi-1, mi-2, mi-3).

[**step 6**] Choose one branch in the normal form of M_1 to generate state s_1 and then reduce the normal form of M_1'' according to the assignments in the chosen branch. Since M_1 has only one branch, from M_{1c} state s_1 is defined by a set of pairs as follows:

$$s_1 = \{(t_0, 0), (t_1, 0), (result1, 2), (stk1, (< null >, 0)), (current\_st, '0'), (p, \&str),$$
$$(g, 'y'), (str, ``0011")\}$$

Thus, program M_{1f} denoted by M_2 is executed at the next state.

$$M_2 \overset{\text{def}}{=} F \wedge (t_0 = 0 \wedge t_1 = 0 \wedge result1 = 2 \wedge stk1 = (< null >, 0) \wedge current\_st = '0'$$
$$\wedge p = \&str[1] \wedge g = 'y' \wedge str = ``0011" \wedge c = '0' \wedge W; R)$$

With the assignments in state s_1, M_{1c}'' is reduced to *true*, while $\neg M_1''$ is reduced to $\neg \odot M_{1f}'' \equiv \bigcirc \neg M_{1f}''$. Further, M_{1f}'' denoted by M_2'' is executed at the next state.

Due to the limited space, we do not list the reduction process step by step. In the same way as steps 1–3 and 4–6, we can generate states s_2, s_3, s_4, s_5 and s_6 given as follows by steps 7–9, 10–12, 13–15, 16–18, 19–21, respectively.

$$s_2 = \{(t_0, 0), (t_1, 0), (result1, 2), (stk1, (< null >, 0)), (current\_st, '0'), (p, \&str[1]),$$
$$(g, 'y'), (str, ``0011"), (c, '0')\}$$
$$s_3 = \{(t_0, 1), (t_1, 0), (result1, 2), (stk1, (< '0', null >, 1)), (current\_st, '1'), (p, \&str[2]),$$
$$(g, 'y'), (str, ``0011"), (c, '0')\}$$
$$s_4 = \{(t_0, 2), (t_1, 0), (result1, 2), (stk1, (< '1', '0', null >, 2)), (current\_st, '1'), (p, \&str[3]),$$
$$(g, 'y'), (str, ``0011"), (c, '1')\}$$
$$s_5 = \{(t_0, 2), (t_1, 1), (result1, 2), (stk1, (< '0', null >, 1)), (current\_st, '1'), (p, \&str[4]),$$
$$(g, 'y'), (str, ``0011"), (c, '1')\}$$
$$s_6 = \{(t_0, 2), (t_1, 2), (result1, 2), (stk1, (< null >, 0)), (current\_st, '0'), (p, \&str[5]),$$
$$(g, 'y'), (str, ``0011"), (c, ' \backslash 0')\}$$

Then program M_{6f} denoted by M_7 is executed at the next state.

$$M_7 \overset{\text{def}}{=} F \wedge (t_0 = 2 \wedge t_1 = 2 \wedge result1 = 1 \wedge stk1 = (< null >, 0) \wedge current\_st = '0'$$
$$\wedge p = \&str[5] \wedge g = 'y' \wedge str = ``0011" \wedge c =' \backslash 0' \wedge \varepsilon)$$

With the assignments in state s_6, M_{6c}'' is reduced to *true*, while $\neg M_6''$ is reduced to $\neg \odot M_{6f}'' \equiv \bigcirc \neg M_{6f}''$. Further, M_{6f}'' denoted by M_7'' is executed at the next state.

[**step 22**] M_7 is transformed into its normal form as follows.

$$M_7 \overset{\text{def}}{=} M_{7c} \wedge \varepsilon$$
$$M_{7c} \equiv t_0 = 2 \wedge t_1 = 2 \wedge result1 = 1 \wedge stk1 = (< null >, 0) \wedge current\_st = '0'$$
$$\wedge p = \&str[5] \wedge g = 'y' \wedge str = ``0011" \wedge c =' \backslash 0'$$

[step 23] M_7'' is reduced into its normal form as shown in formulas (mi-1, mi-2, mi-3).

[step 24] Choose one branch in the normal form of M_7 to generate state s_7 and then reduce the normal form of M_7'' according to the assignments in the chosen branch. Since M_7 has only one branch, from M_{7c}'' state s_7 is defined by a set of pairs as follows:

$$s_7 = \{(t_0, 2), (t_1, 2), (result1, 1), (stk1, (< null >, 0)), (current\_st,' 0'), (p, \&str[5]),$$
$$(g,' y'), (str, \text{``0011''}), (c,' \backslash 0')\}$$

The program terminates at state s_7. With the assignments in state s_7, M_{7c}'' is reduced to $true$, while $\neg M_7''$ is reduced to $\neg \odot M_{7f}'' \equiv \bigcirc \neg M_{7f}''$. Further, M_7 and $\neg M_7'' \equiv false$, which means that M satisfying Property 1 is valid.

5   Conclusion

This paper presents a novel approach to verifying programs against context free properties by means of executing programs at code level. It allows us not only to model a system but also to specify a temporal or context free property in the same language MSVL. However, the verification inputs of programs have to be generated by employing other methods including dynamic symbolic execution and static analysis. In the future, we will investigate how to generate a reasonable set of verification inputs by means of dynamic symbolic execution using Cloud9 [3] and constraint solver Z3 [6] as well as static analysis tools based on CEGAR [4] and CPAChecker [2]. In addition, we will further improve the verification tool so that more complex systems can be verified in a more efficient way.

References

1. IEC 62531:2012(e) (IEEE Std 1850–2010): Standard for property specification language (PSL). IEC 62531:2012(E) (IEEE Std 1850–2010), pp. 1–184, June 2012
2. Beyer, D., Keremoglu, M.E.: CPACHECKER: a tool for configurable software verification. In: Gopalakrishnan, G., Qadeer, S. (eds.) CAV 2011. LNCS, vol. 6806, pp. 184–190. Springer, Heidelberg (2011). https://doi.org/10.1007/978-3-642-22110-1_16
3. Ciortea, L., Zamfir, C., Bucur, S., Chipounov, V., Candea, G.: Cloud9: a software testing service. SIGOPS Oper. Syst. Rev. **43**(4), 5–10 (2010)
4. Clarke, E.M.: SAT-based counterexample guided abstraction refinement. In: Bošnački, D., Leue, S. (eds.) SPIN 2002. LNCS, vol. 2318, p. 1. Springer, Heidelberg (2002). https://doi.org/10.1007/3-540-46017-9_1
5. Clarke, E.M., Emerson, E.A.: Design and synthesis of synchronization skeletons using branching time temporal logic. In: Kozen, D. (ed.) Logic of Programs 1981. LNCS, vol. 131, pp. 52–71. Springer, Heidelberg (1982). https://doi.org/10.1007/BFb0025774
6. de Moura, L., Bjørner, N.: Z3: an efficient SMT solver. In: Ramakrishnan, C.R., Rehof, J. (eds.) TACAS 2008. LNCS, vol. 4963, pp. 337–340. Springer, Heidelberg (2008). https://doi.org/10.1007/978-3-540-78800-3_24

7. Duan, Z.: Temporal Logic and Temporal Logic Programming. Science Press, Beijing (2005)
8. Duan, Z., Tian, C.: A unified model checking approach with projection temporal logic. In: Liu, S., Maibaum, T., Araki, K. (eds.) ICFEM 2008. LNCS, vol. 5256, pp. 167–186. Springer, Heidelberg (2008). https://doi.org/10.1007/978-3-540-88194-0_12
9. Hao, K., Duan, Z.: Traceable automata. Chin. J. Comput. **5**, 340–348 (1990)
10. Kesten, Y., Pnueli, A.: A complete proof system for QPTL. J. Logic Comput. **12**(5), 701–745 (1995)
11. Kobayashi, N., Luke Ong, C.H.: A type system equivalent to the modal mu-calculus model checking of higher-order recursion schemes. In: IEEE Symposium on Logic in Computer Science, 2009. LICS 2009, pp. 179–188 (2009)
12. Wang, M., Tian, C., Zhang, N., Duan, Z., Yao, C.: Translating C programs to MSVL programs. https://arxiv.org/abs/1809.00959 (2018)
13. Wang, X., Tian, C., Duan, Z., Zhao, L.: MSVL: a typed language for temporal logic programming. Front. Comput. Sci. **11**(5), 762–785 (2017)
14. Yang, K., Duan, Z., Tian, C., Zhang, N.: A compiler for MSVL and its applications. Theoretical Computer Science (2017). https://doi.org/10.1016/j.tcs.2017.07.032
15. Zhang, N., Duan, Z., Tian, C.: Model checking concurrent systems with MSVL. SCI. CHINA Inf. Sci. **59**(11), 118101 (2016)

Determination of Dual Distances
for a Kind of Perfect Mixed Codes

Tianyi Mao[1,2(✉)]

[1] Department of Mathematics, CUNY Graduate Center, New York, NY, USA
maotianyi1@gmail.com
[2] Sam's Club Technology Research Team, Dallas, TX, USA

Abstract. A series of mixed perfect codes with minimal distances $d = 3$ has been constructed in term of the partitions of vector space over finite field $\mathbb{F}_p$ by B.Lindström. In this paper the minimal distance of the dual codes of a certain class of such perfect codes has been determined. As an application of this result we constructed a series of good orthogonal arrays with mixed levels and good inhomogeneous asymmetric quantum codes.

Keywords: Partitions of finite vector spaces · Perfect mixed code
Minimal distance · Dual codes · Array · Asymmetric quantum code

1 Introduction

Let $A_i(1 \leq i \leq s)$ be finite abelian groups and $|A_i| = N_i$, $2 \leq N_1 \leq N_2 \leq \ldots \leq N_s$, $A = A_1 \oplus A_2 \oplus \ldots \oplus A_s$, so each element of A can be expressed by $v = (v_1, \ldots, v_s)(v_i \in A_i)$. The Hamming weight of v is defined by

$$wt(v) = \#\{i : 1 \leq i \leq s, v_i \neq 0\}.$$

For $v = (v_1, \ldots, v_s)$ and $u = (u_1, \ldots, u_s)$ in A, the Hamming distance between v and u is defined by

$$d_H(v, u) = \#\{i : 1 \leq i \leq s, v_i \neq u_i\} = wt(v - u).$$

A mixed code C over A is a subset of A with size $K = |C| \geq 2$. The minimal distance of C is defined by

$$d = d(C) = \min\{d_H(c, c') : c, c' \in C, c \neq c'\}.$$

If C is a subgroup of A, then $d(C) = \min\{wt(c) : 0 \neq c \in C\}$. We denote (A, K, d) as the parameters of the mixed code C.

Mixed codes are one of the generalizations of classical codes where all $A_i(1 \leq i \leq s)$ are the same finite group or finite field. One of the fundamental problems is the same as in classical case: to construct mixed codes over a fixed group A with larger size K and larger $d(C)$.

To analogue the classical case, the following bounds are established in order to judge the goodness of mixed codes.

© Springer Nature Switzerland AG 2018
S. Tang et al. (Eds.): AAIM 2018, LNCS 11343, pp. 88–97, 2018.
https://doi.org/10.1007/978-3-030-04618-7_8

Lemma 1 ([1]). *Suppose we have a mixed code C with parameters (A, K, d). Then*
(1) (Hamming bound)

$$|A|(= N_1 \ldots N_s) \geq K \cdot f([\frac{d-1}{2}]).$$

where for integer $l \geq 0$,

$$f(l) = 1 + \sum_{\lambda=1}^{l} \sum_{1 \leq i_1 < \ldots < i_\lambda \leq s} (N_{i_1} - 1) \ldots (N_{i_\lambda} - 1)$$

is the size of a closed ball with radius l in A. The code C is called perfect if it reaches the Hamming bound.
(2) (Singleton bound)

$$K \leq N_1 N_2 \ldots N_{s-d+1}.$$

The code C is called MDS code if it reaches the Singleton bound.

A series of MDS mixed codes are constructed as algebraic-geometry codes in [1] when $A_i = \mathbb{F}_q^{d_i} (1 \leq i \leq s)$ for a fixed finite field $\mathbb{F}_q$. On the other hand, there is an algebraic method to get perfect mixed codes with minimal distances $d = 3$ by so-called partition of finite vector spaces (see [4,5]).

Let p be a fixed prime number, $n \geq 1$, $V_i(1 \leq i \leq s)$ be subspaces of $V = \mathbb{F}_p^n$, $dim_{\mathbb{F}_p} V_i = n_i \geq 1(1 \leq i \leq s)$. We say that $\{V_1, \ldots, V_s\}$ is a partition of V if $V \setminus \{0\}$ is a disjoint union of $V_i \setminus \{0\}(1 \leq i \leq s)$. (One can define a partition into subgroups for any finite abelian group G, but it is proved in [3] that if a finite abelian group G have a partition into subgroups, then G should be an elementary p-group so that $G = \mathbb{F}_p^n$.) From the definition we know that if $\{V_1, \ldots, V_s\}$ is a partition of $V = \mathbb{F}_p^n$, $V_i = \mathbb{F}_p^{n_i}(1 \leq i \leq s)$, then

$$p^n - 1 = \sum_{i=1}^{s} (p^{n_i} - 1), \text{ namely, } p^n + s - 1 = \sum_{i=1}^{s} p^{n_i}. \tag{1}$$

Let $A_i = \mathbb{F}_p^{n_i}(1 \leq i \leq s)$ be viewed as the subspaces V_i of V and $A = A_1 \oplus \ldots \oplus A_s$. Now we consider the following $\mathbb{F}_p$-linear mapping:

$$\phi : A = A_1 \oplus A_2 \oplus \ldots A_s \longrightarrow V = \mathbb{F}_p^n$$

$$(v_1, v_2, \ldots, v_s) \longmapsto v_1 + v_2 + \ldots + v_s.$$

The kernel

$$C = ker(\phi) = \{(v_1, \ldots, v_s) \in A : v_1 + \ldots + v_s = 0\}$$

is a linear code over A. It is easy to see that $\phi(A) = V$, $d(C) = 3$, and $K = |C| = p^k$ where

$$k = \dim A - \dim V = \sum_{i=1}^{s} n_i - n. \tag{2}$$

Since

$$K \cdot \left(1 + \sum_{i=1}^{s}(p^{n_i} - 1)\right) = p^{k+n} \qquad \text{(by (1))}$$
$$= p^{\dim A} = |A|, \qquad \text{(by (2))}$$

we know that C is a perfect mixed code with parameters $(A, K, 3)$. We denote this perfect code by $C = C(V, \pi)$ where $V = \mathbb{F}_p^n$ and $\pi = \{V_1, \ldots, V_s\}$ is the partition of V.

For any fixed basis $\{v_1, \ldots, v_n\}$ of $V = \mathbb{F}_p^n$, an element v of V can be uniquely expressed as $v = a_1 v_1 + \ldots + a_n v_n (a_i \in \mathbb{F}_p)$. We identify v as the vector $v = (a_1, \ldots, a_n)$. In this way we have the normal inner product on V: for $v = (a_1, \ldots, a_n)$ and $u = (b_1, \ldots, b_n)$ in V, the inner product of v and u is

$$\langle v, u \rangle = v u^T = \sum_{\lambda=1}^{n} a_\lambda b_\lambda \in \mathbb{F}_p.$$

For subspaces $V_i = \mathbb{F}_p^{n_i} (1 \leq i \leq s)$ of V, the inner product on V_i is the restriction of $\langle \cdot, \cdot \rangle$ on V_i. Then we can define the inner product on $A = A_1 \oplus A_2 \oplus \ldots \oplus A_s (A_i = \mathbb{F}_p^{n_i})$ as: for $v = (v_1, \ldots, v_s)$ and $u = (u_1, \ldots, u_s)(v_i, u_i \in A_i)$,

$$\langle v, u \rangle = \sum_{i=1}^{s}(v_i, u_i) \in \mathbb{F}_p.$$

Let C be a linear code over A, $K = |C| = p^k (k \geq 1)$. The dual (linear) code $C^\perp$ of C is defined by

$$C^\perp = \{v \in A : (v, c) = 0 \text{ for all } c \in C\}.$$

It is easy to see that $|C^\perp| = \frac{|A|}{|C|} = p^{\dim A - k}$. To determine the minimal distance of $C^\perp$ from the structure of C is one of the important problems in coding theory which has not only theoretical interests, but also practical applications in combinatorial designs (orthogonal arrays with different levels) and communication (quantum codes). It seems that there are only few mixed linear codes C such that $d(C^\perp)$ have been determined.

In Sect. 2 of this paper we determine $d(C^\perp)$ for a class of perfect mixed codes C with $d(C) = 3$ derived from a kind of specific partition of $V = \mathbb{F}_p^n$. Then we will show an application of our result in combinatorial design theory and quantum code theory in Sect. 3.

2 Determination of $d(C^\perp)$

In Sect. 1 we have shown that perfect mixed codes can be derived from partitions of a vector space $V = \mathbb{F}_p^n$. Many partitions of V have been constructed in past 30 years ([3–6] and the references therein). In this paper we consider a general construction given in [6]. We will introduce this construction briefly.

Let V be the finite field $\mathbb{F}_p^n$. For a fixed basis $\{v_1,\ldots,v_n\}$ over $\mathbb{F}_p$. We consider $\mathbb{F}_{p^n}$ as $\mathbb{F}_p^n$ by identifying $v = a_1v_1+\ldots+a_nv_n \in \mathbb{F}_{p^n}\,(a_i \in \mathbb{F}_p)$ with $(a_1,\ldots,a_n) \in \mathbb{F}_p^n$. Then we view any subspace W of V as a subset of the finite field $\mathbb{F}_{p^n}$, so that for w_1 and w_2 in W we have their product w_1w_2 in $V = \mathbb{F}_{p^n}$.

Theorem 1 ([6], Theorem 3.1). Let $V = \mathbb{F}_p^n (= \mathbb{F}_{p^n})$ and $V = U \oplus W$ where U and W are subspaces of V. Suppose that

(1) $\dim W = s \geq 1$, $W = \mathbb{F}_{p^s}$
(2) U has a subspace partition $\{U_1,\ldots,U_t\}$ and $\dim U_i = d_i \leq s$ for $1 \leq i \leq t$

Then for each $i(1 \leq i \leq t)$ and $\gamma \in W \setminus \{0\}$ we can define a d_i-dimensional subspace $U_{i\gamma}$ of V such that $\pi = \{W, U, U_{i\gamma}(1 \leq i \leq t, \gamma \in W \setminus \{0\})\}$ form a partition of V.

Proof. Let $\{w_1, w_2,\ldots, w_s\}$ be a basis of $W = \mathbb{F}_{p^s}$, $\{u_{i1},\ldots, u_{id_i}\}$ be a basis of $U_i(1 \leq i \leq t)$. Then for each $\gamma \in W \setminus \{0\}$ and each $i(1 \leq i \leq t)$, the d_i elements $B_i = \{u_{ij} + \gamma w_j : 1 \leq j \leq d_i\}$ of V are $\mathbb{F}_p$-linear independent, where γw_j be the multiplication in $W = \mathbb{F}_{p^s}$. Let $U_{i\gamma}$ be the subspace of V spanned by B_i so that $\dim U_{i\gamma} = d_i$. It can be seen that $\pi = \{W, U, U_{i\gamma}(1 \leq i \leq t, \gamma \in W \setminus \{0\})\}$ is a partition of V (See the proof in [6]).

Now we have a perfect mixed linear code $C(V, \pi)$ over

$$A = W \oplus U \oplus \left(\bigoplus_{1 \leq i \leq t, \gamma \in W \setminus \{0\}} U_{i\gamma} \right)$$

with size $K = p^{\dim A - \dim V}$, where

$$\dim C = \dim A - \dim V$$

$$= \sum_{1 \leq i \leq t, \gamma \in W \setminus \{0\}} \dim U_{i\gamma} = (|W| - 1) \sum_{1 \leq i \leq t} d_i = (p^s - 1) \sum_{i=1}^t d_i$$

and

$$C = C(V, \pi)$$

$$= \left\{ (c_w, c_u, c_{i\gamma}; 1 \leq i \leq t, \gamma \in W \setminus \{0\}) \in A : c_w + c_u + \sum_{1 \leq i \leq t, \gamma \in W \setminus \{0\}} c_{i\gamma} = 0 \right\}.$$

In this section we show the following result.

Theorem 2. Let π be the partition of $V = \mathbb{F}_{p^n}$ given by Theorem 1, $C^\perp$ be the dual code of $C = C(V, \pi)$. Then

$$d(C^\perp) = \min \left\{ p^s, tp^s - \sum_{i=1}^t p^{s-d_i} + 1 \right\}$$

$$= \begin{cases} p^s, & \text{if } t \geq 2, \\ p^s - \sum_{i=1}^t p^{s-d_i} + 1, & \text{if } t = 1. \end{cases} \tag{3}$$

Proof. Firstly we show that.

Lemma 2. *For each i, j, γ where $1 \leq i \leq t$, $1 \leq j \leq d_i$, $0 \neq \gamma \in W$, let*

$$v_{ij\gamma} = (-\gamma w_j, -u_{ij}, 0, \ldots, 0, u_{ij} + \gamma w_j, 0, \ldots, 0)$$

Subspaces	W	U	$\ldots$	$U_{i\gamma}$		$\ldots$
Components	$-\gamma w_j$	$-u_{ij}$	0	$u_{ij} + \gamma w_j$	0	

where $u_{ij} \in U_i \subset U$, $\gamma \in W \setminus \{0\}$ so that $u_{ij} + \gamma w_j \in U_{i,\gamma}$ and $-\gamma w_j \in W$, as shown in the table above. Then $\{v_{ij\gamma} : \gamma \in W \setminus \{0\}, 1 \leq i \leq t, 1 \leq j \leq d_i\}$ is a basis of $C = C(V, \pi)$.

Proof. Since the number of $\{v_{ij\gamma}\}$ is the same as $\dim C = (p^s - 1) \sum_{i=1}^{t} d_i$, we only need to show that $\{v_{ij\gamma}\}$ are linear independent.

Suppose that $v = \sum_{i,j,\gamma} c_{ij\gamma} v_{ij\gamma} = 0$ for $c_{ij\gamma} \in \mathbb{F}_p$. For any fixed i and γ, the $U_{i\gamma}$-component of v is $0 = \sum_{j=1}^{d_i} c_{ij\gamma}(u_{ij} + \gamma w_j)$. Since $\{u_{ij} + \gamma w_j : 1 \leq j \leq d_i\}$ is a basis of $U_{i\gamma}$, we have $c_{ij\gamma} = 0$ for any $i(1 \leq i \leq t), j(1 \leq j \leq d_i)$ and $\gamma \in W \setminus \{0\}$. This completes the proof of Lemma 2.3.

Now we prove Theorem 2.2. Let $c = (c_w, c_u, c_{i\gamma}; 1 \leq i \leq t, \gamma \in W \setminus \{0\})$ be a nonzero codeword in $C^{\perp}$. From Lemma 2.2 we have

$$0 = \langle c, v_{ij\gamma} \rangle = \langle c_w, -\gamma w_j \rangle + \langle c_u, -u_{ij} \rangle + \langle c_{i\gamma}, u_{ij} + \gamma w_j \rangle.$$

Namely, for each $\gamma \in W \setminus \{0\}$, $1 \leq i \leq t$ and $1 \leq j \leq d_i$ we have

$$\langle c_w + \gamma w_j \rangle + \langle c_u, u_{ij} \rangle = \langle c_{i\gamma}, u_{ij} + \gamma w_j \rangle. \tag{4}$$

(I) Assume $c_w = 0$. Then

$$\langle c_u, u_{ij} \rangle = \langle c_{i\gamma}, u_{ij} + \gamma w_j \rangle (\gamma \in W \setminus \{0\}, 1 \leq i \leq t, 1 \leq j \leq d_i). \tag{5}$$

If $c_{i\gamma} = 0$ for some i and $\gamma \in W \setminus \{0\}$, by (5) we know that $\langle c_u, u_{ij} \rangle = 0$ for all j, $1 \leq j \leq d_i$. And then $\langle c_{i\gamma}, u_{ij} + \gamma w_j \rangle = 0$ for all $\gamma \in W \setminus \{0\}$ and $1 \leq j \leq d_i$. Since $\{u_{ij} + \gamma w_j : 1 \leq j \leq d_i\}$ is a basis of $U_{i\gamma}$, we get $c_{i\gamma} = 0$ for all $\gamma \in W \setminus \{0\}$.

If $c_u = 0$, by (5) we get all $c_{i\gamma} = 0$ and c would be zero. Therefore $c_u \neq 0$ and there exists some i such that $c_{i\gamma} \neq 0$ for all $\gamma \in W \setminus \{0\}$. This means that $wt(c) \geq |W| = p^s$.

It is easy to construct $c \in C^{\perp}$ such that $wt(c) = p^s$. We can choose a fixed i, and let $c_w = 0$ and $c_{k\gamma} = 0$ for all $k \neq i$ and $\gamma \in W \setminus \{0\}$. Choose arbitrary $c_u \neq 0$. For each fixed $\gamma \in W \setminus \{0\}$, the equations in (5) uniquely determine $c_{i\gamma} \neq 0$. Such codewords are of Hamming weight p^s.

(II) Assume that $c_w \neq 0$. From (4) we know that if $c_{i\gamma} = c_{i\gamma'} = 0$ for $\gamma, \gamma' \in W \setminus \{0\}$, $\gamma \neq \gamma'$ and some i, then

$$\langle c_w, (\gamma - \gamma') w_j \rangle = 0, \forall 1 \leq j \leq d_i, \tag{6}$$

$$\langle c_u, u_{ij} \rangle = -\langle c_w, \gamma w_j \rangle = -\langle c_w, \gamma' w_j \rangle, \forall 1 \leq j \leq d_i.$$

Let $\pi = \gamma - \gamma' \in W \setminus \{0\} = \mathbb{F}_{p^s}^*$. The "times by π" operator

$$\phi_\pi : \mathbb{F}_{p^s} \longrightarrow \mathbb{F}_{p^s}$$
$$x \longmapsto \pi x$$

is invertible $\mathbb{F}_p$-linear operator. We view c_w and w_j as vectors in $W = \mathbb{F}_p^s$, then $\pi x = x A_\pi$ where A_π is $s \times s$ invertible matrix over $\mathbb{F}_p$. By (6) we have

$$0 = \langle c_w, w_j A_\pi \rangle = c_w A_\pi^T w_j^T = \langle c_w A_\pi^T, w_j \rangle, \forall 1 \leq j \leq d_i.$$

Since $\{w_j : 1 \leq j \leq d_i\}$ are linear independent, there are exactly p^{s-d_i} elements $c_w A_\pi^T$ in W such that $\langle c_w A_\pi^T, w_j \rangle = 0$. Thus there are at most p^{s-d_i} elements $\gamma \in W \setminus \{0\}$ such that $c_{i\gamma} = 0$. Therefore for each i, there are at least $p^s - p^{s-d_i}$ elements $\gamma \in W \setminus \{0\}$ such that $c_{i\gamma} \neq 0$. Then by $c_w \neq 0$ we get $wt(c) \geq 1 + \sum_{i=1}^t (p^s - p^{s-d_i}) = t p^s - \sum_{i=1}^t p^{s-d_i} + 1$.

From the details in the proof we can construct a codeword c with weight $t p^s - \sum_{i=1}^t p^{s-d_i} + 1$. In fact, we can choose arbitrary $c_w \neq 0$, and let $c_u = 0$. Now we have

$$\langle c_w, \gamma w_j \rangle = \langle c_{i\gamma}, u_{ij} + \gamma w_j \rangle.$$

For fixed i and γ, using the fact that $\{u_{ij} + \gamma w_j\}$ is a basis of $U_{i\gamma}$, we conclude that the equation above uniquely determines the value of $c_{i\gamma}$. And $c_{i\gamma} = 0$ is equivalent to $\langle c_w, \gamma w_j \rangle = 0$ for all j, $1 \leq j \leq d_i$. The number of such γ is p^{s-d_i}. This gives a construction of a codeword c with weight $t p^s - \sum_{i=1}^t p^{s-d_i} + 1$.

From (I) and (II) we conclude $d(C^\perp) = \min\{p^s, t p^s - \sum_{i=1}^t p^{s-d_i} + 1\}$.

For any partition $W_1, \ldots, W_r$ of W, we have a corresponding perfect mixed linear code C_0 over W. This induces a refinement of the partition π over V, say $\pi' : V = (\cup_{1 \leq i \leq r} W_r) \bigcup U \bigcup (\cup_{1 \leq i \leq t, \gamma \in W \setminus \{0\}} U_{i\gamma})$, and a new perfect mixed linear code C'. The following theorem determines the minimal distance of C':

Theorem 3. *Let π' be the partition of $V = \mathbb{F}_{p^n}$ given above, $C'^\perp$ be the dual code of $C' = C(V, \pi')$. Then*

$$d(C'^\perp) = \min\{p^s, t p^s - \sum_{i=1}^t p^{s-d_i} + d^\perp\}$$

where $d^\perp = d(C_0^\perp)$.

Proof. Let

$$v_{ij\gamma} = (0, \ldots, 0, -\gamma w_j, 0, \ldots, 0, -u_{ij}, 0, \ldots, 0, u_{ij} + \gamma w_j, 0, \ldots, 0)$$

Subspaces	$W_1 \ldots W_i$	$\ldots W_r$	U	$U_{i\gamma}$	$\ldots$
Components	$0 \quad \ldots -\gamma w_j$	$\ldots 0$	$-u_{ij}$	$u_{ij} + \gamma w_j$	0

where $-\gamma w_j$ belongs to some $W_i(1 \leq i \leq r)$ as in the table above, and

$$v_k = (v_k^0, 0, \ldots, 0)$$

where $v_k^0(k = 1, 2, \ldots)$ is a basis of C_0. Then $v_{ij\gamma}$ and v_k form a basis of C'.

For a codeword $c \in C'^{\perp}$, let c_H be the first r components of c, and write $c = (c_H, c_T)$.

(I) If $c_H = 0$, then $\langle c, v_{ij\gamma} \rangle = 0$. By the proof in Theorem 2.3 we have $wt(c) \geq p^s$.

(II) If $c_H \neq 0$, then by $(c, v_k) = 0$ we have $wt(c_H) \geq d^{\perp}$. By the same method in the proof of Theorem 2.3, we can easily derive that $wt(c_T) \geq tp^s - \sum_{i=1}^{t} p^{s-d_i}$. So $wt(c) \geq tp^s - \sum_{i=1}^{t} p^{s-d_i} + d^{\perp}$.

By the methods in the proof of Theorem 3, we can construct examples showing that the equality can be reached.

3   Applications

In this section we present two applications of Theorem 2.2. The first application is in combinatorial design theory, to construct orthogonal array with mixed level.

Definition 1. *Let A_j $(1 \leq j \leq s)$ be finite sets, $|A_j| = N_j$, $A = A_1 \times A_2 \times \ldots \times A_s$, $M = (a_{ij})_{1 \leq i \leq N, 1 \leq j \leq s}$ be an $N \times s$ matrix where $a_{ij} \in A_j$. For each subset I of $\{1, 2, \ldots, s\}$, let $M_I = (a_{ij})_{1 \leq i \leq N, j \in I}$ be the $N \times |I|$ submatrix of M. The matrix M is called an orthogonal array with parameters $(A, N, l)(l \geq 1)$ if each element $c = (c_j)_{j \in I}$ in $A_I = \prod_{j \in I} A_j$ appears in the N rows of M_I with exact $\frac{N}{|A_I|} = \frac{N}{\prod_{j \in I} N_j}$ times.*

It is known in [2] that if $A_j(1 \leq j \leq s)$ are abelian groups and C is an additive code in $A = A_1 \oplus \ldots \oplus A_s$ with parameters (A, K, d), then the $K \times s$ matrix M consisted by all K codewords of C as rows, is an orthogonal array with parameters $(A, K, d^{\perp} - 1)$ where $d^{\perp}$ is the minimal distance of $C^{\perp}$, the dual code of C. From Theorems 1 and 2 we get the following result:

Theorem 4. *Let $V = \mathbb{F}_p^n = U \oplus W$ where $U = \mathbb{F}_p^{n-s}$, $W = \mathbb{F}_p^s$ and U has a subspace partition $\{U_1, \ldots, U_t\}$, $\dim U_i = d_i \leq s(1 \leq i \leq t)$. Then there exist orthogonal arrays with parameters $(A, K, 2)$ and $(A, p^n, d^{\perp} - 1)$ where*

$$A = \mathbb{F}_p^s \oplus \mathbb{F}_p^{n-s} \oplus \underbrace{G \oplus \ldots \oplus G}_{p^s - 1 \ times}, G = \mathbb{F}_p^{d_1} \oplus \ldots \oplus \mathbb{F}_p^{d_t} \qquad (7)$$

$$K = p^{(d_1 + \ldots + d_t)(p^s - 1)}$$

$$d^{\perp} = \begin{cases} p^s, & if \ t \geq 2 \\ p^s - \sum_{i=1}^{t} p^{s-d_i} + 1, & if \ t = 1 \end{cases}$$

The second application of Theorem 2 is in quantum code theory, to construct asymmetric inhomogenous quantum codes (AIQC). Firstly we introduce basic definitions and results on inhomogeneous quantum codes, see [8] for the detail.

Let $A_i (1 \le i \le n)$ be finite abelian groups, $|A_i| = q_i$ and

$$A = A_1 \oplus A_2 \oplus \ldots \oplus A_n$$

be an abelian group with $N = q_1 \ldots q_n$ elements. We consider the N-dimensional complex vector space

$$V = V_1 \oplus V_2 \oplus \ldots \oplus V_n, V_i = \mathbb{C}^{q_i}$$

For each A_i, let $\{|c\rangle : c \in A_i\}$ be a fixed orthonormal basis of V_i. Namely, for $c, c' \in A_i$,

$$\langle c|c' \rangle = \begin{cases} 1, \text{if } c = c' \\ 0, \text{otherwise} \end{cases}$$

where $\langle , \rangle$ denotes the Hermitian inner product on complex vector space. Then V has the orthonormal basis

$$\{|c\rangle = |c_1 c_2 \ldots c_n\rangle = |c_1\rangle \otimes |c_2\rangle \otimes \ldots \otimes |c_n\rangle : c = (c_1, c_2, \ldots, c_n) \in A = A_1 \oplus A_2 \oplus \ldots \oplus A_n\} \tag{8}$$

and $|c_i\rangle$ is called the ith qudit (quantum digit) of $|c\rangle$. An inhomogeneous quantum state is a non-zero vector in V which is uniquely expressed by

$$|v\rangle = \sum_{c \in A} \phi(c)|c\rangle \ (\phi(c) \in \mathbb{C})$$

Let $\hat{A}_i$ be the character group of A_i. It is well-known that there exists an isomorphic $A_i \to \hat{A}_i$, $b_i \mapsto \chi_{b_i}$ so that $\hat{A}_i = \{\chi_{b_i} : b_i \in A_i\}$.

Now we introduce quantum errors. Each quantum error is an unitary operation acting on the complex vector space V. At each qudit we have two types of errors $X(a_i)$ and $Z(b_i)$, $a_i, b_i \in A_i$ acting on $V_i = \mathbb{C}^{q_i}$ defined by their action on the basis $\{|c\rangle : c \in A_i\}$ of V_i as

$$X(a_i)|c\rangle = |a_i + c\rangle, Z(b_i)|c\rangle = \chi_{b_i}(c)|c\rangle$$

On the quantum state space $V = V_1 \otimes V_2 \otimes \ldots \otimes V_n$, we have quantum error operators $X(a)$ and $Z(b)$, $a = (a_1, a_2, \ldots, a_n)$, $b = (b_1, b_2, \ldots, b_n) \in A$ defined by their actions on the basis (8) as

$$\begin{aligned} X(a)|c\rangle &= X(a_1)|c_1\rangle \otimes X(a_2)|c_2\rangle \otimes \ldots \otimes X(a_n)|c_n\rangle \\ &= |a_1 + c_1\rangle \otimes |a_2 + c_2\rangle \otimes \ldots \otimes |a_n + c_n\rangle = |a + c\rangle \\ Z(b)|c\rangle &= Z(b_1)|c\rangle \otimes Z(b_2)|c_2\rangle \otimes \ldots \otimes Z(b_n)|c_n\rangle \\ &= \chi_{b_1}(c_1)|c_1\rangle \otimes \chi_{b_2}(c_2)|c_2\rangle \otimes \ldots \otimes \chi_{b_n}(c_n)|c_n\rangle = \chi_b(c)|c\rangle \end{aligned}$$

Let m be the exponent of A, namely, m is the smallest positive integer such that $ma = 0$ for all $a \in A$. Let $\omega = e^{\frac{2\pi i}{m}}$. Then the set

$$E_n = \{w^\lambda X(a)Z(b) : \lambda \in \{0,1,\ldots,m\}, a,b \in A\}$$

is a nonabelian group with $m|A|^2 = mN^2$ elements, called the quantum error group of V.

For each quantum error $e = w^\lambda X(a)Z(b)$ we define the X-weight $W_X(e)$ and Z-weight $W_Z(e)$ by

$$W_X(e) = W_H(a) = \#\{i : 1 \le i \le n, a_i \ne 0\}$$

$$W_Z(e) = W_H(b) = \#\{i : 1 \le i \le n, b_i \ne 0\}$$

and for positive integers d_X and d_Z we define the subset $S(d_X, d_Z)$ of E_n by

$$S(d_X, d_Z) = \{e \in E_n : W_X(e) \le d_X - 1, \text{ and } W_Z(e) \le d_Z - 1\}$$

Definition 2. *An subgroup Q of A is called an asymmetric inhomogenous quantum code (AIQC) with parameters $((A, K, d_X/d_Z))$ if $K = |Q|$ and for each $e \in S(d_X, d_Z)$ and $|v\rangle, |v'\rangle \in Q$ such that $\langle v|v'\rangle = 0$, we have $\langle v|e|v'\rangle = 0$.*

One important method to construct good AIQCs is the stablizer method (see [7] for homogenous case, [8] for symmetric case and [9] for general asymmetric and inhomogenous case). With this method, an AIQC can be constructed by a classical additive mixed code as shown in the following result:

Lemma 3 *([9], Theorem 3.2). Let $A_i(1 \le i \le s)$ be finite abelian groups, $A = A_1 \oplus A_2 \oplus \ldots \oplus A_s$. If there exist mixed additive codes C_1 and C_2 with parameters (A, K_1, d_1) and (A, K_2, d_2) respectively and $C_2^\perp \subset C_1$, then there exists an asymmetric inhomogenous quantum code Q with parameters $((A, \frac{K_1 K_2}{|A|}, d_X/d_Z))$ where*

$$d_X = \min\{w_H(c) : c \in C_2 \setminus C_1^\perp\}$$

$$d_Z = \min\{w_H(c) : c \in C_1 \setminus C_2^\perp\}$$

From Theorems 1, 2 and Lemma 3 we constructed the following AIQC by taking $C_1 = C$, $C_2 = C^\perp$:

Theorem 5. *Let $V = \mathbb{F}_p^n = U \oplus W$ where $U = \mathbb{F}_p^{n-s}$, $W = \mathbb{F}_p^s$ and U has a subspace partition $\{U_1, \ldots, U_t\}$, $\dim U_i = d_i \le s(1 \le i \le t)$. Let A be the additive group defined by 8. Then there exists an asymmetric inhomogenous quantum code with parameters $((A, 1, d_X/d_Z))$ where $d_X = 3$ and*

$$d_Z = \begin{cases} p^s, & \text{if } t \ge 2 \\ p^s - \sum_{i=1}^t p^{s-d_i} + 1, & \text{if } t = 1 \end{cases} \tag{9}$$

By using the results on partition of finite vector space given in [3,6], many series of good AIQCs can be obtained by Theorem 5.

References

1. Feng, K., Xu, L., Hickernell, F.: Linear error-block codes. Finite Fields Their Appl. **12**, 638–652 (2006)
2. Hedayet, A.S., Sloane, N.J.A., Stufken, J.: Orthogonal Arrays: Theory and Applications. Springer Series in Statistics. Springer, New York (1999). https://doi.org/10.1007/978-1-4612-1478-6
3. Heden, O.: Partitions of finite Abelian groups. Eur. J. Comb. **7**, 11–25 (1986)
4. Herzog, M., Schönheim, J.: Linear and nonlinear single error-correcting perfect mixed codes. Inf. Control. **18**, 364–368 (1971)
5. Lindström, B.: Group partitions and mixed perfect codes. Can. Math. Bull. **18**, 57–60 (1975)
6. El-Zanati, S.I., Seelinger, G.F., Sissokho, P.A., Spence, L.E., Vanden Eynden, C.: Partitions of finite vector spaces into subspaces. J. Comb. Des. **16**, 329–341 (2008)
7. Wang, L., Feng, K., Ling, S., Xing, C.: Asymmetric quantum codes: characterization and constructions. IEEE Trans. Inf. Theory **56**(6), 2938–2945 (2010)
8. Wang, W., Feng, R., Feng, K.: Inhomogenous quantum codes (I): additive case. Sci. China: Math. **53**, 2501–2510 (2010)
9. Wang, W., Feng, K.: Inhomogenous quantum codes (III): asymmetric case (2010, preprint)

Approximation and Competitive Algorithms for Single-Minded Selling Problem

Francis Y. L. Chin[1], Sheung-Hung Poon[2], Hing-Fung Ting[1], Dachuan Xu[3], Dongxiao Yu[4], and Yong Zhang[5(✉)]

[1] Department of Computer Science, The University of Hong Kong, Pok Fu Lam, Hong Kong
{chin,hfting}@cs.hku.hk
[2] School of Computer Science, University of Nottingham Ningbo China, Ningbo, People's Republic of China
Sheung-Hung.Poon@nottingham.edu.cn
[3] Beijing Institute for Scientific and Engineering Computing, Beijing University of Technology, Beijing, People's Republic of China
xudc@bjut.edu.cn
[4] Institute of Intelligent Computing, School of Computer Science and Technology, Shandong University, Qingdao 266237, People's Republic of China
dxyu@sdu.edu.cn
[5] Research Center for High Performance Computing, Joint Engineering Research Center for Health Big Data Intelligent Analysis Technology, Shenzhen Institutes of Advanced Technology, Chinese Academy of Sciences, Shenzhen, People's Republic of China
zhangyong@siat.ac.cn

Abstract. The problem of item selling with the objective of maximizing the revenue is studied. Given a seller with k types of items and n single-minded buyers, i.e., each buyer is only interested in a particular bundle of items, to maximize the revenue, the seller must carefully assign some amount of bundles to each buyer with respect to the buyer's accepted price. Each buyer b_i is associated with a value function $v_i(\cdot)$ such that $v_i(x)$ is the accepted unit bundle price b_i is willing to pay for x bundles. We show that the single-minded item selling problem is NP-hard. Moreover, we give an $O(\sqrt{k})$-approximation algorithm. For the online version, i.e., the buyers come one by one and the decision on each buyer must be made before the arrival of the next buyer, an $O(\sqrt{k} \cdot (\log h + \log k))$-competitive algorithm is achieved, where h is the highest unit item price among all buyers.

1 Introduction

The selling problems is a common scenario in financial markets and is motivated by certain real goods selling situations in economics. For instance, an investor wants to sell his shares in order to maximize the revenue. The prices of shares

S. Tang et al. (Eds.): AAIM 2018, LNCS 11343, pp. 98–110, 2018.
https://doi.org/10.1007/978-3-030-04618-7_9

fluctuate over time, and one may only know the shares' prices in history but not their future prices. At a specific time point, an investor needs to decide whether he should sell some shares and if so, the number of shares to sell.

Item selling involves two parties, sellers and buyers. The sellers want to sell a set of items, at some specific prices, and the buyers will buy the items if their prices are acceptable. The objective is to maximize the total revenue of the seller. To meet such an objective, prices of the items may need to be carefully adjusted to suit the different expectations of the buyers who have different required amounts, and/or at different occasions, etc. If a buyer's accepted price is lower than the designated price of an item, the buyer will reject the deal; otherwise, he will accept the deal.

Problem Statement

The seller has k types of items $\mathcal{I} = \{1, 2, \ldots, k\}$ and the amount of items in type i is m_i. There is a group of n buyers $B = \{b_1, b_2, \ldots, b_n\}$ who are interested in buying these items. Assume that the buyers are *single-minded*, i.e., each buyer $b_i \in B$ is only interested in a particular subset (bundle) $B_i \subseteq \mathcal{I}$ of items. When buyer b_i buys x bundles of B_i, the amount of each item $j \in B_i$ sold to b_i is x. Each buyer b_i is associated with a value function $v_i(\cdot)$ such that $v_i(x)$ is the accepted unit price buyer b_i is willing to pay for buying x bundles of B_i. Note that the interested bundles of different buyers may be different and the unit bundle prices of different bundles may not be comparable. Let $w_i(x)$ be the unit item price of buyer b_i, i.e., $w_i(x) = v_i(x)/|B_i|$, where $|B_i|$ is the size of the bundle B_i. Via $w_i(.)$, bundles sold to different buyers are comparable. Denote $\mathcal{A} = \{x_1, x_2, \ldots, x_n\}$ be an assignment such that x_i bundles of B_i is assigned to buyer b_i with the unit bundle price $v_i(x_i)$. The total revenue of the assignment $\mathcal{A}$ is

$$\sum_{i=1}^{n} x_i \cdot v_i(x_i) = \sum_{i=1}^{n} x_i \cdot |B_i| \cdot w_i(x_i).$$

The objective of the *item selling problem* is to maximize the revenue by assigning a certain amount of bundles to each buyer at his accepted unit price.

Related Works

Item selling is a realistic and important problem in computational economics, and the aim is to devise fast algorithms to give the optimal, or a reasonably good approximate, selling scheme.

On the seller's side, both multiple types of items and single type of items have been considered. For each type of item, the amount may be unlimited [1,5,10], or bounded [2,6,12]. As for the buyers' behaviors, there are mainly three models that have been considered: *single-minded* [3,4,6,13] (each user is interested only in a particular set of items), *unit-demand* [1,2,5] (each user will buy at most one item in total) and *envy-free* [6,10] (after the assignment, no user would prefer to be assigned a different set of items with the designated prices; loosely speaking, each user is happy with his/her purchase). Previous works have considered different combinations of the behavior of the seller and that of the buyer (e.g., envy-free pricing for single-minded buyers when there is unlimited supply [6]).

For the online version of the selling problem, a fundamental problem called *one-way trading* has been very well studied over the years. The one-way trading problem was introduced by El-Yaniv et al. [9], and the objective is to maximize the seller's revenue by selling some amount of a product to a sequence of buyers. Given the upper bound M and lower bound m of the accepted price, an $O(\log(M/m))$-competitive algorithm was given in [9]; moreover, the algorithm was proved to be optimal by a derived matching lower bound. Fujiwara et al. [11] have studied the one-way trading problem under the assumption that the input prices follow some given probability distribution. If the given distribution satisfies the monotone hazard rate, Chin et al. [8] showed that one-way trading can be approximated within a constant factor. Without the knowledge of the upper bound and lower bound of the accepted price, Chin et al. [7] gave an $O(\log r^*(\log^{(2)} r^*) \ldots (\log^{(i-1)} r^*)(\log^{(i)} r^*)^{1+\epsilon})$-competitive algorithm, where r^* is the ratio between the highest price and the lowest price in the buyer's sequence and i is any positive integer. Moreover, the lower bound of such unbounded one-way trading was proved to be $\Omega(\log r^*(\log^{(2)} r^*) \ldots (\log^{(i-1)} r^*)(\log^{(i)} r^*))$. For multiple types of items and single-minded buyers, Zhang et al. gave an $O(\sqrt{k} \cdot \log h \cdot \log k)$-competitive algorithm in [13], where k is the number of item types and h is the highest unit price among all buyers.

Our Contributions

In this paper, we consider both the offline version and the online version of the single-minded item selling problem. For the offline version, the item selling problem is proved to be NP-hard in Sect. 2. Furthermore, we show in Sect. 3 that this selling problem can be approximated within a factor of $O(\sqrt{k})$, where k is the number of item types. For the online version, we present an algorithm with the competitive ratio $O(\sqrt{k} \cdot (\log h + \log k))$ in Sect. 4, where h is the highest unit item price among all buyers, which improves the previous $O(\sqrt{k} \cdot \log h \cdot \log k)$-competitive algorithm in [13].

2   Hardness of the Selling Problem

In this section, we show that the item selling problem studied in this paper is NP-hard.

Theorem 1. *The item selling problem is NP-complete.*

Proof. Clearly, the item selling problem is in NP. The NP-hardness of the item selling problem is via the reduction from the 3-partition problem, which is described as follows. Given a multiset S of $n = 3p$ positive integers, $a_1, a_2, \ldots, a_{3p}$ with a total sum of pB, where $B/4 < a_j < B/2$ for $j = 1, \ldots, 3p$, can S be partitioned into p triplets $S_1, S_2, \ldots, S_p$ such that the sum of the numbers in each subset is equal to B? The subsets $S_1, S_2, \ldots, S_p$ must form a partition of S in the sense that they are disjoint and they cover S.

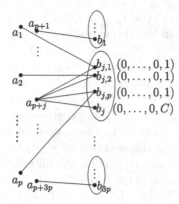

Fig. 1. The construction in the reduction from 3-partition problem to item selling problem.

For the reduction, we first form one type of item i_t with amount B for each triplet S_t for $t = 1, \ldots, p$. For each element a_j $(j = 1, \ldots, 3p)$, we form a set of p users $u_{j,t}$ for all $t = 1, \ldots, p$. See Fig. 1. For each user $u_{j,t}$, we set its value functions as $v_{j,t}(1) = \ldots = v_{j,t}(a_j - 1) = 0$ and $v_{j,t}(a_j) = 1$. We further form a new type of item i_{p+j} with amount B, and set the user $u_{j,t}$'s bundle of interest to be $I_{u_{j,t}} = \{i_t, i_{p+j}\}$. In addition, we form a new user u_j, and set its bundle of interest to be just $I_{u_j} = \{i_{p+j}\}$, and its value functions as $v_j(1) = \ldots = v_j(B - a_j - 1) = 0$ and $v_j(B - a_j) = B$. We note that in total, there are $4p$ items and $3p(p + 1)$ users in the construction. Note that all these value functions are step value functions, i.e., $v_i(x) = v_i(\lceil x \rceil)$. So, fractional bundles are allowed to be assigned to users.

Now suppose that there is a partition of p triplets $S_1, S_2, \ldots, S_m$ such that the sum of the numbers in each subset is equal to B. For the type of item i_{p+j} $(j = 1, \ldots, 3p)$, we assign $B - a_j$ bundles to user u_j, and assign the other a_j bundles to the corresponding user $u_{j,t}$, where the triplet S_t contains a_j. In such a way, the sale of the type of item i_{p+j} contributes $(B - a_j)B + a_j$. For each triplet $S_t = \{a_{j_1}, a_{j_2}, a_{j_3}\}$ in the partition, we assign a_{j_1} bundles to user $u_{j_1,t}$, assign zero bundles to user $u_{j_1,\bar{t}}$ where $\bar{t}$ is any value of $1, \ldots, p$ not equal to t, assign a_{j_2} bundles to user $u_{j_2,t}$, assign zero bundles to user $u_{j_2,\bar{t}}$, assign a_{j_3} bundles to user $u_{j_3,t}$, and assign zero bundles to user $u_{j_3,\bar{t}}$, respectively. We also set item i_t to be sold to exactly these three users $u_{j_1,t}$, $u_{j_2,t}$, and $u_{j_3,t}$. In such a way, the sale of each type of item i_t contributes B values to the revenue. Consequently, the total revenue obtained is $pB + \sum_{j=1}^{3p}((B - a_j)B + a_j) = pB + \sum_{j=1}^{3p}(B^2 - a_jB + a_j) = pB + 3pB^2 - (pB)B + pB = 2pB^2 + 2pB$. Then we have a solution for the item selling problem with total revenue at least pB^2.

For the proof of the reverse direction, we suppose that the total revenue of the item selling problem is at least $2pB^2 + 2pB$. For the type of item i_{p+j} $(j = 1, \ldots, 3p)$, to maximize the revenue, we need to assign $B - a_j$ bundles to user u_j, and assign the other a_j bundles to certain user $u_{j,t}$ so that the maximum revenue obtained from item type i_{p+j} is $(B - a_j)B + a_j{}^1$. Consequently, the maximum revenue which can be attained from item types $i_{p+1}, i_{p+2}, \ldots, i_{p+3p}$ is $\sum_{j=1}^{3p}((B-a_j)B+a_j) = \sum_{j=1}^{3p}(B^2-a_jB+a_j) = 3pB^2-(pB)B+pB = 2pB^2+pB$. Moreover, the maximum revenue obtained from each item type i_t $(t = 1, \ldots, p)$ is at most B, which occurs when each unit of item i_t contribute one in value. To make this happen, the B units of item i_t has to be partitioned by a subset of S. Since each item i_{p+j} $(j = 1, \ldots, 3p)$ corresponds to one a_j for a specific subscript j for obtaining revenue of a_j, the maximum revenue for all item types i_{p+j} $(j = 1, \ldots, 3p)$ results in the selection of all a_j for $j = 1, \ldots, 3p$. These $3p$ numbers of a_j's have to be used to partitioned all amounts for item types i_t for $t = 1, \ldots, p$. Thus the partition in the capacities of all item types i_t $(t = 1, \ldots, p)$ forms a partition for S such that the sum of those a_j's in the capacity of each item type i_t is equal to B. Since $B/4 < a_j < B/2$ for each a_j, the capacity amount B of each item type i_t is partitioned by exactly three a_j's. Thus the partition we obtained above is a 3-partition. This completes our proof.   □

3   Approximation Algorithms

Let $h = \max\{w_i(x)\}$ be the highest unit item price among all buyers b_i and amounts x of bundles. We assume that h is known to the algorithm. Let $\bar{w}_i(x)$ be the step unit item price such that

$$\bar{w}_i(x) = \frac{h}{2^j} \quad \text{if} \quad \frac{h}{2^j} \le w_i(x) < \frac{h}{2^{j-1}}.$$

Fact 1. $w_i(x)/2 < \bar{w}_i(x) \le w_i(x)$ for any buyer b_i and any amount x.

Let OPT be the optimal solution w.r.t. the original value function $v_i(\cdot)$ (or the original unit item value function $w_i(\cdot)$). Let $\overline{OPT}$ be the optimal solution w.r.t. the step unit item value function $\bar{w}_i(\cdot)$.

Lemma 1. $OPT/2 < \overline{OPT} \le OPT$.

Proof. If we directly implement the optimal assignment w.r.t. $w_i(\cdot)$, a feasible assignment $\mathcal{A}$ for the step unit item price function $\bar{w}_i(\cdot)$ can be achieved. From Fact 1, the revenue achieved from such feasible assignment $\mathcal{A}$ is at least $OPT/2$ and at most OPT. Let A denote the revenue of the assignment $\mathcal{A}$, we have $OPT/2 < A \le OPT$. On the other hand, A is upper bounded by the optimal revenue on $\overline{OPT}$ since $\mathcal{A}$ is a feasible assignment w.r.t. $\bar{w}_i(\cdot)$. So, $A \le \overline{OPT}$. Note that $w_i(x)/2 < \bar{w}_i(x) \le w_i(x)$ for any i and x, we have $\overline{OPT} \le OPT$.

Combining the above statements, this lemma is correct.   □

---

[1] Although fractional value is allowed, however, when the value of B is large enough, to maximize the revenue, the number of bundles assigned to users must be integers.

Lemma 1 gives us a heuristic that if considering just the step value function, the decrease of the optimal revenue is not too much, which is at least half of the optimal revenue for the original value function. Therefore, in the next part, we consider the step unit item value function $\bar{w}_i(\cdot)$, which can be easily achieved from the value function $v_i(\cdot)$.

Given the highest unit item value h among all interested bundles, the number of different values of $\bar{w}_i(\cdot)$ is at most $\lceil \log h \rceil$. Let $C_{i,j} = \{x | \bar{w}_i(x) = h/2^j\}$ denote the set of amount of bundles B_i with step unit item price $h/2^j$. If the unit item price is non-increasing or non-decreasing, which are reasonable assumptions in reality, $C_{i,j}$ is continuous. In this work, any kind of value function is allowed and thus, $C_{i,j}$ might not be continuous.

Let (i,j) be the pair of bundle B_i with the unit item price $h/2^j$ and denote (i,j,ℓ) be the assignment such that selling ℓ bundles B_i with the unit item price $h/2^j$.

Given the buyer set B and the value functions for each buyer, a naïve idea is to assign some bundles to the buyer with the highest revenue. By this idea, even with smaller unit item price, a bundle with larger size may have a higher revenue. For example, suppose that the amount of each item is 1, buyer $b_{i>k}$'s bundle $B_i = \mathcal{I}$ and its value function $v_i(x) = k$ for any x. Note that k is the number of item types in $\mathcal{I}$. Buyer set B contains other buyers b_j such that $B_j = \{j\}$ for $j \in [1,..,k]$ and $v_j(x) = k - \epsilon$. In this case, the revenue from this idea is k while the optimal revenue is $k \cdot (k - \epsilon)$. The approximation ratio of this idea is quite bad and close to k.

We may consider instead assigning bundles according to their unit item price. The performance of this idea however is still not good. For example, suppose that the amount of each item is 1, buyer b_1's bundle $B_1 = \{1\}$ and its unit item price is $w_1(x) = k$ for any x, while another buyer b_2's bundle $B_2 = \mathcal{I}$ and its unit item price is $w_2(x) = k - \epsilon$ for any x. Hence, the revenue of such idea is k while the optimal revenue is $k \cdot (k - \epsilon)$. The approximation of this idea still approaches k.

The above two greedy approaches can be regarded as two extreme ideas and do not consider the size of the assigned bundles. We combine them together and derive a new selection criterion, the product of the revenue and the unit item price. Formally speaking, for buyer b_i, consider the values of

$$\bar{w}_i(x) \cdot v_i(x) = (\bar{w}_i(x))^2 \cdot |B_i|.$$

The algorithm is described as follows.

Theorem 2. *Given the step value function $\bar{w}_i(\cdot)$, the approximation ratio of the above algorithm is $O(\sqrt{k})$.*

Proof. Let ALG and $\overline{OPT}$ be the revenue from Algorithm 1 and the optimal solution w.r.t. the step value function $\bar{w}_i(\cdot)$.

The optimal solution can be seen as assigning bundles to buyers one by one following the ordered sequence after line 1 of the algorithm. For two buyers b_i and $b_{i'}$, if $B_i \cap B_{i'} \neq \emptyset$, assigning item to buyer b_i may block assigning items to buyer

---

Algorithm 1. Selling bundles to single-minded buyers

1: Sort the pairs (i,j) according to the values $(\bar{w}_i(x))^2 \cdot |B_i|$.   ▷ The order is arbitrary for the same values from different buyers.
2: Let r_t be the remaining amount of item t. Initially, $r_t = m_t$ for all $t \in \mathcal{I}$.
3: **while** the sorted sequence is not empty **do**
4:    Select the highest one from the sequence, let (i,j) be such pair.
5:    **if** B_i is not assigned in previous steps **then**
6:      **if** $\min\{r_t | t \in B_i\} \geq \min\{x | x \in C_{i,j}\}$ **then**
7:        Set the assignment to be $(i, j, \max\{x | x \in C_{i,j}, x \leq \min\{r_t | t \in B_i\}\})$
8:        Update the remaining amount of items in B_i.
9:      **end if**
10:    **else**
11:      Let (i, j', ℓ') be the assignment on B_i in previous steps.
12:      **if** $\min\{r_t | t \in B_i\} + \ell' \geq \min\{x | x \in C_{i,j}\}$ and
13:      $h/2^j \cdot \max\{x | x \in C_{i,j}, x \leq \min\{r_t | t \in B_i\} + \ell'\} \geq \sqrt{k} \cdot h/2^{j'} \cdot \ell'$ **then**
14:        Re-assign B_i by $(i, j, \max\{x | x \in C_{i,j}, x \leq \min\{r_t | t \in B_i\} + \ell'\})$.
15:        Update the remaining amount of items in B_i.
16:      **end if**
17:    **end if**
18: **end while**

$b_{i'}$. Thus, an assignment (i, j, ℓ) from the algorithm may block some assignments in the optimal solution, which are behind (i, j, ℓ) in the sorted sequence.

Since each tiny piece of item can block at most one bundle, a tiny amount of B_i may block at most $|B_i|$ assignments. W.l.o.g., suppose (i_p, j_p) $(p = 1, \ldots, t)$ are the blocked assignments. and thus, $t \leq |B_i|$. Let $w_i = h/2^j$ and $w_{i_p} = h/2^{j_p}$, which are the step unit item prices. The unit revenue on (i, j) is $w_i \cdot |B_i|$ and the unit revenue on blocked bundles is at most $\sum_{i=1}^{t} w_{i_p} \cdot |B_{i_p}|$. The ratio between $\sum_{i=1}^{t} w_{i_p} \cdot |B_{i_p}|$ and $w_i \cdot |B_i|$ can be roughly regarded as the approximation ratio of the algorithm. Note that (i_p, j_p) is behind (i, j) in the sorted sequence, thus, $w_i^2 \cdot |B_i| \geq (w_{i_p})^2 \cdot |B_{i_p}|$. Let $w'_{i_p} \geq w_{i_p}$ satisfyingt $(w_i)^2 \cdot |B_i| = (w'_{i_p})^2 \cdot |B_{i_p}|$.

$$\frac{\sum_{p=1}^{t} w_{i_p} \cdot |B_{i_p}|}{w_i \cdot |B_i|} \leq \frac{\sum_{p=1}^{t} w'_{i_p} \cdot |B_{i_p}|}{w_i \cdot |B_i|} = \sum_{p=1}^{t} \frac{w'_{i_p} \cdot |B_{i_p}|}{w_i \cdot |B_i|} = \sum_{p=1}^{t} \frac{w_i}{w'_{i_p}} \tag{1}$$

$$\leq \sqrt{t \cdot \sum_{p=1}^{t} \left(\frac{w_i}{w'_{i_p}}\right)^2} = \sqrt{t \cdot \sum_{p=1}^{t} \frac{|B_{i_p}|}{|B_i|}} \tag{2}$$

$$\leq \sqrt{\sum_{p=1}^{t} |B_{i_p}|} \tag{3}$$

$$\leq \sqrt{k} \tag{4}$$

Inequality (2) holds due to the Cauchy-Schwartz inequality, inequality (3) holds since each item in B_i may block at most one bundle in the optimal solution and thus $|B_i| \geq t$, inequality (4) holds since the total size of disjoint bundles in the optimal solution is at most k.

From the above analysis, we can see that the ratio between the revenue from the blocked bundles of the optimal solution and the revenue from the algorithm is upper bounded by $\sqrt{k}$. However, for a bundle, say B_i, the previous assignment may be re-assigned by some latter assignment with higher revenue but lower unit item value. According to the algorithm, this case happens only when the revenue from the latter assignment is at least $\sqrt{k}$ times the former one.

Let (i, j_1, ℓ_1), $(i, j_2, \ell_2), \ldots, (i, j_t, \ell_t)$ be the assignments for buyer b_i by the algorithm during the execution. Let $y_1, y_2, \ldots y_t$ be the revenues received from these assignments. As mentioned before, revenues from the assigned bundles in the optimal solution and blocked by the assignment (i, j_s, ℓ_s) between (i, j_s, ℓ_s) and (i, j_{s+1}, ℓ_{s+1}) is at most $\sqrt{k} \cdot y_s$, which is upper bounded by y_{s+1}.

Therefore, revenues from the bundles in the optimal solution and blocked by the assignment on buyer b_i before (i, j_t, ℓ_t) is at most y_t, while the optimal revenue blocked by (i, j_t, ℓ_t) is at most $\sqrt{k} \cdot y_t$. Thus, the optimal revenue blocked by the assignment on buyer b_i is at most $O(\sqrt{k}) \cdot y_t$. Note that the optimal solution may assign some bundles to buyer b_i. From the previous analysis, the optimal revenue on buyer b_i is also upper bounded by $\sqrt{k} \cdot y_t$.

Therefore, the approximation ratio of the algorithm is $O(\sqrt{k})$. □

Combining Lemma 1 and Theorem 2, we have the following conclusion.

Theorem 3. *The approximation ratio of Algorithm 1 is $O(\sqrt{k})$.*

4   Online Algorithm

In the online version of the selling problem, buyers come to the seller one by one and each buyer's information is only known to the seller upon his arrival. When a buyer comes, a decision of selling some amount of bundles with designated price must be made immediately before the arrival of the next buyer. Suppose the highest unit item value is known to the seller. This assumption is reasonable in reality where the seller knows the possible highest value but has no idea about the information of each buyer.

The approximation algorithm in the last section gives us a heuristic to design an online algorithm by considering the value of $w_i(x) \cdot v_i(x)$ for each coming buyer b_i. Given the highest unit item price h, the maximal possible value is $h^2 \cdot k$, where k is the number of item types.

According to Lemma 1, the optimal revenue w.r.t. the step value function is at least half of the original optimal revenue. Similar to the approximation algorithm in the last section, when a buyer comes, his value function will be modified to be a 'step value function' according to the highest unit value h. Formally speaking,

$$\bar{w}_i(x) = \frac{h}{2^j} \quad \text{if} \quad \frac{h}{2^j} \leq w_i(x) < \frac{h}{2^{j-1}}.$$

Such modification can be done easily and in the latter part, we assume the value functions are 'step value functions'.

For each item i, its amount m_i is evenly partitioned into $\lceil 2 \log h + \log k \rceil$ levels, say level 1, level 2, $\ldots$, etc. Let $r_{i,j}$ be the available amount of item i in level j. Initially, $r_{i,j} = \frac{m_i}{\lceil 2 \log h + \log k \rceil}$. In the algorithm, items in level j can be only assigned to bundles with the value no less than $\frac{k \cdot h^2}{2^j}$. Thus, the higher the value, the more levels can be assigned. Let $\overline{r_{i,j}} = \sum_{\ell=j}^{\lceil 2 \log h + \log k \rceil} r_{i,\ell}$ be the available amount of item i which can be assigned to bundles with value $h^2 \cdot k / 2^j$. Let $R_{i,j} = \min\{\overline{r_{i',j}} | i' \in B_i\}$ be the maximal amount of bundles which can be assigned to buyer b_i with value $h^2 \cdot k / 2^j$.

For each buyer b_i, any value $(h/2^j)^2 \cdot |B_i|$ belongs to one range $(h^2 \cdot k / 2^j, h^2 \cdot k / 2^{j-1}]$, and each range contains at most one value. Thus, given a unit price $h/2^{j'}$ of buyer b_i, its corresponding level j can be immediately achieved. Let $p_i(j)$ be buyer b_i's unit price w.r.t. level j, in the above case, $p_i(j) = h/2^{j'}$. Denote $\overline{C_{i,j}}$ to be the set of amount of bundles B_i at level j, i.e., $p_i(j) \in (h^2 \cdot k / 2^j, h^2 \cdot k / 2^{j-1}]$.

For each coming buyer, Algorithm 2 greedily determines the price and amount so as to maximize the revenue in the partition framework.

Algorithm 2. Assigning bundles for a coming buyer b_i

1: if $\bigcup_j \{x | x \in \overline{C_{i,j}}, x \le R_{i,j}\} \ne \emptyset$ then
2:      Let $\ell = \arg\max_j \max\{x | x \in \overline{C_{i,j}}, x \le R_{i,j}\} \cdot p_i(j)$.
3:      Assign $\max\{x | x \in \overline{C_{i,\ell}}, x \le R_{i,\ell}\}$ bundles with the unit item price $p_i(\ell)$.
4:      Modify Available Amount of Items.
5: end if

As mentioned before, the available amount of items in level ℓ can be used for bundles with value no less than $h^2 \cdot k / 2^\ell$. In other words, the assigned bundle with value no less than $h^2 \cdot k / 2^\ell$ can use the available amount in level $\ell' \ge \ell$. Thus, the amount $\max\{x | x \in \overline{C_{i,\ell}}, x \le R_{i,\ell}\}$ of bundles is justified. When assigning bundles, the available amounts at lower numbered levels will be used first and when the amounts at lower numbered levels run out, the amounts at higher numbered levels will be used. More precisely, we use level ℓ first, and then level $\ell + 1$, level $\ell + 2$, and so on.

The procedure of modifying the amount at each affected levels is shown in Algorithm 3.

After handling the sequence of buyers, let ALG and OPT be the revenue received from the online algorithm and the optimal algorithm respectively. There are two cases in the final configuration after handling all buyers.

– $R_{i,j} = 0$. For buyer b_i with unit price $p_i(j)$, levels higher than j are full.
– $R_{i,j} > 0$. For buyer b_i with unit price $p_i(j)$, levels higher than j are not full.

Let ALG_1 be the assignments w.r.t. the first case.

Algorithm 3. Modify Available Amount of Items

1: **for** each item $j \in B_i$ **do**
2:      Let $s = \arg\min_t \{\overline{r_{j,\ell}} - \overline{r_{j,t}} \geq \max\{x | x \in \overline{C_{i,\ell}}, x \leq R_{i,\ell}\}\}$.
3:      **for** $p = \ell$ to $s - 1$ **do**
4:         $r_{j,p} = 0$.
5:      **end for**
6:      $r_{j,s} = \overline{r_{j,\ell}} - \overline{r_{j,t}} - \max\{x | x \in \overline{C_{i,\ell}}, x \leq R_{i,\ell}\}$.
7:      **for** $p = \ell$ to s **do**
8:         $\overline{r_{j,p}} = \overline{r_{j,\ell}} - \max\{x | x \in \overline{C_{i,\ell}}, x \leq R_{i,\ell}\}$.
9:      **end for**
10: **end for**

In the optimal assignment, buyer b_i gets some bundles B_i with the unit item price w_i. Suppose it belongs to level j. Consider the case that there exists an item $\ell \in B_i$ such that $\overline{r_{\ell,j}} = 0$, i.e., the amount of bundles in B_i at level j is full. Note that $\overline{r_{\ell,j}}$ is the value from the online algorithm. Let OPT_1 be the revenue of the optimal assignment w.r.t. this case, and let OPT_2 be the revenue of the remaining optimal assignment. Thus, $OPT = OPT_1 + OPT_2$.

If $B_i \bigcap B_j \neq \emptyset$, the assignment on B_i by the online algorithm may block the assignment on B_j by the optimal solution, where $i = j$ or $i \neq j$. Denote (i, j, ℓ) be an assignment on buyer b_i at level j with amount ℓ. W.l.o.g., let $f = (i, j, \ell) \in ALG_1$, and let F be the blocked bundles by the optimal assignment at the same level j. Note that the value of f may not be the largest in this level, however, other values in this level are at most twice of the value of f, i.e., $w_q^2 \cdot |B_q| \leq 2 \cdot (p_i(j))^2 \cdot |B_i|$ for any $(i_q, j_q, \ell_q) \in F$ at this level. Similar to the analysis in Sect. 3, we enlarge the unit item price of other bundles such that the value is equal to twice the value on f, i.e.,

$$(w_q)^2 \cdot |B_q| \leq (w_q')^2 \cdot |B_q| = 2 \cdot (p_i(j))^2 \cdot |B_i|.$$

The ratio between the revenues in F and f is

$$\frac{\sum_{q \in F} w_q \cdot |B_q|}{p_i(j) \cdot |B_i|} \leq \frac{\sum_{q \in F} w_q' \cdot |B_q|}{p_i(j) \cdot |B_i|} = \sum_{q \in F} \frac{w_q' \cdot |B_q|}{p_i(j) \cdot |B_i|} = 2 \cdot \sum_{p \in F} \frac{p_i(j)}{w_q'} \quad (5)$$

$$\leq 2 \cdot \sqrt{t \cdot \sum_{q \in F} \left(\frac{p_i(j)}{w_q'}\right)^2} = 2 \cdot \sqrt{t \cdot \sum_{q \in F} \frac{|B_q|}{2 \cdot |B_i|}} \quad (6)$$

$$\leq \sqrt{2 \cdot \sum_{q \in F} |B_q|} \quad (7)$$

$$\leq \sqrt{2k} \quad (8)$$

Let f_i be the assigned bundle by the online algorithm at some level j and let F_i be the assigned bundles by the optimal algorithm at the same level but blocked by f_i. From the above analysis, the revenue on $\bigcup f_i$ is at least $O(1/\sqrt{k})$ of the optimal revenue on $\bigcup F_i$.

In the optimal solution, each assignment may belonging to some level j' ($j' > j$). Similar to the above analysis, we may also increase the unit item price of these bundles and prove that the revenue at level j' and belongs to OPT_1 is at most $O(\sqrt{k})$ times the revenue at level j by the online algorithm. Since there are $O(\log h + \log k)$ levels, we have the following conclusion.

Theorem 4. $OPT_1 = O(\sqrt{k} \cdot (\log h + \log k)) \cdot ALG_1$.

Now we analyze the remaining assignment in OPT_2. Let $f \in OPT_2$ be an assignment for buyer b_i in the optimal solution and belongs to level j.

Fact 2. If $f \in OPT_2$, some amount of bundles B_i must be assigned by the online algorithm.

Proof. In the final configuration, level j is not full w.r.t. the bundle B_i. Thus, when buyer b_i comes, the online algorithm must assign some amount of B_i to buyer b_i.   □

- Suppose the corresponding level after the assignment of B_i by the online algorithm is not full, i.e., $R_{i,j} \neq 0$. According to the greedy approach, the revenue received on B_i by the online algorithm is no less than the revenued received by the optimal solution.
- Suppose the corresponding level, say level j', after the assignment of B_i by the online algorithm is full. In this case, such assignment is the last one at level j' for bundle B_i. From the online algorithm, assigning bundles at level j is also a candidate assignment. In this case, assigning at level j makes $R_{i,j} = 0$, but the received revenue is no more than the assignment at level j'. According to the algorithm, such assignment at level j does not have a higher revenue on B_i. Thus, if we replace the assignment on B_i and make level j full, the revenue does not increase but the assignment f in the optimal solution will be switched to OPT_1.
  Let $TEMP$ be the configuration that replace all such assignments. It can be seen that $TEMP \leq ALG$. Some optimal assignments will be switched to OPT_1. From the analysis of the previous case, the revenues of the remaining assignments which belong to OPT_2 are no more than the revenues from the same buyers by the online algorithm.

Now, we conclude that

Theorem 5. $OPT = O(\sqrt{k} \cdot (\log h + \log k)) \cdot ALG$.

Proof. $OPT = OPT_1 + OPT_2 = OPT_1' + OPT_2'$, where OPT_1' contains the assignments after switching some assignment from OPT_2 to OPT_1 and OPT_2' is the remaining assignment in OPT_2. From the previous analysis,

$$OPT_2' \leq ALG.$$

According to Theorem 4,

$$OPT_1' = O(\sqrt{k} \cdot (\log h + \log k)) \cdot TEMP \leq O(\sqrt{k} \cdot (\log h + \log k)) \cdot ALG.$$

Combining the above inequalities, we have

$$OPT = OPT_1' + OPT_2' = O(\sqrt{k} \cdot (\log h + \log k)) \cdot ALG.$$

$\square$

Acknowledgements. This research is supported by China's NSFC grants (No. 61433012, U1435215, 11871081, 61602195), Hong Kong GRF grant (17210017, HKU 7114/13E), and Shenzhen research grant JCYJ20160229195940462 and GGFW2017073114031767.

References

1. Bansal, N., Chen, N., Cherniavsky, N., Rurda, A., Schieber, B., Sviridenko, M.: Dynamic pricing for impatient bidders. ACM Trans. Algorithms **6**(2), 35 (2010)
2. Briest, P., Krysta, P.: Buying cheap is expensive: hardness of non-parametric multi-product pricing. In: Proceedings of the Eighteenth Annual ACM-SIAM Symposium on Discrete Algorithms (SODA 2007), pp. 716–725, 07–09 January 2007, New Orleans, Louisiana
3. Chalermsook, P., Chuzhoy, J., Kannan, S., Khanna, S.: Improved hardness results for profit maximization pricing problems with unlimited supply. In: Gupta, A., Jansen, K., Rolim, J., Servedio, R. (eds.) APPROX/RANDOM -2012. LNCS, vol. 7408, pp. 73–84. Springer, Heidelberg (2012). https://doi.org/10.1007/978-3-642-32512-0_7
4. Chalermsook, P., Laekhanukit, B., Nanongkai, D.: Independent set, induced matching, and pricing: connections and tight (subexponential time) approximation hardnesses. In: Proceedings of FOCS 2013, pp. 370–379 (2013)
5. Chen, N., Ghosh, A., Vassilvitskii, S.: Optimal envy-free pricing with metric substitutability. In: Proceedings of the 9th ACM conference on Electronic commerce (EC 2008), pp. 60–69 (2008)
6. Cheung, M., Swamy, C.: Approximation algorithms for single-minded envy-free profit-maximization problems with limited supply. In: Proceedings of 49th Annual IEEE Symposium on Foundations of Computer Science (FOCS 2008), pp. 35–44 (2008)
7. Chin, F.Y.L., et al.: Competitive algorithms for unbounded one-way trading. Theor. Comput. Sci. **607**, 35–48 (2015)
8. Chin, F.Y.L., Lau, F.C.M., Tan, H., Ting, H.-F., Zhang, Y.: Unbounded one-way trading on distributions with monotone hazard rate. In: Gao, X., Du, H., Han, M. (eds.) COCOA 2017. LNCS, vol. 10627, pp. 439–449. Springer, Cham (2017). https://doi.org/10.1007/978-3-319-71150-8_36
9. El-Yaniv, R., Fiat, A., Karp, R., Turpin, G.: Optimal search and one-way trading online algorithms. Algorithmica **30**(1), 101–139 (2001)
10. Fiat, A., Wingarten, A.: Envy, multi envy, and revenue maximization. In: Leonardi, S. (ed.) WINE 2009. LNCS, vol. 5929, pp. 498–504. Springer, Heidelberg (2009). https://doi.org/10.1007/978-3-642-10841-9_48

11. Fujiwara, H., Iwama, K., Sekiguchi, Y.: Average-case competitive analyses for one-way trading. J. Comb. Optim. **21**(1), 83–107 (2011)
12. Im, S., Lu, P., Wang, Y.: Envy-free pricing with general supply constraints. In: Saberi, A. (ed.) WINE 2010. LNCS, vol. 6484, pp. 483–491. Springer, Heidelberg (2010). https://doi.org/10.1007/978-3-642-17572-5_41
13. Zhang, Y., Chin, F., Ting, H.-F.: Online pricing for bundles of multiple items. J. Glob. Optim. **58**(2), 377–387 (2014)

An Empirical Analysis of Feasibility Checking Algorithms for UTVPI Constraints

K. Subramani[1], Piotr Wojciechowski[1(✉)], Zachary Santer[1], and Matthew Anderson[2]

[1] LDCSEE, West Virginia University, Morgantown, WV, USA
k.subramani@mail.wvu.edu, {pwojciec,zsanter}@mix.wvu.edu
[2] AFRL, RITA, Rome, NY, USA
matthew.anderson.37@us.af.mil

Abstract. In this paper, we document the results of a detailed implementation study of two different algorithms for checking real (linear) and integer feasibility in a conjunction of Unit Two Variable per Inequality (UTVPI) constraints. Recall that a UTVPI constraint is a linear relationship of the form: $a \cdot x_i + b \cdot x_j \leq c_{ij}$, where $a, b \in \{-1, 0, 1\}$. A conjunction of UTVPI constraints is called a UTVPI constraint system (UCS). UTVPI constraints subsume difference constraints. Unlike difference constraints, the linear and integer feasibilities for UCSs do not coincide. UCSs find applications in a number of different domains such as abstract interpretation, packing, and covering. There exist several algorithms for UCS linear feasibility and integer feasibility with various running times. We will focus on the linear feasibility algorithms in [19] (LF_1) and [13] (LF_2). We also focus on the integer feasibility algorithms in [18] (IF_1) and [13] (IF_2). We compare our implementations to the Yices SMT solver [17] running linear real arithmetic (QF_LRA) and linear integer arithmetic (QF_LIA). Our experiments indicate that LF_1 is moderately superior to LF_2 in terms of time, and that IF_1 is **vastly** superior to IF_2 in terms of time. Additionally on small inputs the Yices Solver performs better than the implemented algorithms, however the implemented algorithms perform much better on larger inputs.

1 Introduction

This paper is concerned with an empirical analysis of algorithms for checking linear and integer feasibilities of a Unit Two Variable per Inequality (UTVPI) constraint system (UCS). Briefly, a UTVPI constraint is a linear relationship of the form: $a \cdot x_i + b \cdot x_j \leq c_{ij}$, where $a, b \in \{-1, 0, 1\}$. A conjunction of such constraints is called a UCS and can be represented in matrix form as: $\mathbf{A} \cdot \mathbf{x} \leq \mathbf{c}$. Let n represent the number of variables in $\mathbf{A} \cdot \mathbf{x} \leq \mathbf{c}$, and let m represent the number of constraints.

If either a or b is 0, then the constraint is said to be an absolute constraint; otherwise, it is said to be a relative constraint. If $a = -b$, then the constraint is

© Springer Nature Switzerland AG 2018
S. Tang et al. (Eds.): AAIM 2018, LNCS 11343, pp. 111–123, 2018.
https://doi.org/10.1007/978-3-030-04618-7_10

said to be a difference constraint. Difference constraints are widely used to model specifications in real-time scheduling [9,10], image processing [8], and program verification [6].

Definition 1. *The linear feasibility (**LF**) problem for UTVPI constraints is: Given a UCS* $\mathbf{A} \cdot \mathbf{x} \leq \mathbf{c}$*, does it have a real (linear) solution?*

Definition 2. *The integer feasibility (**IF**) problem for UTVPI constraints is: Given a UCS* $\mathbf{A} \cdot \mathbf{x} \leq \mathbf{c}$*, does it have an integer solution?*

The following example illustrates why the **LF** and **IF** problems are distinct for UCSs.

Example 1. Integer feasibility in a UCS immediately implies linear feasibility; the converse is not true. For instance, consider the UCS defined by the following constraints:

$$
\begin{aligned}
x_1 + x_2 &\leq 1 & -x_1 + x_2 &\leq 0 \\
x_1 - x_2 &\leq 0 & -x_1 - x_2 &\leq -1
\end{aligned}
\tag{1}
$$

Note that these are equivalent to $x_1 = x_2$ and $x_1 + x_2 = 1$. Thus, it is clear that System (1) has no lattice point (integer) solution. However, it contains the fractional point $(\frac{1}{2}, \frac{1}{2})$ and is thus non-empty.

Definition 3. *We divide UCSs into the following categories:*

1. *F0: UCSs that are not linearly feasible.*
2. *F1: UCSs that are linearly feasible but not integrally feasible.*
3. *F2: UCSs that are integrally feasible.*

UTVPI constraints occur in a number of problem domains including, but not limited to, program verification [13], abstract interpretation [7,14], real-time scheduling [9], and operations research. Indeed, many software and hardware verification queries are naturally expressed using this fragment of integer linear arithmetic, i.e., the case in which the solution of a UTVPI system is required to be integral. We note that when the goal is to model indices of an array or queues in hardware or software, rational solutions are unacceptable [13]. Other application areas include spatial databases [16] and theorem proving.

Both the **LF** and **IF** problems have been widely studied in the literature. For the **LF** problem, we refer the interested reader to [13,14,19] and for the **IF** problem we refer to [4,11–13,15].

The chief contributions of this paper are a detailed implementation profile of the LF_1 [19] and LF_2 [13] algorithms for the **LF** problem and the IF_1 [18] and IF_2 [13] algorithms for the **IF** problem. We also compare these implementations to the Yices SMT Solver [17]. We note that we used the flags QF_LIA and QF_RIA for our experiments. *Note that QF_RDL and QF_IDL apply to difference logic only, i.e., UTVPI constraints are not permitted.* This paper assumes familiarity with the detailed constructions and algorithms in these papers.

The rest of this paper is organized as follows: In Sect. 2, we discuss the two linear feasibility algorithms analyzed in this paper. Section 3 describes our experimental setup. Section 4 documents our observations. We conclude in Sect. 5 by summarizing our contributions and identifying avenues for future research.

2   Linear Feasibility Algorithms

In this section, we describe the linear feasibility algorithms compared in this paper.

2.1   Linear Feasibility Algorithm LF_1

We now briefly describe the linear feasibility algorithm from [19].

Corresponding to the input UCS, Algorithm LF_1 constructs a constraint network $\mathbf{N}$ as defined in [19].

This constraint network utilizes different types of edges:

1. White edges ($x_i \overset{c_{ij}}{\square} x_j$): these correspond to constraints of the form $x_i + x_j \leq c_{ij}$.
2. Black edges ($x_i \overset{c_{ij}}{\blacksquare} x_j$): these correspond to constraints of the form $-x_i - x_j \leq c_{ij}$.
3. Gray edges ($x_i \overset{c_{ij}}{\blacksquare} x_j$) and ($x_i \overset{c_{ij}}{\square} x_j$): these correspond to constraints of the form $x_i - x_j \leq c_{ij}$ and $-x_i + x_j \leq c_{ij}$ respectively.

Example 2. The UCS (2) corresponds to the constraint network in Fig. 1.

Note that the edges of weight 63 in Fig. 1 are to ensure that every vertex is reachable from x_0 by a path of any type. The weight of these edges is $(2 \cdot n + 1) \cdot C$

$$x_1 + x_3 \leq 0$$
$$x_2 - x_3 \leq -7$$
$$x_4 - x_2 \leq 3$$
$$-x_1 - x_4 \leq 5$$
$$x_1 \leq 6 \qquad (2)$$

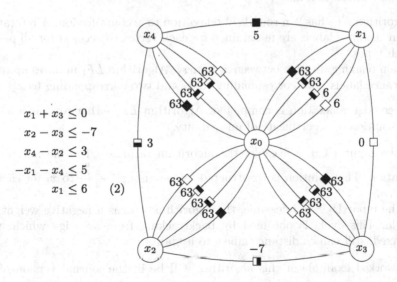

Fig. 1. Example constraint network.

where C is the largest absolute weight of any edge in the original graph. Thus, for Fig. 1 this is $(2 \cdot 4 + 1) \cdot |-7| = 63$. This large weight ensures that no negative weight gray cycles are added to the graph by the introduction of these edges.

Each valid path in $\mathbf{N}$ has type $t \in \{ \square, \blacksquare, \blacksquare, \blacksquare \}$, if it can be reduced to a single edge of type t.

Definition 4. *A* shortest white path *between* x_i *and* x_j *is a white path between* x_i *and* x_j *with minimum weight.*

We define shortest black path and shortest gray path similarly.
Algorithm LF_1 maintains four distance labels for each vertex, x_i, viz.,

1. $d_i^{\blacksquare}$: This label corresponds to a path that reduces to an edge of the form $(x_0 \overset{c}{\blacksquare} x_i)$. In fact, it is the weight of the current shortest gray path from x_0 to x_i.
2. $d_i^{\blacksquare}$: This label corresponds to a path that reduces to an edge of the form $(x_i \overset{c}{\blacksquare} x_0)$. In fact, it is the weight of the current shortest gray path from x_i to x_0.
3. $d_i^{\square}$: This label corresponds to a path that reduces to an edge of the form $(x_0 \overset{c}{\square} x_i)$. In fact, it is the weight of the current shortest white path from x_0 to x_i.
4. $d_i^{\blacksquare}$: This label corresponds to a path that reduces to an edge of the form $(x_0 \overset{c}{\blacksquare} x_i)$. In fact, it is the weight of the current shortest black path from x_0 to x_i.

Algorithm LF_1 is based on a modified version of Bellman-Ford. The key differences are as follows:

1. Algorithm LF_1 has $2 \cdot n$ rounds of relaxation to account for longer refutations.
2. Four distance labels are maintained for each vertex to account for all possible path types.
3. When relaxing an edge between x_i and x_j, Algorithm LF_1 updates up to four distance labels (two corresponding to x_i and two corresponding to x_j).

Once edge relaxation is completed, Algorithm LF_1 either returns a feasible linear solution or a certificate of infeasibility.

1. If the input UCS is feasible, the algorithm returns a feasible half-integral solution. This solution is constructed by assigning $\frac{d_i^{\square} - d_i^{\blacksquare}}{2}$ to x_i for each $i = 1 \ldots n$.
2. If the input UCS is infeasible, the algorithm returns a negative weight gray cycle. This cycle is obtained by backtracking from an edge which, when relaxed, still causes distance labels to update.

A worked example of this algorithm will be in the journal version of this paper.

2.2   Linear Feasibility Algorithm from LF_2

We now briefly describe the linear feasibility algorithm from [13].

Algorithm LF_2 converts the UCS into an equivalent system of difference constraints. Each constraint is converted as follows:

1. Each constraint of the form $x_i + x_j \leq c_{ij}$ becomes $x_i^+ - x_j^- \leq c_{ij}$ and $-x_i^- + x_j^+ \leq c_{ij}$.
2. Each constraint of the form $x_i - x_j \leq c_{ij}$ becomes $x_i^+ - x_j^+ \leq c_{ij}$ and $-x_i^- + x_j^- \leq c_{ij}$.
3. Each constraint of the form $-x_i + x_j \leq c_{ij}$ becomes $x_i^- - x_j^- \leq c_{ij}$ and $-x_i^+ + x_j^+ \leq c_{ij}$.
4. Each constraint of the form $-x_i - x_j \leq c_{ij}$ becomes $x_i^- - x_j^+ \leq c_{ij}$ and $-x_i^+ + x_j^- \leq c_{ij}$.
5. Each constraint of the form $x_i \leq c_{ij}$ becomes $x_i^+ - x_i^- \leq 2 \cdot c_{ij}$.
6. Each constraint of the form $-x_i \leq c_{ij}$ becomes $x_i^- - x_i^+ \leq 2 \cdot c_{ij}$.

Algorithm LF_2 then runs the Bellman-Ford algorithm on the directed graph **G** corresponding to the constructed system of difference constraints. The algorithm maintains a single distance label for each vertex. The distance label for the vertex x_i^+ is d_i^+, and for the vertex x_i^-, it is d_i^-.

Example 3. Figure 2 shows the graph corresponding to System (3).

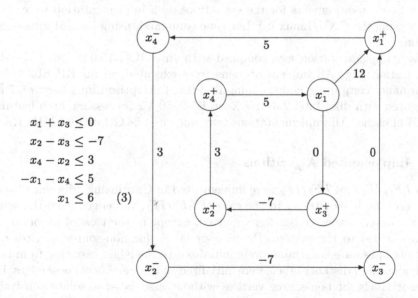

$$x_1 + x_3 \leq 0$$
$$x_2 - x_3 \leq -7$$
$$x_4 - x_2 \leq 3$$
$$-x_1 - x_4 \leq 5$$
$$x_1 \leq 6 \quad (3)$$

Fig. 2. Example graph.

Once edge relaxation is completed, Algorithm LF_2 either returns a feasible linear solution or a certificate of infeasibility.

1. If the input UCS is feasible, the algorithm returns a feasible half-integral solution. This solution is constructed by assigning $\frac{d_i^+ - d_i^-}{2}$ to x_i for each $i = 1 \ldots n$.
2. If the input UCS is infeasible, the algorithm returns a negative cycle. This cycle is obtained by backtracking from an edge which, when relaxed, still causes distance labels to update.

A worked example for this algorithm will be in the journal version of this paper.

3   Empirical Study

In this section, we describe the implementation of the algorithms, and the specifications of the system used to obtain our results. Details on the Algorithms can be found in Sect. 2.

3.1   Experimental Setup

Algorithm timing was performed within the implementations themselves, via calls to the GNU/Linux clock_gettime() function, specifying use of the CLOCK_MONOTONIC_RAW system clock. In this way, the running time requirements of each portion of each implementation could be analyzed separately. Space requirements for the entirety of each implementation were determined using the GNU/Linux usr/bin/time command, using its %M format conversion.

Each implementation was compiled with Intel ICC 15.0.2, using the -fast optimization flag. All implementations were executed on an HP SL230 high performance computing node running Red Hat Enterprise Linux Server 6.7 and configured with dual Intel 2.6 GHz Xeon E5-2650 V2 processors, each featuring 20 MB of cache. All implementations were allocated 64 GB of 1866 MHz RAM.

3.2   Implemented Algorithms

Both LF_1/IF_1 and LF_2/IF_2 were implemented in C, utilizing adjacency lists to represent graph structures. In the case of LF_1/IF_1, the edges from the source vertex to every other vertex were omitted, except in the case of absolute constraints added to the system. Predecessor labels for non-source vertices with associated absolute constraints were initialized to the edges resulting from these constraints, and distance labels were initialized to the weights of these edges. Predecessor labels for non-source vertices without associated absolute constraints were initially set null, and distance labels were initialized to the weight that the omitted edges would have had. This poses no danger, as the weight of source node edges not associated with absolute constraints is set in such a way that these edges cannot be part of a negative cost cycle.

In the case of LF_2/IF_2, a Bellman-Ford implementation was used to determine linear feasibility, since LF_1, as presented in [19], is organized analogously to this algorithm. The Johnson All-Pairs algorithm is used to determine an integral solution when IF_2 determines that such a solution exists [5].

The model generation algorithm within IF_2 was integrated directly into the Johnson All-Pairs implementation. Thus, the transitive and tight closure $\mathbf{C}*$ did not need to be represented in the form of a data structure in order to generate an integral solution. This eliminates IF_2's $O(n^2)$ space complexity term, leaving a space complexity of $O(m + n)$. This was accomplished by organizing the Johnson All-Pairs implementation such that it would process the original set of constraints $\mathbf{C}$ in the same order as the model generation algorithm processes $\mathbf{C}*$. Whenever an edge within $\mathbf{C}*$ is discovered, it is immediately used to modify bounds, then forgotten.

A number of different means by which the **LF** portions of these algorithms could be further optimized have been explored. Implementations of both algorithms were made where the relaxation loop in the **LF** portion of each implementation follows the FIFO Label-Correcting Algorithm given in [1]. These implementations also check the entire predecessor structure for negative cost cycles every $\sqrt{2 \cdot n}$ passes through the graph.

Where LF_2 uses vertices that each contain only one distance label, these modifications follow the approach given in [1] directly. In the case of LF_1, each vertex contains four distance labels. Thus, the queue structure stores a combination of vertex and distance label type updated. As a result, only edge relaxations that could lead to further distance label updates are conducted in the next pass through the graph.

A couple of variations of another mechanism were also implemented. "Cycle-originator" tokens were added to each vertex data structure. These tokens are initialized to a vertex's first predecessor edge and then propagated along edges during the relaxation process. Two different methods of propagation were used (CO1 and CO2) and these are described in detail later in this section. In this way, each cycle-originator can move through the predecessor structure as it develops. If a given cycle-originator token makes its way back to its original vertex, then we have a possible negative cost cycle. The algorithm then backtracks through the predecessor structure from that point. This determines if a negative cost cycle has actually been detected.

The first variation on the cycle-originator mechanism, here referred to as CO1, sets all vertex cycle-originators equal to predecessors after the first pass through the graph. In subsequent passes, this variation passes a cycle-originator down a relaxed edge in the predecessor structure whenever that edge relaxation leads to a distance label update. This variation works well with the linearly infeasible UCSs used for profiling in this work. However, it has not been proven to work in any particular situation.

The second variation, here referred to as CO2, sets a vertex's cycle-originator edge equal to its predecessor edge after an edge relaxation leads to a change in a vertex's predecessor edge. Whenever an edge relaxation leads to a distance label update without a change in the vertex's predecessor edge, the cycle-originator at the head of the relaxed edge is set equal to that at its tail. Since all predecessor edges are initially set null, this does not require any special handling at the beginning of system relaxation.

If a negative cost cycle is formed in the predecessor structure, and if the cycle stays constant as the distance labels of constituent vertices continue to decrease, then the CO2 mechanism is guaranteed to detect this cycle. At least the final edge added to the predecessor structure cycle will be represented as a cycle-originator and will then be passed through the predecessor structure, back to where it started, as the edges within the cycle continue to be relaxed.

This scenario of a negative cost cycle being formed within the predecessor structure, then remaining unchanged, can at least be guaranteed to occur in the case that there is only one possible negative cost cycle within a system of inequalities. However, the CO2 mechanism failed in the vast majority of linearly infeasible input systems used for profiling in this work.

There was little difference between how the cycle-originator mechanisms were implemented in the two algorithm implementations. Since cycle-originator tokens must be compared against predecessor labels, there must be equal numbers of each associated with each vertex. As such, the vertices in the LF_1 implementation each have four, and the vertices in the LF_2 implementation each have one. In both variations, the BACKTRACK() implementation used with each algorithm can detect when it backtracks to the source vertex and will then undo its work, as these mechanisms may lead to false positives.

The code used to implement these algorithms can be found at [3].

4   Empirical Results

In this section, we provide the results of our empirical analysis.

We tested multiple variations of the algorithms. These variations are as follows:

1. IF1: the integer feasibility algorithm from [19].
2. IF2: the integer feasibility algorithm from [13].
3. IF3: The Yices SMT Solver using linear integer arithmetic (QF_LIA) [17].
4. LF1: the linear feasibility algorithm from [19].
5. LF2: the linear feasibility algorithm from [13].
6. LF3: The Yices SMT Solver using linear real arithmetic (QF_LRA) [17].
7. LF1C: the linear feasibility algorithm from [19] optimized using cycle originators.
8. LF1F: the linear feasibility algorithm from [19] optimized using the FIFO label correcting algorithm.
9. LF2C: the linear feasibility algorithm from [13] optimized using cycle originators.
10. LF2F: the linear feasibility algorithm from [13] optimized using the FIFO label correcting algorithm.

Since a UCS can contain non-difference constraints, for example $x_1 + x_2 \leq 4$, using real difference logic (QF_RDL) or integer difference logic (QF_IDL) is not sufficient to solve the constraint systems used.

These algorithms were tested on both sparse and dense constraint networks. These networks were randomly generated using the UTVPI generator found at [2]. The following types of networks were generated:

1. Constraint networks with no negative gray cycles.
2. Constraint networks with a single negative gray 3-cycle.
3. Constraint networks with a single long negative gray cycle.
4. Constraint networks with many negative gray 3-cycles.
5. Constraint networks with 10 medium length gray cycles.

The sizes of the generated networks are detailed in Table 1.

Table 1. Size of generated constraint networks.

# of variables	256	1024	4096	8192	16384
# of constraints (sparse)	20,000	30,000	40,000	50,000	60,000
# of constraints (dense)	80,000	120,000	160,000	200,000	240,000
Long cycle length	256	1024	4096	8192	16384
# of short cycles	16	32	64	90	128
Medium cycle length	16	32	64	90	128

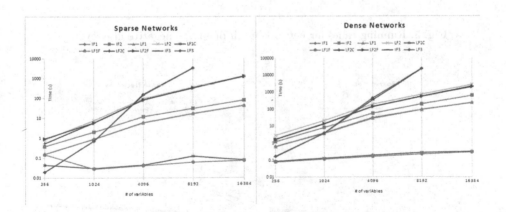

Fig. 3. Running times for networks with no negative gray cycles.

Figures 3 through 7 show the running times of the algorithms examined in this paper both with and without optimizations. Since Figs. 4 through 7 deal with

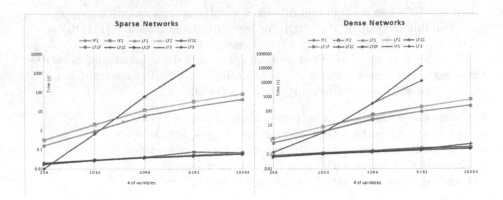

Fig. 4. Running times for networks with one negative gray 3-cycle.

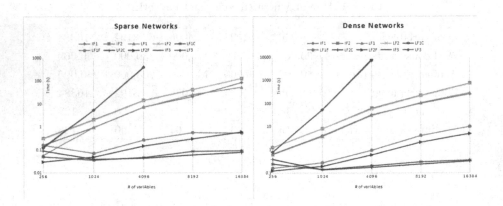

Fig. 5. Running times for networks with one large negative gray cycle.

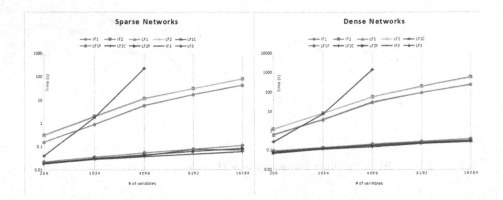

Fig. 6. Running times for networks with many negative gray 3-cycles.

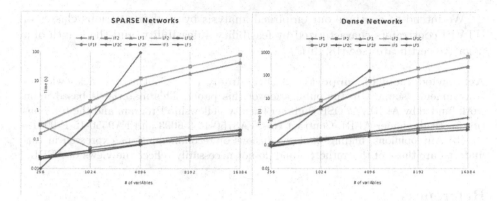

Fig. 7. Running times for networks with ten medium length negative gray cycles.

linearly infeasible systems we do not include the running time of the integer feasibility algorithms in those graphs.

Based on the data in this section, we can draw the following conclusions:

1. Without optimization, LF_1 runs faster than LF_2, on average, in all cases.
2. Both optimizations drastically reduced the running time of both algorithms and resulted in much closer running times.
3. In most cases both optimizations resulted in similar running times. However, cycles originators yielded noticeably faster running times on graphs with a single large negative cycle.
4. On small networks (256 variables), Yices runs faster than either LF_1 or LF_2. However, performance quickly drops off. On networks with 8,192 variables Yices often fails to complete, and Yices never completes computation on networks with 16,384 variables.

5   Conclusion

In this paper, we documented our observations regarding empirical analyses of algorithms for the **LF** and **IF** problems in UTVPI constraints. In particular, we profiled the algorithms in [19] (LF_1) and [18] (IF_1) vis-a-vis the algorithms in [13] (LF_2 and IF_2) and the Yices SMT solver [17]. Our experiments indicated that for the **LF** problem, LF_1 outperforms LF_2 in all cases. In the case of the **IF** problem, the IF_1 and IF_2 outperform each other in different situations. Including time to read the input file and construct each algorithm's UCS representation, then combining time requirements of the **LF** and **IF** algorithms, LF_1/IF_1 and LF_2/IF_2 run in very similar amounts of time. The Yices SMT solver outperformed both algorithms on small networks, however on larger networks LF_1/IF_1 and LF_2/IF_2 outperformed Yices with Yices failing to complete computation on the largest networks.

We intend to continue our empirical analysis by studying various classes of UTVPI constraints characterized by feasibility, infeasibility, and the length of a negative certificate (see [18,19]).

Acknowledgment of Support and Disclaimer. (a)   Contractor   acknowledges Government's support in the publication of this paper. This material is based upon work funded by AFRL/AFOSR Summer Faculty Fellowship Program and the Information Institute, under AFRL Contract Nos. FA8750-16-3-6003 and FA9550-15-F-0001.

(b) Any opinions, findings and conclusions or recommendations expressed in this material are those of the author(s) and do not necessarily reflect the views of AFRL.

References

1. Ahuja, R.K., Magnanti, T.L., Orlin, J.B.: Network Flows: Theory, Algorithms and Applications. Prentice-Hall, Upper Saddle River (1993)
2. Anderson, M., Wojciechowski, P., Subramani, K.: Generator for systems of UTVPI constraints. https://www.dropbox.com/sh/qjcyey8mtyd8mkh/AABPktep24zf6vzw00TRFLjHa?dl=0
3. Anderson, M., Wojciechowski, P., Subramani, K.: Implementaions of UTVPI algorithms. https://www.dropbox.com/sh/6x4pzwm98a7djq8/AACKiGVNjfjHU1p23IAXDyS1a?dl=0
4. Bagnara, R., Hill, P.M., Zaffanella, E.: Weakly-relational shapes for numeric abstractions: improved algorithms and proofs of correctness. Formal Methods Syst. Des. **35**(3), 279–323 (2009)
5. Cormen, T.H., Leiserson, C.E., Rivest, R.L., Stein, C.: Introduction to Algorithms, 3rd edn. The MIT Press, Cambridge (2009)
6. Cotton, S., Maler, O.: Fast and flexible difference constraint propagation for DPLL(T). In: Biere, A., Gomes, C.P. (eds.) SAT 2006. LNCS, vol. 4121, pp. 170–183. Springer, Heidelberg (2006). https://doi.org/10.1007/11814948_19
7. Cousot, P., Cousot, R.: Abstract interpretation: a unified lattice model for static analysis of programs by construction or approximation of fixpoints. In: POPL, pp. 238–252 (1977)
8. Cox, I.J., Rao, S.B., Zhong, Y.: Ratio regions: a technique for image segmentation. In: Proceedings of the International Conference on Pattern Recognition, pp. 557–564. IEEE, August 1996
9. Gerber, R., Pugh, W., Saksena, M.: Parametric dispatching of hard real-time tasks. IEEE Trans. Comput. **44**(3), 471–479 (1995)
10. Han, C.C., Lin, K.J.: Job scheduling with temporal distance constraints. Technical report UIUCDCS-R-89-1560, University of Illinois at Urbana-Champaign, Department of Computer Science (1989)
11. Harvey, W., Stuckey, P.J.: A unit two variable per inequality integer constraint solver for constraint logic programming. In: Proceedings of the 20th Australasian Computer Science Conference, pp. 102–111 (1997)
12. Jaffar, J., Maher, M.J., Stuckey, P.J., Yap, H.C.: Beyond finite domains. In: Proceedings of the Second International Workshop on Principles and Practice of Constraint Programming (1994)
13. Lahiri, S.K., Musuvathi, M.: An efficient decision procedure for UTVPI constraints. In: Gramlich, B. (ed.) FroCoS 2005. LNCS (LNAI), vol. 3717, pp. 168–183. Springer, Heidelberg (2005). https://doi.org/10.1007/11559306_9

14. Miné, A.: The octagon abstract domain. Higher-Order Symb. Comput. **19**(1), 31–100 (2006)
15. Revesz, P.Z.: Tightened transitive closure of integer addition constraints. In: SARA (2009)
16. Sitzmann, I., Stuckey, P.J.: O-trees: a constraint-based index structure. In: Australasian Database Conference, pp. 127–134 (2000)
17. SRI International. Yices: An SMT solver. http://yices.csl.sri.com/
18. Subramani, K., Wojciechowski, P.: Analyzing lattice point feasibility in UTVPI constraints. In: Beck, J.C. (ed.) CP 2017. LNCS, vol. 10416, pp. 615–629. Springer, Cham (2017). https://doi.org/10.1007/978-3-319-66158-2_39
19. Subramani, K., Wojciechowski, P.J.: A combinatorial certifying algorithm for linear feasibility in UTVPI constraints. Algorithmica **78**(1), 166–208 (2017)

Quality-Aware Online Task Assignment Using Latent Topic Model

Yang Du[1], Yu-E Sun[2(✉)], He Huang[3(✉)], Liusheng Huang[1], Hongli Xu[1],
and Xiaocan Wu[3]

[1] School of Computer Science and Technology,
University of Science and Technology of China, Hefei, China
jannr@mail.ustc.edu.cn, {lshuang,xuhongli}@ustc.edu.cn
[2] School of Rail Transportation, Soochow University, Suzhou, China
sunye12@suda.edu.cn
[3] School of Computer Science and Technology, Soochow University, Suzhou, China
huangh@suda.edu.cn, xiaocanwu9577@gmail.com

Abstract. Crowdsourcing has been proven to be a useful tool for solving the tasks hard for computers. Due to workers' uneven qualities, it is crucial to model their reliabilities for computing effective task assignment plans and producing accurate estimations of truths. However, existing reliability models either cannot accurately estimate workers' fine-grained reliabilities or require external information like text description. In this paper, we consider dividing tasks into clusters (*i.e., topics*) based on workers' behaviors, then propose a Bayesian latent topic model for describing the topic distributions and workers' topical-level expertise. We further present an online task assignment scheme which incorporates the latent topic model to dynamically assign each incoming worker a set of tasks with the Maximum Expected Gain (MEG). The experimental results demonstrate that our method can significantly decrease the number of task assignments and achieve higher *accuracy* than the state-of-the-art approaches.

Keywords: Crowdsourcing · Latent topic model
Online task assignment · Truth discovery

1 Introduction

In recent years, crowdsourcing has emerged as a promising paradigm which leverages the wisdom of crowds to solve the tasks that are hard for computers, such as image tagging [19] and sentiment analysis [11]. The popular platforms, like AMT [5] and CrowdFlower [4], also enable the tasks requester to access the crowds conveniently. However, crowdsourcing still suffers from critical data quality problem since workers are not always trustable [9]. In fact, Vuurens et al. pointed out that only 55% of workers are proper workers whose average precision is only 75%, and 39% of workers are spammers [18]. Thus, the quality control problem has attracted much attention from both academia and industry.

S. Tang et al. (Eds.): AAIM 2018, LNCS 11343, pp. 124–135, 2018.
https://doi.org/10.1007/978-3-030-04618-7_11

To improve the quality of crowdsourced data, a widely adopted strategy is adding redundancy by distributing each task to multiple workers and aggregating noisy answers to infer correct answers (*i.e., truths*) [6]. In this standard workflow, an accurate reliability model that describes workers' expertise levels on different tasks can significantly benefit the task assignment process and the truth discovery process. Existing work often treats workers' reliabilities as single values (worker probability, WP) but overlooks workers' fine-grained reliabilities. Given WP's limitation, some research employs the confusion matrix (CM) that models a worker's probability of answering a when the truth is d. Also, the latent domain model (LDM) tries to capture the domain-level reliability by dividing them into latent domains based on their text description. A significant drawback of CM and LDM is that the reliability model they learned may not be accurate since a worker can show uneven quality on a set of tasks with the same truth or same task description. For example, the BlueBird dataset is collected by asking 39 workers to provide judgments for 108 images that whether a given image contains a duck. Suppose the correct judgments of the images are known, we can use 39 workers' correctness on an image as this image's features. Then we perform dimension reduction on the images' features to reduce the dimensionality from 39 to 2. In Fig. 1, we plot the images based on their reduced features and use *red diamonds/blue dots* to represent the images with truths *YES/NO*. We further run the k-means algorithm with parameter $k = 4$ on the reduced features and plot the cluster boundaries with different background colors. By this means, we can easily distinguish that there exists finer clustering structure among the tasks with the same truth or text description.

Intuitively, we can utilize workers' behaviors to divide tasks into latent topics and estimate workers' topical-level expertise. Therefore, we present a novel latent topic model which assumes there exist hidden topics among the tasks, and a worker shows the same reliability on the tasks belonging to the same topic. In this paper, we consider a crowdsourcing system where there exists a crowdsourcing platform, a set of task requesters, and a set of crowdsourcing workers (as shown in Fig. 2). We focus on the online task assignment problem where workers arrive online, and the platform can only assign tasks to a worker upon arrival. Based on the proposed latent topic model, we can estimate a task's topic distribution from

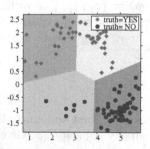

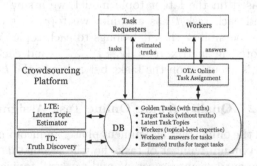

Fig. 1. Task clusters in Bluebird   **Fig. 2.** Framework of QAOTA

currently collected answers and further predict a worker's answering behavior on this task. Then we propose an adaptive quality-aware task assignment scheme that dynamically assigns each incoming worker a set of tasks with Maximum Expected Gain (MEG). In summary, three main contributions of this work are as follows:

(1) To the best of our knowledge, we are the first to utilize the latent topic model to design online task assignment scheme that dynamically estimates the topic distribution of tasks and makes predictions for workers' answering behaviours.
(2) We employ a Bayesian latent topic model that utilizes workers' answering behaviors to divide tasks into topics, estimates workers' topical-level expertise, and propose an online task assignment scheme that dynamically assigns each incoming worker a set of tasks with Maximum Expected Gain.
(3) We empirically show that the proposed online task assignment scheme can significantly outperform the state-of-the-art approaches by achieving higher accuracy with fewer number of assignments.

2   Problem Statement

2.1   Data Model

We assume that each target/golden task t has B_t candidate answers $C_t = \{c_{t,b}\}_{b=1}^{B_t}$ and one correct answer d_t (*i.e., truth*). Let $\mathcal{T} = \{t_1, t_2, \cdots, t_M\}$ be a set of M target tasks published by the task requesters and let $\mathcal{T}' = \{t'_1, t'_2, \cdots, t'_{M'}\}$ be the M' golden tasks with known truths that the platform maintains. Given the truths of the target tasks are unknown a priori, we record the estimated truths of the target tasks with $\{d_t^*|t \in \mathcal{T}\}$. As in literature [15], the platform can estimate the reliabilities of workers by assigning golden tasks to them in the forms of qualification tests or hidden tests. Assuming that there exists a set of N workers $\mathcal{W} = \{w_1, w_2, \cdots, w_N\}$, we use notation A to store the currently collected answers for the target tasks and use notation A' to record workers' answers for the golden tasks. Also, we use notation $\pi_{a_i}^t(\pi_{a'_i}^t)$, $\pi_{a_i}^w(\pi_{a'_i}^w)$ to represent the associated task and the provider of a given answer $a_i(a'_i) \in A(A')$. Based on the latent topic model, we assume that there exist K latent topics and each task t has an unknown topic distribution $\phi_t = \{\phi_{t,k}\}_{k=1}^K$ recording the possibility that it belongs to each topic. Workers' topical-level expertise is stored in matrix $R = \{r_{n,k}\}_{n=1,k=1}^{N,K}$ and each element $r_{n,k}$ represents worker w_n's accuracy on the tasks belonging to the k-th topic.

2.2   Quality-Aware Online Task Assignment

In the online task assignment phase, an incoming worker w will first announce the maximum number of tasks he will accept this time, namely capacity s_w. Given the current answer set A^c and workers' topical-level expertise, we can compute the *current distribution matrix* of task t_m, denoted by $Q_t^c = \{q_{m,k,b}^c\}_{k=1,b=1}^{K,B_{t_m}}$. In this matrix, an element $q^c_{m,k,b}$ represents the possibility that the truth of task t_m is b and this task belongs to the k-th topic. Thus, we can compute the possibility that the truth of task t_m is b by summarizing all the elements associated with the subscript b, formally: $\sum_{k=1}^K q^c_{m,k,b}$. Then we output the estimated truths by selecting the answers with the maximum possibilities of being the truths. Letting $Q^c = \{Q_1^c, Q_2^c, \cdots, Q_M^c\}$ denote the current distribution matrices of all tasks, the accuracy of the estimated truths can be computed by:

$$F(Q^c) = \frac{\sum_{m=1}^M \max_{b \in [1, B_{t_m}]} \left(\sum_{k=1}^K q^c_{m,k,b} \right)}{M} \tag{1}$$

In this paper, we focus on the online task assignment problem. Let V_w denote the tasks assigned for worker w and let Q^{V_w} denote the updated distribution matrix after collecting worker w's answers for the tasks in V_w. The goal of online task assignment is to choose an optimal set of unfinished tasks, namely V_w^*, that maximizes the accuracy of estimated truths after collecting w's answers for these tasks. We formally define the quality-aware online task assignment problem as:

Definition 1 (Quality-aware Online Task Assignment). *When a worker w announces his capacity s_w, given the workers' topical-level expertise R, the current answer set A^c, and the current distribution matrices Q^c, the problem of quality-aware online task assignment is to choose the optimal task set V_w^* with no more than s_w tasks such that $V_w^* = \arg \max_{V_w} F(Q^{V_w})$.*

3   Latent Topic Estimation

In this section, we first demonstrate our parameter estimation method based on the Gibbs-EM algorithm, then propose our model selection criterion.

3.1   Bayesian Latent Topic Model

The inputs of the Bayesian latent topic model (BLTM) are M' golden tasks $\mathcal{T}' = \{t_1', t_2', \cdots, t_{M'}'\}$, N workers $\mathcal{W} = \{w_1, w_2, \cdots, w_N\}$, topic number K, and workers' correctness on the golden tasks X. The outputs of BLTM are tasks' topic distribution $\{\phi_t\}_{t=1}^{t=M'}$, workers' topical-level expertise $R = \{r_{n,k}\}_{n=1,k=1}^{N,K}$. We use α to denote the Dirichlet hyper-parameters of latent topic distributions and use z_i to denote the latent topics of the tasks. In the following, we will describe the generative processes of the latent topics and the topical-level expertise.

Latent Topics. We assume that there exist K task clusters, namely topics, in the crowdsourcing system. Letting $\{\phi_t\}_{t=1}^{t=M'}$ denote the topic distributions of tasks, we assume that for each task t, its topic distribution ϕ_t is drawn from a Dirichlet distribution with parameters $\alpha = \{\alpha_k\}_{k=1}^K$.

Topical-Level Expertise. By assuming that a worker maintains the same reliability level on the tasks belonging to the same topic, we model each worker w's expertise as a K-dimensional vector $\{r_{w,k}\}_{k=1}^{K}$, where each element $r_{w,k}$ denotes worker w's reliability level on the k-th topic. By assuming that the wrong answers are uniformly distributed, the possibility that worker w submits answer a for a task t belonging to the k-th topic is:

$$p(a|z_t = k, d_t = b, r_{w,k}) = (r_{w,k})^{\delta(a-b)} \cdot \left(\frac{1 - r_{w,k}}{B_t - 1}\right)^{1-\delta(a-b)} \tag{2}$$

In Eq. 2, we use notation $\delta(\cdot)$ to represent the Kronecker function which equals 1 if the input is 0, and 0 otherwise.

3.2 Parameter Estimation

Given the observations X, the objective of the proposed model is to learn the optimal model parameters (α, R) which maximize the likelihood of observing X. However, it is intractable to directly compute the likelihood $p(X|\alpha, R)$. Thus, we employ Gibbs-EM [13] to find the optimal model parameters by iteratively sampling latent variables and updating the parameters.

In the E-step, we employ a Gibbs sampler to sequentially sample each latent variables z_i from the distribution over this variable given the observations and all other latent variables. Let the subscript $\neg i$ denote a set of data with the i-th element being excluded. The conditional posterior of latent variable z_i can be computed as:

$$p(z_i = k|Z_{\neg i}, X, \alpha, R) = \frac{p(Z, X|\alpha, R))}{p(Z_{\neg i}, X|\alpha, R)} \propto \frac{p(Z, X|\alpha, R))}{p(Z_{\neg i}, X_{\neg i}|\alpha, R)} \tag{3}$$

Based on the chain rule, we can factorize the term $p(Z, X|\alpha, R)$ into the product of two factors, formally: $p(Z, X|\alpha, R) = p(Z|\alpha) \cdot p(X|Z, R)$. Given the Dirichlet-Multinomial conjugacy, we can sequentially sample z_i by:

$$p(z_i = k|Z_{\neg i}, X, \alpha, R) \propto (n_{\pi_{a_i}^t, zi} + \alpha_{z_i}) \cdot r_{\pi_{a_i}^w, z_i}^{x_i} \cdot (1 - r_{\pi_{a_i}^w, z_i})^{1-x_i} \tag{4}$$

In the M-step, we update the model parameters (α, R) with sampled latent variables. Letting $\{Z^{(s)}\}_{s=1}^{S}$ be the generated latent variables, we can use Newton's method to update the Dirichlet hyper-parameters α. Letting α_0 be the sum of all elements in α, letting $n_{t_m}^{(r)}$ be the sum of topic labels assigned to task t_m, and letting $n_{t_m,k}^{(s)}$ be the times that a task t_m was assigned with k-th topic labels in the s-th round samples, we can compute the gradient of the log-likelihood $p(Z|\alpha)$ by:

$$g_{\alpha_k} = \sum_{s=1}^{S} \sum_{m=1}^{M} \Psi(\alpha_0) - \Psi(n_{t_m}^{(s)} + \alpha_0) + \Psi(n_{t_m,k}^{(s)} + \alpha_k) - \Psi(\alpha_k) \tag{5}$$

With the following Hessian matrix:

$$h_{jk} = \sum_{s=1}^{S} \sum_{m=1}^{M} \Psi'(\alpha_0) - \Psi'(n_{t_m}^{(s)} + \alpha_0) + \delta(j-k)\left(\Psi'(n_{t_m,k}^{(s)} + \alpha_k) - \Psi'(\alpha_k)\right), \tag{6}$$

we can finally update the hyper-parameters α by $\alpha_{new} = \alpha_{old} - H^{-1}g_\alpha$. For the model parameters R, we update a worker w's topical-level expertise on the k-th topic by: $r_{w,k} = \frac{n_{w,k,1}+1}{n_{w,k,0}+n_{w,k,1}+2}$.

3.3   Model Selection

In the following, we describe our criterion for determining the number of topics. Based on the integrated completed likelihood (ICL) [7], the proposed metric computes the expectation of the joint probability of the observations and latent variables. Formally, the ICL value of a fixed K can be computed by:

$$\text{ICL}(K) = \log \int_{(\alpha,R)} p(X, Z|\alpha, R)\, p(\alpha, R)\, d\alpha\, dR \qquad (7)$$

We follow an approximation strategy as described in [1] that replaces the latent variables by their MAP estimations, which is: $\hat{Z} = \arg\max_Z p(Z|X, \hat{\alpha}, \hat{R})$. Given that $p(X, Z|\alpha, R)$ can be factorized into the product of $p(X|Z, R)$ and $p(Z|\alpha)$, we employ the Bayesian information criterion (BIC) to approximate each term and obtain a BIC-like approximation for ICL:

$$\text{ICL}(K) \simeq \log p(\hat{Z}|\hat{\alpha}) + \log p(X|\hat{Z}, \hat{R}) - \frac{K}{2}\log|X| - \frac{K}{2}\sum_{n=1}^{N}\log|X_{w_n}| \qquad (8)$$

In Eq. 8, we use notation $|X_{w_n}|$ to denote the number of answers that worker w_n has submitted.

4   Online Task Assignment

In this section, we first describe the computation process of the current distribution matrix, then propose a quality-aware online task assignment (OTA) scheme based on LTM.

4.1   Computing Current Distribution Matrix

Given the workers' topical-level expertise and their answers for task t, we can compute the current distribution matrix $Q_t^c = \{q_{t,k,b}^c\}_{k=1,b=1}^{K,B_t}$ where a matrix element $q_{t,k,b}^c$ represents the posterior probability that topic is k and truth is b, formally $q_{m,k,b}^c = p(z_t = k, d_t = b|A^{(t)}, R)$. Given the fact that $p(A^{(t)}|R)$ is a constant term, the formal equation can be transformed to $q_{m,k,b}^c \propto p(z_t = k, d_t = b, A^{(t)}|R)$. Letting $X_{t|b}$ denote the correctness of answers $A^{(t)}$ given truth b, $q_{m,k,b}^c$ can be obtained by:

$$q_{m,k,b}^c = p(d_t = b, z_t = k)\, p(X_{t|b}|R, z_t = k)\, p(A^{(t)}|d_t = b, z_t = k, X_{t|b}) \qquad (9)$$

In Eq. 9, factors $p(d_t = b, z_t = k)$ represents the prior probability that a task with truth b belongs to the k-th topic and can be estimated from workers' answers for golden tasks. Combining the Eq. 2 and the definition of topical-level expertise, we can compute the matrix element by:

$$q^c_{t,k,b} \propto p(d_t = b, z_t = k) \prod_{a_i \in A^{(t)}} (r_{\pi^w_{a_i},k})^{\delta(a_i - b)} \cdot \left(\frac{1 - r_{\pi^w_{a_i},k}}{B_t - 1} \right)^{1 - \delta(a_i - b)} \quad (10)$$

Truth Inference. Given the current distribution matrix Q^c_t, we select the answer with the highest probability to be the truth as the estimated truth d_t. Formally, the estimated truth is chosen by:

$$d^*_t = \arg \max_{b \in [1, B_t]} p(d_t = b | A^{(t)}, R) = \arg \max_{b \in [1, B_t]} \sum_{k=1}^{K} q^c_{t,k,b} \quad (11)$$

4.2 Adaptive Task Assignment

In this following, we will demonstrate our online task assign assignment scheme that follows the principle of Maximum Expected Gain (MEG) with a block parameter *Thres*.

Maximum Expected Gain. The MEG principle first estimates the expected gain in accuracy if a worker is assigned to complete a given task, then allocates the tasks with maximum expected gains ($\geq Thres$) to this worker. Assuming that worker w has been assigned with a target task t, the worker is required to provide an answer $a_{t,w}$ for this task. Letting $Q^{a_{t,w}}$ denote the updated distribution matrix when worker w provides answer $a_{t,w}$, the expected accuracy $G(t, w)$ which represents the weighted sum of the estimated truth's accuracy when worker w providing different answers can be computed by:

$$G(t, w) = \sum_{a_{t,w} \in [1, B_t]} p(a_{t,w} | A^{(t)}, \{r_{w,k}\}^K_{k=1}) \cdot F(Q^{a_{t,w}}) \quad (12)$$

In Eq. 12, the first term $p(a_{t,w} | A^{(t)}, \{r_{w,k}\}^K_{k=1})$ can be obtained by enumerating all possible combinations (z_t, d_t) and summarizing the joint probabilities $p(a_{t,w}, z_t, d_t | A^{(t)}, \{r_{w,k}\}^K_{k=1})$, which is:

$$p(a_{t,w} | rest) = \sum_{k=1}^{K} \sum_{b=1}^{B_t} p(a_{t,w} | z_t = k, d_t = b, r_{w,k}) \cdot q^c_{t,k,b} \quad (13)$$

With the previous defined topical-level expertise model, we can further replace the term $p(a_{t,w} | z_t = k, d_t = b, r_{w,k})$ with $(r_{\pi^w_{a_i},k})^{\delta(a_{t,w} - d_t)} \cdot (\frac{1 - r_{\pi^w_{a_i},k}}{B_t - 1})^{1 - \delta(a_{t,w} - d_t)}$.

In Eq. 10, we show that an element $q^c_{t,k,b}$ is in proportion to the product of the $(z_t = k, d_t = b)$'s prior and workers' conditional probabilities of providing their answer when task t belongs to the k-th topic. Based on this, we can update the distribution matrix by:

$$q_{t,k,b}^{a_{t,w}} \propto q_{t,k,b}^c \cdot p(a_{t,w}|z_t = k, d_t = b, r_{w,k}) \tag{14}$$

Recall that the sum of the elements in a distribution matrix Q_t^c equals to 1, we can normalize the updated distribution matrix with the following equation:

$$q_{t,k,b}^{a_{t,w}} = \frac{q_{t,k,b}^c \cdot p(a_{t,w}|z_t = k, d_t = b, r_{w,k})}{\sum_{k=1}^K \sum_{b=1}^{B_t} q_{t,k,b}^c \cdot p(a_{t,w}|z_t = k, d_t = b, r_{w,k})}, \tag{15}$$

where the denominator is the term $p(a_{t,w}|A^{(t)}, \{r_{w,k}\}_{k=1}^K)$. Given Eqs. 1 and 11, the expected accuracy can be computed by:

$$G(t, w) = \sum_{a_{t,w}} p(a_{t,w}|A^{(t)}, \{r_{w,k}\}_{k=1}^K) \cdot \arg\max_b \sum_{k=1}^K q_{t,k,b}^{a_{t,w}} \tag{16}$$

Based on Eq. 15, we can integrate out the term $p(a_{t,w}|A^{(t)}, \{r_{w,k}\}_{k=1}^K)$ and obtain the following equation:

$$G(t, w) = \sum_{a_{t,w}} \max_b \sum_{k=1}^K q_{t,k,b}^c \cdot (r_{\pi_{a_i}^w, k})^{\delta(a_{t,w}-b)} \cdot \left(\frac{1 - r_{\pi_{a_i}^w, k}}{B_t - 1}\right)^{1-\delta(a_{t,w}-b)} \tag{17}$$

5   Experiments

In this section, we first describe our experimental setup, then evaluate our online task assignment scheme in two aspects: accuracy and the number of assignments.

5.1   Experimental Setup

Datasets. To evaluate the performance of our model, we run experiments on four real-world datasets: Fact[1], BlueBird [19], Valence [17], and WSD [17]. As shown in Table 1, we list the model selection results on the original datasets. The model selection process on the original BlueBird dataset is shown in Fig. 3. By plotting the ICL values associated with different K, we can see that the ICL value first grows when K increases, then falls after reaching the peak value, i.e., -2221. Therefore, we can say that the $K = 4$ is the best choice for the Bluebird dataset. Likewise, we compare the ICL values on the other three datasets and select their optimal topic numbers. In our experiments, we employ a bootstrapping method as described in [22] to simulate the golden tasks and workers' answers on the golden tasks. We assume that each worker has answered $M' = 40$ golden tasks in the past. Thus, we simulate 1000 training sets for each dataset, where each contains workers' answers for 40 simulated tasks. By employing the random permutation model as in [2] to permutate the arrival order, we run online

Table 1. Description of the datasets

Dataset	Properties					
	#Tasks	#Users	#Per-task answers	Sparsity	#Topics	ICL
BlueBird	108	39	39	0	4	−2221
Fact	550	91	5.08	0.9440	2	−647
Valence	100	38	10	0.7368	2	−549
WSD	177	34	10	0.7058	2	−213

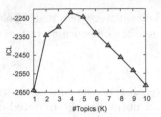

Fig. 3. ICL values on Bluebird

task assignment experiments for our methods and baselines, then evaluate the schemes by their average performance over 1000 rounds.

Baseline Methods. For the evaluation of our OTA scheme, we compared MEG-LTM with three baselines: (1) RR-MV, (2) CrowdDQS, and (3) QASCA. The simple baseline RR-MV treats workers as equal and distributes tasks evenly among the workers. CrowdDQS employs WP to model workers' reliabilities and computes the maximum potential gain that an incoming worker can achieve on each task to dynamically assign tasks for this workers. In our experiments, we run QASCA with CM model since it can better capture workers' fine-grained reliabilities than the WP model.

Evaluation Metrics. We evaluate our OTA scheme and three baselines in terms of *accuracy* and the *number of assignments*. Accuracy denotes the proportion of correctly inferred tasks among all the target tasks, and we use the number of assignments to record the number of votes that an OTA scheme collects in an experiment.

Experiment Workflow. To evaluate an OTA scheme on a dataset with fixed worker capacity, we first run its reliability estimation module on the boot-strapped training set, then perform OTA schemes under the worker capacity constraint, and finally evaluate the OTA schemes in terms of accuracy and the number of assignments. In the experiments, we set the block parameter $Thres$ to 0.001 for the schemes which dynamically allocate tasks based on estimated gains.

5.2   Performance of Online Task Assignment

We plot the accuracy of each algorithm in Fig. 4 and plot the number of assignments in Fig. 5. In Figs. 4(a) and 5(a), we compare the performance of the OTA schemes on the BlueBird dataset when the worker capacity varies from 10 to 100. In Fig. 4(a), we notice that the accuracy of MEG-LTM first increases rapidly when capacity increases and then slowly increases when capacity exceeds 30. From Fig. 5(a), we can observe that the assignment number of MEG-LTM first increases when workers' capacity is less than 30 and then become stable. One

---

[1] https://sites.google.com/site/crowdscale2013/shared-task/task-fact-eval.

possible reason is that when the capacity grows from 10 to 30, the number of answers that collected for each task is still small. In that case, enabling workers to contribute more answers can significantly improve the overall accuracy of aggregated answers, which means the proposed method MEG-LTM will only block a few assignments. However, when workers' capacity exceeds 30, the high-quality workers are allocated with more tasks which significantly improves the overall accuracy. In this case, low-quality workers are blocked from taking assignments. Thus, the overall accuracy and the assignment number of MEG-LTM becomes stable when the capacity exceeds 30. We notice that the proposed method MEG-LTM shows obvious advantages than two baselines CrowdDQS and RR-MV. MEG-LTM can also utilize significantly fewer assignments to achieve slightly lower accuracy than QASCA. In particular, MEG-LTM uses only 505.5 assignments (70.8% fewer than QASCA) and achieves 89.72% overall accuracy (0.89% lower than QASCA) when the worker capacity is 100. When comparing MEG-LTM with the second best baseline CrowdDQS, MEG-LTM shows significant advantages in both accuracy and the assignment number. In detail, MEG-LTM achieves 4.15% higher accuracy than CrowdDQS with 26.80% assignments and achieves 14.28% higher accuracy than RR-MV with only 12.96% assignments when the worker capacity is 30. While comparing the proposed method MEG-LTM with three baselines on the other three datasets, we notice that MEG-LTM outperforms all baselines in both accuracy and the number of assignments.

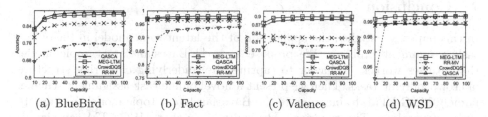

(a) BlueBird   (b) Fact   (c) Valence   (d) WSD

Fig. 4. Accuracy w.r.t. capacity

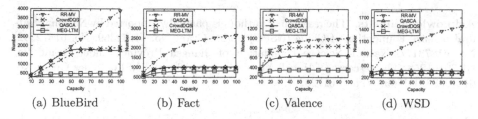

(a) BlueBird   (b) Fact   (c) Valence   (d) WSD

Fig. 5. Number of assignments w.r.t. capacity

6   Related Work

There has been extensive work that incorporates worker expertise into task assignment and truth discovery. In the following, we will begin with current task assignment schemes that employ different reliability models, then discuss the application of reliability model in truth discovery. The idea of task assignment is to assign appropriate tasks to workers and improve the quality of estimated truths. CDAS [11] employes the WP model and dynamically estimates workers' quality. CrowdDQS [8] also employs the WP model, and it dynamically chooses to issue more golden tasks to a worker for reliability estimation or to assign target tasks to collect data. QASCA [23] conducts the task assignment by allocating the tasks with maximum accuracy (or f1-score) gains to each incoming worker. Unlike previous OTA schemes, DOCS [21] utilizes the existing knowledge base to divide tasks into latent domains, and further conducts task assignments based on workers' domain-level expertise. Researchers have also proposed various truth discovery approaches that utilize different reliability models. Based on the WP model, Yin et al. [20] employed an EM algorithm to learn workers' reliabilities and the truths. Dong et al. [10] conducted truth discovery by exploiting the confidence interval. To capture the fine-grained expertise, some work [14,16] employed the CM models to describe workers' behaviors under different truths. Some other work [12,19] utilized the task information (e.g., text description) to divide tasks into latent domains and further estimate workers' domain-level expertise. Also, Du et. al. [3] considered the co-clustering structure in a crowdsourcing system and tried to estimate workers' expertise on different task clusters.

7   Conclusion

In this paper, we studied the application of the latent topic model in the online task assignment problem. We propose a novel Bayesian latent topic model that utilizes workers' behavior to capture the underlying clustering structure of crowdsourcing tasks. Then we present a novel online task assignment scheme, namely MEG-LTM, to incorporate the Bayesian latent topic model to compute assignment plans. The experimental results show that MEG-LTM can significantly outperform the state-of-the-art approaches by achieving higher accuracy with fewer number of assignments.

Acknowledgements. The research of authors is partially supported by National Natural Science Foundation of China (NSFC) under Grant No. 61672369, No. 61572342, No. 61873177, Natural Science Foundation of Jiangsu Province under Grant No. BK20161258. The research of Hongli Xu is supported by the NSFC under Grant No. 61472383, U1709217, and 61728207.

References

1. Biernacki, C., Celeux, G., Govaert, G.: Assessing a mixture model for clustering with the integrated completed likelihood. IEEE Trans. Pattern Anal. Mach. Intell. **22**(7), 719–725 (2000)
2. Devanur, N.R., Hayes, T.P.: The adwords problem: online keyword matching with budgeted bidders under random permutations. In: Proceedings of the EC, pp. 71–78. ACM (2009)

3. Du, Y., Xu, H., Sun, Y.-E., Huang, L.: A general fine-grained truth discovery app-roach for crowdsourced data aggregation. In: Candan, S., Chen, L., Pedersen, T.B., Chang, L., Hua, W. (eds.) DASFAA 2017. LNCS, vol. 10177, pp. 3–18. Springer, Cham (2017). https://doi.org/10.1007/978-3-319-55753-3_1
4. Finin, T., Murnane, W., Karandikar, A., Keller, N., Martineau, J., Dredze, M.: Annotating named entities in Twitter data with crowdsourcing. In: Proceedings of the NAACL HLT 2010 Workshop on Creating Speech and Language Data with Amazon's Mechanical Turk, pp. 80–88. ACL (2010)
5. Gabriele, P., Jesse, C., Ipeirotis, P.G.: Running experiments on amazon mechanical turk. Judgm. Decis. Mak. **5**(5), 411–419 (2010)
6. Gao, J., Li, Q., Zhao, B., Fan, W., Han, J.: Truth discovery and crowdsourcing aggregation: a unified perspective. Proc. VLDB Endow. **8**(12), 2048–2049 (2015)
7. Keribin, C., Brault, V., Celeux, G., Govaert, G.: Estimation and selection for the latent block model on categorical data. Stat. Comput. **25**(6), 1201–1216 (2015)
8. Khan, A.R., Garcia-Molina, H.: CrowdDQS: dynamic question selection in crowd-sourcing systems. In: Proceedings of SIGMOD, pp. 1447–1462. ACM (2017)
9. Li, G., Wang, J., Zheng, Y., Franklin, M.J.: Crowdsourced data management: a survey. IEEE Trans. Knowl. Data Eng. **28**(9), 2296–2319 (2016)
10. Li, Q., et al.: A confidence-aware approach for truth discovery on long-tail data. Proc. VLDB Endow. **8**(4), 425–436 (2014)
11. Liu, X., Lu, M., Ooi, B.C., Shen, Y., Wu, S., Zhang, M.: CDAS: a crowdsourcing data analytics system. Proc. VLDB Endow. **5**(10), 1040–1051 (2012)
12. Ma, F., et al.: Faitcrowd: fine grained truth discovery for crowdsourced data aggre-gation. In: Proceedings of ACM SIGKDD, pp. 745–754. ACM (2015)
13. Minka, T.: Estimating a Dirichlet Distribution (2000)
14. Moreno, P.G., Artés-Rodríguez, A., Teh, Y.W., Perez-Cruz, F.: Bayesian nonpara-metric crowdsourcing. J. Mach. Learn. Res. **16**(1), 1607–1627 (2015)
15. Oleson, D., Sorokin, A., Laughlin, G.P., Hester, V., Le, J., Biewald, L.: Program-matic gold: targeted and scalable quality assurance in crowdsourcing. Hum. Com-put. **11**(11) (2011)
16. Simpson, E., Roberts, S., Psorakis, I., Smith, A.: Dynamic Bayesian combination of multiple imperfect classifiers. In: Guy, T., Karny, M., Wolpert, D. (eds.) Intelligent Systems Reference Library Series: Decision Making and Imperfection, pp. 1–35. Springer, Heidelberg (2013). https://doi.org/10.1007/978-3-642-36406-8_1
17. Snow, R., O'Connor, B., Jurafsky, D., Ng, A.Y.: Cheap and fast–but is it good?: evaluating non-expert annotations for natural language tasks. In: Proceedings of EMNLP, pp. 254–263. Association for Computational Linguistics (2008)
18. Vuurens, J., de Vries, A.P., Eickhoff, C.: How much spam can you take? An anal-ysis of crowdsourcing results to increase accuracy. In: Proceedings ACM SIGIR Workshop on Crowdsourcing for Information Retrieval (CIR11), pp. 21–26 (2011)
19. Welinder, P., Branson, S., Perona, P., Belongie, S.J.: The multidimensional wisdom of crowds. In: Proceedings of NIPS, pp. 2424–2432. Curran Associates, Inc. (2010)
20. Yin, X., Han, J., Yu, P.S.: Truth discovery with multiple conflicting information providers on the web. IEEE Trans. Knowl. Data Eng. **20**(6), 796–808 (2008)
21. Zheng, Y., Li, G., Cheng, R.: DOCS: a domain-aware crowdsourcing system using knowledge bases. Proc. VLDB Endow. **10**(4), 361–372 (2016)
22. Zheng, Y., Li, G., Li, Y., Shan, C., Cheng, R.: Truth inference in crowdsourcing: is the problem solved? Proc. VLDB Endow. **10**(5), 541–552 (2017)
23. Zheng, Y., Wang, J., Li, G., Cheng, R., Feng, J.: QASCA: A quality-aware task assignment system for crowdsourcing applications. In: Proceedings of SIGMOD, pp. 1031–1046. ACM (2015)

Calibration Scheduling with Time Slot Cost

Kai Wang$^{(\boxtimes)}$

Department of Computer Science, City University of Hong Kong,
83 Tat Chee Avenue, Kowloon, Hong Kong SAR, China
kai.wang@my.cityu.edu.hk

Abstract. In this paper we study the scheduling problem with calibration and time slot cost. In this model, the machine has to be calibrated to run a job and the calibration remains valid for a fixed time period of length T, after which it must be recalibrated before running more jobs. On the other hand, a certain cost will be incurred when the machine executes a job and the cost is determined by the time slots occupied by the job in the schedule. We work on the jobs with release times, deadlines and identical processing times. The objective is to schedule the jobs on a single machine and minimize the total cost while calibrating the machine at most K times. We propose dynamic programmings for different scenarios of this problem, as well as a greedy algorithm for the non-calibration version of this problem.

1 Introduction and Related Work

The scheduling with calibrations was originally motivated from the Integrated Stockpile Evaluation (ISE) program which requires expensive calibrations to test nuclear weapons periodically [4]. This motivation can be extended to the scenarios where the machines need to be calibrated periodically to ensure high-quality products, which has many industrial applications, including robotics and digital cameras [3,10,14]. In this calibration model, the machine must be calibrated before it runs a job. When the machine is calibrated at time t, it stays in calibrated status for a fixed time period of length T, after which it must be recalibrated to continue running the jobs. The time interval $[t, t + T]$ is referred to as the *calibration interval*. In the ideal model, calibrating a machine is instantaneous, meaning that the machine could continue running the job immediately after being calibrated and the machine can switch from uncalibrated to calibrated status instantaneously.

K. Wang—We thank Prof. Minming Li and Mr. Yanbin Gong for their helpful discussions. The work described in this paper was supported by a grant from Research Grants Council of the Hong Kong Special Administrative Region, China (Project No. CityU 11268616).

S. Tang et al. (Eds.): AAIM 2018, LNCS 11343, pp. 136–148, 2018.
https://doi.org/10.1007/978-3-030-04618-7_12

There are few theoretical results about scheduling with calibrations. In 2013, Bender et al. [2] proposed a theoretical framework for scheduling with calibrations. They considered jobs of unit processing time with release times and deadlines, aiming at minimizing the total number of calibrations. In the single-machine setting, they proposed a greedy, optimal, polynomial-time algorithm called *Lazy-Binning*, and for the multiple machine setting, they showed that the Lazy-Binning algorithm on multiple machines is 2-approximation, while the complexity status of this problem still remains open. Fineman and Sheridan [7] generalized the problem with resource-augmentation [9] and considered the jobs with non-unit processing times on multiple machines. They showed the relationship of the problem with the classical machine-minimization problem [12]. Angel et al. [1] developed different results on several generalizations of this problem, including many calibration types and calibration activation time. Chau et al. [5] worked on the trade-off between weighted flow time and calibration cost for unit-time jobs and gave both online approximation results and offline optimal result for this problem. They gave several online approximation results on different settings of single or multiple machines for weighted or unweighted jobs and also a dynamic programming for the offline problem.

On the other hand, for the classical scheduling problem without calibration, Wan and Qi [13] introduce time slot cost into the objective function, in which the time is discretized into many time slots by unit length and whenever a job is scheduled during a time slot, a certain cost must be incurred. For example, the price of electricity can be different over time, as well as the availability of electricity. Later research of integrating the objective with time slot cost can be found in [6,15,16].

In this paper, we work on the scheduling problem with the consideration of both calibrations on the machine and the time slot cost. Instead of integrating the objective with these two aspects, we investigated the problem on a single machine aiming to minimize the total time slot cost with a limited number of calibrations. Especially we study the problem of jobs for several cases (Sect. 2). We first propose a dynamic programming approach to solve the problem for the jobs of identical processing time (Sect. 2.1). Then we show an improved dynamic programming approach when the jobs are agreeable and have identical processing time (Sect. 2.2). At last, for the special case where the time slot function is monotonic we show that the running time of the dynamic programming can be further reduced (Sect. 2.3). Moreover, for the problem without the consideration of calibration, we propose an efficient greedy algorithm for jobs of arbitrary processing times (Sect. 3).

The main contribution of this work is to extend the dynamic programming technique from Chau et al. [5]. The foundation of the dynamic programming approach in both our work and their work is that once we fix the schedule of one job, we are able the divide the problem into several sub-problems. And such a job is the job of the smallest weight in their work since the optimal schedule follows Smallest-Weight-First scheduling policy when the available time slots are predetermined. Similarly, in our work such a job is the job of the latest deadline, as the optimal schedule follows EDF (Earliest-Deadline-First) scheduling policy. Moreover, in our work, we extend the model to consider jobs of non-unit identical processing times.

Formulation. We are given a set J of n jobs, where each job $j \in J$ has release time r_j, deadline d_j, processing time p_j. For each job $j \in J$, we call $I_j = (r_j, d_j]$ the time interval of job j. In this paper, we work on the scenario where jobs have identical processing time, i.e. $p_j = p, \forall j \in J$ where p is a constant. We consider the schedule on a single machine which can be trusted to run a job only when it is calibrated. The calibration remains valid for a consecutive time period of length T once started. In this model, time is divided into many unit slots, and we denote the time interval $(t-1, t]$ as *time slot t*. During each time slot, the machine could process a job by one unit of workload, and each time slot t is associated with a non-negative cost $c(t)$, i.e. when the machine executes a job during time slot t, it will incur the cost $c(t)$. A feasible solution includes the schedule of calibrations (i.e. when to start a calibration) and the schedule of jobs (i.e. when to start a job). The objective is to schedule all the jobs on a single machine such that each job is scheduled during its time interval non-preemptively, so as to minimize the total time slot cost while calibrating the machine at most K times where K is input. The problem is denoted as $1|r_j, d_j, p_j = p, T, K| \sum_t c(t)$, according to the classical three-field notation [8].

Let interval $I_0 = (\min_{j \in J} r_j, \max_{j \in J} d_j]$ be the time horizon of the schedule and let $L = \max_{j \in J} d_j - \min_{j \in J} r_j$ be the length of the time horizon. Without loss of generality we assume $L \geq P$ where $P = \sum_{j \in J} p_j$. We assume all the inputs are integers and in any feasible solution both the calibration and a job should start at a time which is an integer.

As a matter of fact, once the available time slots are predetermined, the schedule of jobs can be obtained by applying *Earliest-Deadline-First* (EDF) scheduling algorithm, which always assigns the job of earliest deadline from the pending jobs to the current available time slot.

2   Slot Cost with Calibration

In this section, we work on the problem on a single machine with the consideration of calibrations on the machine. We consider the jobs of identical processing time p and aim to find a schedule on single machine so as to minimize the total time slot cost while calibrating the machine at most K times where K is input. Note that, depending on the input, it might happen that $p \ll T$ or $p \gg T$ where in the latter case we would have to start multiple calibrations in order to finish one job, and hence it might happen that $K \gg n$.

Let $\Psi = \{t \mid \min_{j \in J} r_j \leq t \leq \max_{j \in J} d_j\}$ be the set of possible starting time or completion time of the jobs. In the following, we assume jobs are sorted by non-decreasing order of their deadlines and assume that the index of the jobs starts from 1.

Lemma 1 (EDF). *There exists a feasible schedule such that for any two jobs* i, j *with* $i < j$ *(i.e.* $d_i \leq d_j$*), if* $r_i \leq t_j$ *then* $t_i \leq t_j$ *where* t_i, t_j *are the corresponding starting time of jobs* i *and* j *in the schedule respectively.*

Lemma 1 relies on the swap of the (partial) schedule of job i and j, which works in the preemptive setting when we swap the execution of jobs i and j by one unit, and requires that $p_i = p_j$ in the non-preemptive setting. The proof of Lemma 1 can be found in the full version (Fig. 1).

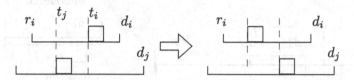

Fig. 1. An illustration for the proof of Lemma 1. When we swap the schedule of the two jobs, the new schedule follows *Earliest-Deadline-First* rule

Distinct Release Time and Deadline Property on Single Machine. As we consider the schedule on a single machine, any two jobs cannot be scheduled at the same time slot. Hence, without loss of generality, we assume that job release times are distinct and that job deadlines are distinct. Formally, for any pair of jobs $i, j \in J$, if $r_i = r_j$ and $d_i \leq d_j$, then we increase r_j by 1, i.e. $r_j \leftarrow r_j + 1$. A similar argument applies to job deadlines.

2.1 Jobs of Identical Processing Times

We first focus on the jobs of identical processing times where the problem is denoted as $1|r_j, d_j, p_j = p, T, K| \sum_t c(t)$. In the following, we propose a dynamic programming approach.

Definition 1. *Let* $J(j, t_1, t_2) = \{i \mid t_1 - p \leq r_i < t_2, i \leq j, i \in J\}$ *be the set of jobs that are released during time interval* $[t_1 - p, t_2)$ *whose index is at most* j *where* $t_1, t_2 \in \Psi$ *and* $j \in J$. *We define* $f(j, t_1, t_2, u_1, u_2, k)$ *to be the minimum total time slot cost in the partial optimal schedule that completes jobs* $J(j, t_1, t_2)$ *during interval* $(t_1, t_2]$ *with at most* k *calibrations providing that the machine has already been calibrated during time intervals* $(t_1, t_1 + u_1]$ *and* $(t_2 - u_2, t_2]$ *where* $u_1, u_2 \in [0, T]$, $k \in [0, K]$.

We look for the optimal schedule in which no two calibrations overlap with each other, while in the proposed dynamic programming approach, we allow the overlap of the calibrations. Note that Lemma 1 works in this model. In the dynamic programming, we test every possibility of the schedule of job j, and assume that job $j \in J(j, t_1, t_2)$ starts at time t in the optimal schedule, then, we divide the remaining jobs into two subsets, $J(j - 1, t_1, t)$ and $J(j - 1, t + p, t_2)$.

For any job i from $J(j-1, t_1, t)$, it must be scheduled before job j in the optimal schedule, because otherwise scheduling job i after job j will violate Lemma 1. Moreover, all jobs of $J(j-1, t+p, t_2)$ must be scheduled after job j in the optimal schedule because they are released as late as t. Therefore, the problem can be divided into two sub-problems: scheduling jobs $J(j-1, t_1, t)$ during time interval $(t_1, t]$ and scheduling jobs $J(j-1, t+p, t_2)$ during time interval $(t+p, t_2]$ (Fig. 2).

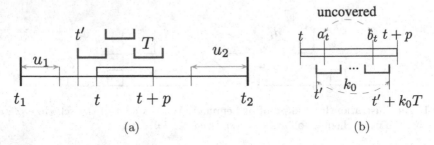

(a)   (b)

Fig. 2. Illustration of the dynamic programming in Proposition 1. (a) Shows the partition of the sub-problems while (b) shows a series of k_0 calibrations to cover the interval $(a_t, b_t]$.

Proposition 1. *Let $F = f(j, t_1, t_2, u_1, u_2, k)$. For the base cases, if $J(j, t_1, t_2) = \emptyset$, we have $F \leftarrow 0$ and if $j \notin J(j, t_1, t_2)$ we have $F \leftarrow f(j-1, t_1, t_2, u_1, u_2, k)$. Let $\mathcal{Q} = \{t \mid t \in [t_1, t_2 - p] \cap [r_j, d_j - p]\}$ be the set of possible starting times of job j. If $\mathcal{Q} = \emptyset$ we have $F \leftarrow \infty$, and if $t_1 + u_1 \geq t_2 - u_2$ we have $F = \min_{t \in \mathcal{Q}} f(j-1, t_1, t, t-t_1, 0, 0) + f(j-1, t+p, t_2, t_2-t-p, 0, 0) + \sum_{z=1}^{p} c(t+z)$. Otherwise, we have $F = \min_{t \in \mathcal{Q}} \sum_{z=1}^{p} c(t+z) +$*

$$
\begin{cases}
\begin{aligned}
&f(j-1, t_1, t, t-t_1, 0, 0) \\
&+ f(j-1, t+p, t_2, t_1 + u_1 - t - p, u_2, k), && \text{if } t \in [t_1, t_1 + u_1 - p] \\
&f(j-1, t_1, t, u_1, t+u_2 - t_2, k) \\
&+ f(j-1, t+p, t_2, t_2 - t - p, 0, 0), && \text{if } t \in [t_2 - u_2, t_2 - p] \\
&\min_{cond.} \; f(j-1, t_1, t, u_1, \max\{t-t', 0\}, k') \\
&+ f(j-1, t+p, t_2, \max\{t^*, 0\}, u_2, k - k_0 - k'), \text{if } t \in (t_1 + u_1 - p, t_2 - u_2)
\end{aligned}
\end{cases}
$$

*where $t^* = t' + k_0 T - t - p$ and **cond.** stands for $k' \in [0, k - k_0]$, $t' \in [t_1, t_2) \cap (a_t - T, a_t]$ and $k_0 = \lceil \frac{b_t - t'}{T} \rceil$, $a_t = \max\{t, t_1 + u_1\}$, $b_t = \min\{t+p, t_2 - u_2\}$.*

Proof. Note that we allow the overlap of the calibrations in the dynamic programming. If $J(j, t_1, t_2) = \emptyset$, nothing needs to be done. If $j \notin J(j, t_1, t_2)$, we would have $J(j, t_1, t_2) = J(j-1, t_1, t_2)$.

The main focus of the dynamic programming is to try every possibility of the starting time of job j, as well as the starting time of the calibrations that job j occupies. Suppose job j starts at time t in the optimal schedule. Combined with the fact that job j is scheduled during time interval $(t_1, t_2]$ in the optimal schedule, we have $t \in [t_1, t_2 - p]$ and $t \in [r_j, d_j - p]$. Therefore set $\mathcal{Q}$ indeed contains all such possible starting times of job j. If $\mathcal{Q} = \emptyset$, it is impossible to schedule job j, hence we have $F \leftarrow \infty$. If $t_1 + u_1 \geq t_2 - u_2$, we would have $(t_1, t_2] \subseteq (t_1, t_1 + u_1] \cup (t_2 - u_2, t_2]$, in other words the machine is completely calibrated during time interval $(t_1, t_2]$. Therefore, we only need to schedule job j at time t and reduce the problem to sub-problems: scheduling jobs $J(j - 1, t_1, t)$ in time interval $(t_1, t]$ and jobs $J(j - 1, t + p, t_2)$ in time interval $(t + p, t_2]$ (as argued earlier). And for both intervals, the machine is completely calibrated.

Case 1. $t \in [t_1, t_1 + u_1 - p]$. In this case the interval $[t, t + p]$ is completely contained in the interval $[t_1, t_1 + u_1]$. Therefore, it is not necessary to start a new calibration to cover the interval $[t, t + p]$ and we just reduce to sub-problems. For the sub-problem of jobs $J(j - 1, t_1, t)$ during interval $(t_1, t]$, the machine is completely calibrated during interval $(t_1, t]$, hence we allocate all the k calibrations to the other sub-problem. For the sub-problem of jobs $J(j - 1, t + p, t_2)$ during interval $(t + p, t_2]$, the calibrated intervals are $(t + p, t_1 + u_1]$ and $(t_2 - u_2, t_2]$.

Case 2. $t \in [t_2 - u_2, t_2 - p]$. This case is symmetric with *Case 1.* in the sense that the interval $[t, t + p]$ is completely contained in the interval $[t_2 - u_2, t_2]$. Therefore, we just reduce to sub-problems and reserve all the k calibrations to the sub-problem of jobs $J(j - 1, t_1, t)$, in which the calibrated intervals are $(t_1, t_1 + u_1]$ and $(t_2 - u_2, t]$. For the sub-problem of jobs $J(j - 1, t + p, t_2)$, the machine is completely calibrated during interval $(t + p, t_2]$.

Case 3. $t \in (t_1 + u_1 - p, t_2 - u_2)$. In this case we have to start new calibrations to schedule job j at time t because $(t, t + p] \setminus ((t_1, t_1 + u_1] \cup (t_2 - u_2, t_2]) \neq \emptyset$ due to $t_1 + u_1 < t_2 - u_2$. By definition, we have $(a_t, b_t] = (t, t + p] \setminus ((t_1, t_1 + u_1] \cup (t_2 - u_2, t_2]) \neq \emptyset$, i.e. $(a_t, b_t]$ denotes the interval in which the machine has not (but have to) been calibrated. In case the length of the interval $(a_t, b_t]$ is larger than T, we would have to start more than one calibration to cover interval $(a_t, b_t]$. Suppose in the optimal schedule, job j starts at time t and set $\mathcal{O}$ contains all the calibrations that intersect with interval $(a_t, b_t]$. We sort the calibrations in $\mathcal{O}$ by the increasing order of their starting time. Let t' be the starting time of the first calibration in $\mathcal{O}$. Then we have $t' \in (a_t - T, a_t]$. If $t' + T < b_t$, we would have $|\mathcal{O}| > 1$. Then, for the second calibration in $\mathcal{O}$, it must start at time $t' + T$ because otherwise time slot $t' + T + 1$ will not be calibrated due to the fact that no two calibrations in the optimal schedule overlap with each other. Therefore, the calibrations in $\mathcal{O}$ must be consecutive, i.e. there is no gap between any two consecutive calibrations. As a result, we would have $|\mathcal{O}| = k_0$ where $k_0 = \lceil \frac{b_t - t'}{T} \rceil$. Moreover, besides interval $(a_t, b_t]$, the machine is also calibrated in intervals $(t', a_t]$ and $(b_t, t' + k_0 T]$ by the calibrations in $\mathcal{O}$. In the dynamic programming, we try every possibility of t' from $[t_1, t_2) \cap (a_t - T, a_t]$ to start the calibrations and then reduce to sub-problems. For the sub-problem of jobs $J(j - 1, t_1, t)$, we try every possibility of the number of calibrations that are allocated to cover the interval $(t_1, t]$, i.e. $k' \in [0, k - k_0]$. And we allocate the remaining $k - k_0 - k'$ calibrations to the other sub-problem of jobs $J(j-1, t+p, t_2)$. Finally, we would reach the optimal schedule since we have tried every possibility of the values t' and k'.

If none of the above three cases is possible, it is impossible to feasibly schedule job j, hence we would set F to be infinity. $\square$

Time Complexity. Note that the computation of $\sum_{z=1}^{p} c(t+z)$ for a fixed value t will takes $O(1)$ time if we use the date structure of prefix sum array. The table size of the dynamic programming is $O(nKT^2L^2)$ and computation for each sub-problem takes $O(KTL)$ steps for enumerating the values t, t', k' in the last case. In total, the time complexity is $O(nK^2T^3L^3)$.

2.2   Agreeable Jobs of Identical Processing Times

In this section, we restrict the problem to a special set of jobs, agreeable jobs. Job set J is called *agreeable* if for any two jobs $i, j \in J$ it satisfies the property that $r_i < r_j$ implies $d_i \leq d_j$. In other words, earlier released jobs have earlier deadlines. Lemma 1 implies that the optimal schedule preserves the order of jobs (non-decreasing order of deadlines) as the jobs are agreeable. We still assume all jobs have identical processing times and further propose an improved dynamic programming approach to solve the problem.

Definition 2. *We define $F(j, k, t)$ to be the minimum total time slot cost of the partial schedule that completes jobs $\{1, 2, \ldots, j\}$ before time t with at most k calibrations, where $j \in J, k \in [0, K], t \in \Psi$. Let $f(i, j, t_1, t_2)$ be the minimum total time slot cost to schedule jobs $\{i, i+1, \ldots, j\}$ during time interval $(t_1, t_2]$ given that the machine has been calibrated during interval $(t_1, t_2]$, where $i, j \in J, t_1, t_2 \in \Psi$.*

Proposition 2. *If $i > j$ we have $f(i, j, t_1, t_2) \leftarrow 0$, and if $(j+1-i)p > t_2 - t_1$ or $[t_1, t_2 - p] \cap [r_j, d_j - p] = \emptyset$ we have $f(i, j, t_1, t_2) \leftarrow \infty$. Otherwise,*

$$f(i, j, t_1, t_2) = \min \begin{cases} f(i+1, j, t_1 + p, t_2) + \sum_{z=1}^{p} c(t_1 + z), & \text{if } t_1 \in [r_i, d_i - p] \\ f(i, j, t_1 + 1, t_2) \end{cases}$$

Proof. If $i > j$, no job needs to be considered by definition, hence we have $f(i, j, t_1, t_2) \leftarrow 0$. If $(j+1-i)p > t_2 - t_1$, time slots in interval $(t_1, t_2]$ is insufficient to schedule jobs $\{i, i+1, \ldots, j\}$, and if $[t_1, t_2-p] \cap [r_j, d_j-p] = \emptyset$ job j cannot be scheduled during time interval $(t_1, t_2]$, therefore we have $f(i, j, t_1, t_2) \leftarrow \infty$. The main focus of the dynamic programming is to determine the starting time of job i in the optimal schedule. In the optimal schedule, job i either starts at time t_1 or after time t_1. If job i starts at time t_1, we would have $t_1 \in [r_i, d_i - p]$. Since the optimal schedule preserves the order of jobs, the remaining jobs $\{i+1, \ldots, j\}$ have to be finished during interval $(t_1 + p, t_2]$, therefore we reduce to sub-problem $f(i+1, j, t_1 + p, t_2)$. If job i starts after time t_1, we just reduce to sub problem $f(i, j, t_1 + 1, t_2)$ since no job will be scheduled in time slot $t_1 + 1$. The recurrence equation is correct as we have tested every possibility of the schedule of job i. $\square$

Proposition 3. *If $j = 0$ we have $F(j, k, t) \leftarrow 0$, and if $j > kT$ or $r_j + p > t$ we have $F(j, k, t) \leftarrow \infty$. Otherwise,*

$$F(j, k, t) = \min \begin{cases} \min_{i \in [1,j], k' \in [1,k]} F(i - 1, k - k', t - k'T) + f(i, j, t - k'T, t) \\ F(j, k, t - 1) \end{cases}$$

The main focus of the dynamic programming in Proposition 3 is to identify a maximal set of consecutive calibrations (an amount of k' calibrations covering interval $(t - k'T, t]$) satisfying that the last calibration finishes at time t and then identify the first job (job i) that is totally scheduled during these calibrations. We put the proof of Proposition 3 in full version.

Time Complexity. The dynamic programming in Proposition 3 takes $O(n^2 K^2 L)$ in which the table size of $F(j, k, t)$ is bounded by $O(nKL)$ and the recurrence equation takes $O(nK)$ steps. For the dynamic programming in Proposition 2, the table size of $f(i, j, t_1, t_2)$ is bounded by $O(n^2 L^2)$, while it can be further reduced to $O(n^2 LKT)$ because $t_2 - t_1 \leq KT$. Moreover, the recurrence equation takes $O(1)$ steps to compute each value of $f(i, j, t_1, t_2)$, hence the time complexity is $O(n^2 KTL)$. In total the time complexity is $O(n^2 LK(K + T))$.

2.3   Monotonic Time Slot Cost Function

In this part, we investigate the case that the slot cost function is monotonic over time. Without loss of generality, we would assume that the slot cost function is monotonic non-decreasing, as the analysis for the other case (monotonic non-increasing) is symmetric when we consider the time horizon from large to small. In other words, the earlier a job is scheduled, the smaller cost it will incur. We propose an improved dynamic programming approach for the case when the job set J is agreeable and all jobs have identical processing times.

Note that the optimal schedule preserves the order of jobs for agreeable jobs by Lemma 1. For each job $j \in J$, we define $\tau_j = \max_{i \in [1,j]}\{r_i + (j - i)p\}$, then τ_j indicates the earliest time that job j could start in any feasible schedule preserving the order of jobs.

We define *calibration block* to be a set of maximal consecutive calibrations such that there is no gap between any two consecutive calibrations in this set (i.e. the calibration starts immediately after the previous calibration finishes). Then we show that for each calibration block in the optimal schedule, the last job, say job j, that is scheduled in this block starts at τ_j in the optimal schedule. Therefore, we propose a dynamic programming approach which identifies the calibration block in the optimal schedule, as well as the last job in each calibration block.

Lemma 2. *There exists an optimal schedule such that for each calibration block, job j starts at time τ_j where job j is the last job of the calibration block.*

Proof. In the following we show how to transform an optimal schedule into another optimal schedule that meets the statement. Consider an optimal schedule σ and a calibration block $\mathcal{B}$ in which the last job j starts at time t_j such that $t_j > \tau_j$. Let t be the latest idle time slot before time t_j (note that time slot t refers to interval $(t-1, t]$ and time slot t is not necessarily to be calibrated), i.e. there is a job which starts at time t and let job i be the job starting at time t. Then we claim that for each job $j' \in J'$, we have $t_{j'} > \tau_{j'}$ where $t_{j'}$ is the starting time of job j' in the schedule σ and $J' = \{i, i+1, \ldots, j\}$. To prove the claim, if $t_{j'} \le \tau_{j'}$ for some job $j' \in J'$, we would have a contradiction that $t_j = t_{j'} + (j-j')p \le \tau_{j'} + (j-j')p \le \tau_j$, where the first equality holds because jobs $\{j', j'+1, \ldots, j\}$ are scheduled consecutively due to the definition of t and the last inequality holds because the definition of τ_j and also $\tau_{j'} = r_i + (j'-i)p$ for some job $i \in [1, j']$ by definition. Hence, the claim is true.

Regrading the feasibility of jobs, advancing each job in J' by one unit time will result in another feasible schedule because of the above claim. Therefore, if time slot t is calibrated, we would obtain another optimal schedule by applying the above advancing process. Otherwise, time slot t is not calibrated. We would obtain another optimal schedule by advancing both jobs in J' and all the calibrations in the calibration block $\mathcal{B}$ by one unit time. Consequently, by repeating such advancing process we would eventually obtain another optimal schedule which meets the statement.   □

Definition 3. *We define $f(i, j, k)$ to be the minimum total time slot cost to schedule jobs $\{i, i+1, \ldots, j\}$ during time interval I' given that the machine is completely calibrated during interval I' where $I' = (\tau_j + p - kT, \tau_j + p]$ and $i, j \in J$, $k \in [0, K]$. We define $F(j, k)$ to be the minimum total time slot cost of the schedule that completes jobs $\{1, 2, \ldots, j\}$ with at most k calibrations where $k \in [0, K]$ and $j \in J$.*

First, we show the Computation of $f(i, j, k)$. As the optimal schedule preserves the order of jobs, we could calculate the starting time of each job of J' directly where $J' = \{i, i+1, \ldots, j\}$. For each job $j' \in J'$, we define $t_{j'}$ to be the starting time of job j' in the optimal schedule. Then we have $t_i = \max\{\tau_i, t_0\}$ where $t_0 = \tau_j + p - kT$, and for $j' \in J' \setminus \{i\}$ we have $t_{j'} = \max\{\tau_{j'}, t_{j'-1} + p\}$. By definition $\tau_{j'}$ is the earliest possible time that job j' could start. We would have $f(i, j, k) \leftarrow \infty$ if $\exists j' \in J'$, $t_{j'} + p > d_{j'}$ and otherwise $f(i, j, k) \leftarrow \sum_{j' \in J'} \sum_{z=1}^{p} c(t_{j'} + z)$. Note that the computation of $\sum_{z=1}^{p} c(t + z)$ for a fixed value t takes $O(1)$ time if we use the date structure of prefix sum array. The computation of $t_{j'}$ takes $O(n)$ time, and hence the computation of each value $f(i, j, k)$ takes $O(n)$ time.

Proposition 4. *If $j = 0$ we have $F(j, k) \leftarrow 0$, and if $jp > kT$ we have $F(j, k) \leftarrow \infty$. Otherwise,*

$$F(j, k) = \min_{cond.} \{F(i-1, k-k') + f(i, j, k')\}$$

where cond. stands for $i \in [1, j], k' \in [1, k], \tau_{i-1} \le \tau_j - k'T$ and we would regard $\tau_0 = -\infty$.

Proof. In the dynamic programming, we maintain the invariant that job j is the last job of a calibration block. Then job j starts at time τ_j by Lemma 2. Let $\mathcal{B}$ be the calibration block in the optimal schedule in which job j is the last job in this block. Suppose there are k' calibrations in calibration block $\mathcal{B}$, then the block will cover interval $(\tau_j + p - k'T, \tau_j + p]$. Then we test every possibility of the number of jobs that are scheduled during calibration block $\mathcal{B}$. Let job $i \in [1, j]$ be the earliest job that is scheduled in calibration block $\mathcal{B}$ in the optimal schedule. Then for job $i - 1$ (if $i - 1 > 0$), it should finish before time t_0 where $t_0 = \tau_j + p - k'T$ according to the definition of calibration block. Therefore, we would have $\tau_{i-1} + p \leq t_0$, i.e. $\tau_{i-1} \leq \tau_j - k'T$. In the dynamic programming, we test every possible value of k' from $[1, k]$ and every possible value of i from $[1, j]$ such that $\tau_{i-1} \leq \tau_j - k'T$, and then reduce to sub-problems $F(i - 1, k - k')$ and $f(i, j, k')$ by scheduling jobs $\{i, i + 1, \ldots, j\}$ during time interval $(\tau_j + p - k'T, \tau_j + p]$. The dynamic programming is correct because we have tested every possible value of i and k'.   □

Time Complexity. The table size of $f(i, j, k)$ is $O(Kn^2)$ and computation of each value of $f(i, j, k)$ takes $O(n)$ time as argued before. The table size of $F(j, k)$ is $O(Kn)$ and computation of each value $F(j, k)$ in Proposition 4 takes $O(nK)$ steps. Hence in total, the time complexity of the dynamic programming is $O(Kn^2(n + K))$.

3   Slot Cost Without Calibration for General Jobs

In this section, we investigate the problem without the consideration of the calibrations on the machines, i.e. the machine is available at any time. We consider general jobs with preemption (can be interrupted during execution and resumed later) where each job could have arbitrary processing time on the condition that $P = \sum_{j \in J} p_j \leq L = \max_{j \in J} d_j - \min_{j \in J} r_j$. The problem is denoted as $1|r_j, d_j, p_j, pmtn| \sum_t c(t)$. We propose a greedy algorithm to solve this problem.

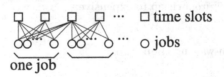

Fig. 3. An illustration

We first show that the problem can be solved by weighted maximum matching over a bipartite graph. Then we propose an improved method. The bipartite graph is constructed as follows. For each job $j \in J$, we create a total amount of p_j copies of vertices that corresponds to job j and let $\mathcal{J}$ be the collection of all job vertices. In total, we have $|\mathcal{J}| = \sum_{j \in J} p_j$ job vertices. Let set $\mathcal{T}$ be the set of all available time slots, and for each time slot we create one vertice, in total we have $T = \max_{j \in J} d_j - \min_{j \in J} r_j$ time slot vertices. Bipartite graph $G = (J, T, E)$ is a graph over vertices J and T where for each vertice j' from J and each vertice t' from T, if the interval of job j covers time slot t' (where j is the job that corresponds to vertice j'), then set E contains an edge $e = (j', t')$ with weight $w_e = c(t')$. An illustration of the graph is shown in Fig. 3.

A maximum matching of graph G that contains P edges corresponds to a feasible schedule to the original problem, as we could assign the time slots to the jobs according to the edges in the matching. Hence, the minimum weighted bipartite matching corresponds to the optimal schedule of the original problem. The minimum weighted bipartite matching can be solved via classical Hungarian algorithm, while in the following we propose a faster algorithm, based on matroid theory.

Definition 4. *Let $M = (T, F)$ be a set system where F is a collection of set $X \subseteq T$ such that there exists a matching of bipartite graph $G = (J, T, E)$ covering the vertices in X.*

By definition, set system M is a transversal matroid [11].

Theorem 1. *Set system M is a transversal matroid.*

Definition 5. *For any matching of bipartite graph $G = (J, T, E)$, the set of time slots X is called* non-wasting *where X are the vertices covered by the matching and $X \subseteq T$.*

By definition, a matching of bipartite graph $G = (J, T, E)$ corresponds to a set of *non-wasting* time slots X, and also corresponds to a partial schedule such that each time slot from X is occupied by some job in the schedule. Since set system M is a matroid, we could obtain the optimal solution by adding the time slots (from cheap to expansive) into a candidate list while maintaining the property that the set of time slots in the candidate list is non-wasting.

The following algorithm gives the optimal solution.

Algorithm 1. Greedy Algorithm

1: Sort the time slots T from cheap to expensive
2: $Q \leftarrow \emptyset$
3: **for** $t \in T$ **do**
4:     **if** $Q \cup \{t\}$ is non-wasting **then**
5:         $Q \leftarrow Q \cup \{t\}$
6:     **end if**
7: **end for**

Lemma 3. *Given a subset $X \subseteq T$ of time slots, the Earliest-Deadline-First (EDF) algorithm can be applied to check whether X is non-wasting in linear time.*

Combined with Lemma 3, we are able to obtain the optimal schedule in $O(nL)$ time where $L = \max_{j \in J} d_j - \min_{j \in J} r_j$.

4   Conclusion

We studied the scheduling problem with the consideration of both calibrations on the machine and time slot cost. We propose dynamic programmings for different scenarios of this problem, as well as a greedy algorithm for the non-calibration version of this problem. For the future work, it is challenging to tackle the open problem about the complexity status of the basic calibration model proposed by Bender et al. [2]. And it is also worth working on the preemptive jobs for arbitrary processing times.

References

1. Angel, E., Bampis, E., Chau, V., Zissimopoulos, V.: On the complexity of minimizing the total calibration cost. In: Xiao, M., Rosamond, F. (eds.) FAW 2017. LNCS, vol. 10336, pp. 1–12. Springer, Cham (2017). https://doi.org/10.1007/978-3-319-59605-1_1
2. Bender, M.A., Bunde, D.P., Leung, V.J., McCauley, S., Phillips, C.A.: Efficient scheduling to minimize calibrations. In: Proceedings of the Twenty-Fifth Annual ACM Symposium on Parallelism in Algorithms and Architectures, SPAA 2013, pp. 280–287. ACM, New York (2013)
3. Bringmann, B., Küng, A., Knapp, W.: A measuring artefact for true 3D machine testing and calibration. CIRP Ann.-Manuf. Technol. **54**(1), 471–474 (2005)
4. Burroughs, C.: New integrated stockpile evaluation program to better ensure weapons stockpile safety, security, reliability (2006)
5. Chau, V., Li, M., McCauley, S., Wang, K.: Minimizing total weighted flow time with calibrations. In: Proceedings of the 29th ACM Symposium on Parallelism in Algorithms and Architectures, SPAA 2017, pp. 67–76. ACM, New York (2017)
6. Chen, L., Megow, N., Rischke, R., Stougie, L., Verschae, J.: Optimal algorithms and a PTAS for cost-aware scheduling. In: Italiano, G.F., Pighizzini, G., Sannella, D.T. (eds.) MFCS 2015. LNCS, vol. 9235, pp. 211–222. Springer, Heidelberg (2015). https://doi.org/10.1007/978-3-662-48054-0_18
7. Fineman, J.T., Sheridan, B.: Scheduling non-unit jobs to minimize calibrations. In: Proceedings of the 27th ACM Symposium on Parallelism in Algorithms and Architectures, SPAA 2015, pp. 161–170. ACM, New York (2015)
8. Graham, R.L., Lawler, E.L., Lenstra, J.K., Kan, A.H.G.R.: Optimization and approximation in deterministic sequencing and scheduling: a survey. Ann. Discrete Math. **5**, 287–326 (1979)
9. Kalyanasundaram, B., Pruhs, K.: Speed is as powerful as clairvoyance. J. ACM **47**(4), 617–643 (2000)
10. Nguyen, H.-N., Zhou, J., Kang, H.-J.: A new full pose measurement method for robot calibration. Sensors **13**(7), 9132–9147 (2013)
11. Oxley, J.G.: Matroid Theory. Oxford University Press, Oxford (1992)
12. Phillips, C.A., Stein, C., Torng, E., Wein, J.: Optimal time-critical scheduling via resource augmentation. In: Proceedings of the 29th ACM Symposium on Theory of Computing, STOC 1997, pp. 140–149. ACM Press, New York (1997)
13. Wan, G., Qi, X.: Scheduling with variable time slot costs. Naval Res. Log. (NRL) **57**(2), 159–171 (2010)
14. Zhang, Z.: A flexible new technique for camera calibration. IEEE Trans. Pattern Anal. Mach. Intell. **22**(11), 1330–1334 (2000)

15. Zhao, Y., Qi, X., Li, M.: On scheduling with non-increasing time slot cost to minimize total weighted completion time. J. Sched. **19**(6), 759–767 (2016)
16. Zhong, W., Liu, X.: A single machine scheduling problem with time slot costs. In: Qian, Z., Cao, L., Su, W., Wang, T., Yang, H. (eds.) Recent Advances in Computer Science and Information Engineering. LNEE, vol. 126, pp. 677–681. Springer, Heidelberg (2012). https://doi.org/10.1007/978-3-642-25766-7_90

The k-power Domination Problem
in Weighted Trees

ChangJie Cheng, Changhong Lu$^{(\boxtimes)}$, and Yu Zhou

School of Mathematical Sciences, Shanghai Key Laboratory of PMMP,
East China Normal University, Shanghai 200241, People's Republic of China
354840621@qq.com, chlu@math.ecnu.edu.cn, 154423228@qq.com

Abstract. The power domination problem of the graph comes from how to choose the node location problem of the least phase measurement units in the electric power system. In the actual electric power system, because of the difference in the cost of phase measurement units at different nodes, it is more practical to study the power domination problem with the weighted graph. In this paper, we present a dynamic programming style linear-time algorithm for k-power domination problem in weighted trees.

Keywords: Power domination · Weighted trees
Linear time algorithm · Dynamic programming

1 Introduction

Let $G = (V, E)$ be a simple graph with vertex set $V = V(G)$ and edge set $E = E(G)$. A simple graph means an undirected graph without multiple edges or loops. The *open neighborhood* of a vertex $v \in V$ is the set $N_G(v) = \{u \in V \mid uv \in E\}$, and the *closed neighborhood* of v is the set $N_G[v] = N_G(v) \cup \{v\}$. A *dominating* set of graph G is a set $D \subseteq V$ such that for each $u \in V \setminus D$, there exists an $x \in D$ adjacent to u. The *domination number* of G, denoted by $\gamma(G)$, is the minimum cardinality amongst all dominating sets of G.

Modern society is inseparable from electricity. Electric power companies need to continuously monitor their systems in the process of delivering electrical energy. One way is to put the phase measurement units at the selected location in their system. How to save the number of phase measurement units is very attractive for the electric power companies, as introduced in [3]. In 2002, Haynes et al. [12] described the power system monitoring as a graph theoretical problem. The original definition of power domination was simplified to the following definition independently in [9–11,13].

Supported in part by National Natural Science Foundation of China (No. 11371008) and Science and Technology Commission of Shanghai Municipality (No. 18dz2271000).

S. Tang et al. (Eds.): AAIM 2018, LNCS 11343, pp. 149–160, 2018.
https://doi.org/10.1007/978-3-030-04618-7_13

Definition 1. *Let $G = (V, E)$ be a graph. A set $D \subseteq V$ is a power dominating set (abbreviated as PDS) of G if and only if all vertices of G have messages either by Observation Rule 1 (abbreviated as OR 1) initially or by Observation Rule 2 (abbreviated as OR 2) recursively.*

OR 1. *A vertex $v \in D$ sends a message to itself and all its neighbors. We say that v observes itself and all its neighbors.*

OR 2. *If an observed vertex v has only one unobserved neighbor u, then v will send a message to u. We say that v observes u.*

Let $G = (V, E)$ be a graph and D be a subset of V. For $i \geq 0$, we define by $P_G^i(D)$ the set of vertices observed by D at step i by the following rules:

(1) $P_G^0(D) = N_G[D]$;

(2) $P_G^{i+1}(D) = \cup\{N_G[v] : v \in P_G^i(D) \text{ such that } |N_G[v] \setminus P_G^i(D)| \leq 1\}$.

Note that for any integer $i \geq 0$, we have $P_G^i(D) \subseteq P_G^{i+1}(D) \subseteq V$; if D is a power domination set of G, then there is a minimal integer i_0 such that $P_G^{i_0}(D) = V$. Hence $P_G^j(D) = P_G^{i_0}(D)$ for every $j \geq i_0$ and we accordingly define $P_G^\infty(D) = P_G^{i_0}(D)$.

Chang et al. [6] generalized the power domination to k-power domination replacing OR 2 with the following observation rule: If an observed vertex v has at most k unobserved neighbors, then v will send a message to all its unobserved neighbors. The definition of k-power dominating set is given below.

Definition 2. *Let $G = (V, E)$ be a graph, $D \subseteq V$, integer $k \geq 0$. We define by $P_{G,k}^i(D)$ the set of vertices observed by D at step i. If $P_{G,k}^\infty(D) = V$, We call D a k-power dominating set (abbreviated as kPDS) of G. The recursive formula is as follows:*

(1) $P_{G,k}^0(D) = N_G[D]$;

(2) $P_{G,k}^{i+1}(D) = \cup\{N_G[v] : v \in P_{G,k}^i(D) \text{ such that } |N_G[v] \setminus P_{G,k}^i(D)| \leq k\}$.

The *k-power domination number* of G, denoted by $\gamma_{p,k}(G)$, is the minimum cardinality of a k-power dominating set of G. When $k = 0$, the k-power domination is usual domination. When $k = 1$, the k-power domination is usual power domination.

Let $G = (V, E, w)$ be a weighted graph, where w is a function from V to positive real numbers. Let $w(D) = \sum_{v \in D} w(v)$ be the weight of D for any subset D of V. The *weighted k-power domination number*, denoted by $\gamma_{kp}^w(G)$, is defined as $\gamma_{kp}^w(G) = \min\{w(D) \mid D \text{ is a } k\text{-power dominating set of } G\}$. The *weighted k-power domination problem* is to determine the k-power domination number of any weighted graph. When the weight of each vertex of the graph G is equal to 1, the problem of weighted k-power domination problem is the k-power domination problem.

Aazami [1,2] proves the NP-hardness of power domination problem on general graphs. Liao and Lee [13] explored an efficient algorithm for power domination problem on interval graphs. Haynes [12] gives a linear time algorithm for k-power domination problem on trees. In [14], a linear-time algorithm for power domination on weighted trees was given. Many kinds of weighted domination problems were studied by some scholars, see [4,5,7,8,15–17,19,20]. In this paper, we will study a linear-time algorithm for k-power domination problem on weighted trees.

Let $T = (V, E, w)$ be a weighted tree with n vertices. It is well known that the vertices of T have an ordering $v_1, v_2, \cdots, v_n$ such that for each $1 \leq i \leq n-1$, v_i is adjacent to exactly one v_j with $j > i$. The ordering is call a *tree ordering* of the tree, where the only neighbor v_j with $j > i$ is called the *father* of v_i and v_i is a *child* of v_j. For each $1 \leq i \leq n-1$, the father of v_i is denoted by $F(v_i) = v_j$. For technical reasons, we assume that $F(v_n) = v_n$.

Suppose that T is a (weighted) tree rooted at r (denoted by (T, r)). We make use of the fact that the class of (weighted) rooted tree can be constructed recursively from copies of the single vertex K_1, using only one rule of composition, which combines two trees (T_1, r_1) and (T_2, r_2) by adding an edge between r_1 and r_2 and calling r_1 the root of the resulting larger tree T. We denote this as follows: $(T, r_1) = (T_1, r_1) \circ (T_2, r_2)$ (Fig. 1).

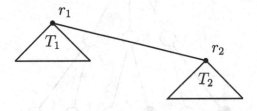

Fig. 1. $(T, r_1) = (T_1, r_1) \circ (T_2, r_2)$

Definition 3. *For a given weighed tree $T = (V, E, w)$ rooted at r, and integer $k \geq 0$. For any $1 \leq p \leq k$, T^p called the extended tree of T if $V(T^p) = V(T) \cup \{v_1, v_2, \cdots, v_p\}$, $E(T^p) = E(T) \cup \{v_1 r, v_2 r, \cdots, v_p r\}$ and $w(v_i) = 1$ for $0 \leq i \leq p$. $v_1, \cdots, v_p$ are called the extended children of r in T^p (Fig. 2).*

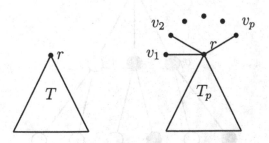

Fig. 2. The tree T and the tree T^p

Definition 4. *For a given weighed tree* $T = (V, E, w)$ *rooted at* r, *and integer* $k \geq 0$, $D \subseteq V(T)$. *Let* p, q *be integers satisfying* $0 \leq p, q \leq k$. *Define the following relationships on tree and vertex subset.*

$[a] = \{(T, D, w) \mid D$ *is a kPDS of* T *and* $r \in D\}$;

$[b]_p = \{(T, D, w) \mid D$ *is a kPDS of* T^p, *but not of* T^{p+1} *and* $r \notin D\}$;

$[c]_q = \{(T, D, w) \mid D$ *is not a kPDS of* T *and* $T - r$, *but if a message is given to* r *in advance, then all vertices of* T *will get message and there are exactly* q *children observed by* $r.\}$

Here are two examples of the above concepts.

Example 1.1. *For the sake of simplicity, we assume that the weight of each vertex is equal to one and* $k = 2$, $p = 1$. $D = \{v_1, v_3\}$. *The structure of the tree* T *is as follows (Fig. 3).*

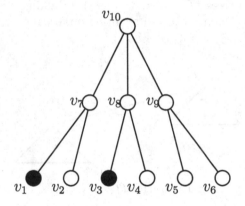

Fig. 3. Example of $[b]_1$ with $k = 2$ and $D = \{v_1, v_3\}$.

By OR 1, v_1, v_3, v_7, v_8 *have message initially (see Fig. 4).*

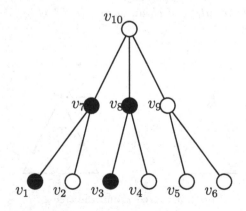

Fig. 4. $P_{T,2}^0(D) = \{v_1, v_3, v_7, v_8\}$.

Then, $v_1, v_2, v_3, v_4, v_7, v_8, v_{10}$ *have message by OR 2 at the first time (see Fig. 5A).*

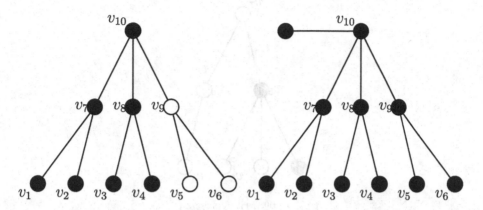

Fig. 5. A (left) $P^1_{T,2}(D) = \{v_1, v_2, v_3, v_4, v_7, v_8, v_{10}\}$ and B (right) $P^3_{T^1,2}(D) = V(T^1)$.

Since $k = 2$, the root v_{10} can send message to its two unobserved neighbors. So, D is a kPDS of the extended tree T^1 (see Fig. 5B), but it is not a kPDS of the extended tree T^2. Hence, $(T, \{v_1, v_3\}, w) \in [b]_1$.

Example 1.2. *For the sake of simplicity, we also assume that the weight of each vertex is equal to one and $k = 2$, $q = 1$. $D = \{v_1\}$. The structure of the tree T is as follows (Fig. 6).*

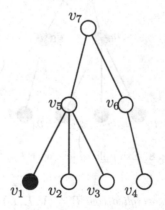

Fig. 6. Example of $[c]_1$ with $k = 2$ and $D = \{v_1\}$.

By OR 1, v_1, v_5 have message initially (Fig. 7).

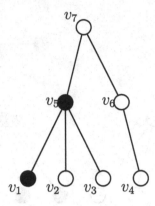

Fig. 7. $P^0_{T,2}(D) = \{v_1, v_5\}$.

As $k = 2$, we can't go further. So, D is not a kPDS of T and $T - v_7$. But if a message is given to v_7 in advance, then $P^2_{T,2}(D) = V(T)$ and v_6 is observed by v_7. (see Fig. 8). Hence, $(T, \{v_1\}, w) \in [c]_1$.

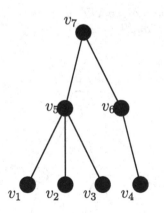

Fig. 8. Example of $[c]_1$ with $k = 2$.

Definition 5. *For a given weighed tree $T = (V, E, w)$, and integer $k \geq 0$, $D \subseteq V(T)$. Let p, q be an integer satisfying $0 \leq p, q \leq k$.*

(1) $r^a(T, r) := min\{w(D) | (T, D) \in [a]\};$
(2) $r^{bp}(T, r) := min\{w(D) | (T, D) \in [b]_p\};$
(3) $r^{cq}(T, r) := min\{w(D) | (T, D) \in [c]_q\}.$

By the above definitions, it i.e. easy to know

Lemma 1. *For a given weighed tree* $T = (V, E, w)$*, and integer* $k \geq 0$*,* $D \subseteq V(T)$*. then*

$$r_{kp}^{w}(T) = min\{r^a(T, r), r^{b0}(T, r), r^{b1}(T, r), \cdots, r^{bk}(T, r)\}.$$

2 Main Theorem and Its Proof

The following lemmas give the key steps for solving the k-power domination problem on the weighted tree. We borrow the method proposed by Wimer [18].

Lemma 2. $[a] = [a] \circ [a] \cup [a] \circ (\cup_{l=0}^{k}[b]_l) \cup [a] \circ (\cup_{h=0}^{k}[c]_h)$.

Proof. Let $(T, D) = (T_1, D_1) \circ (T_2, D_2)$, where $D_1 \subseteq V(T_1), D_2 \subseteq V(T_2)$, $D_1 \cup D_2 = D$. Suppose that $r_1(= r)$ is the root of T_1 and r_2 is the root of T_2 and $r_1 r_2 \in E(T)$. We first show that $[a] \circ [a] \cup [a] \circ (\cup_{l=0}^{k}[b]_l) \cup [a] \circ (\cup_{h=0}^{k}[c]_h) \subseteq [a]$.

(i) If $(T_1, D_1, w) \in [a]$ and $(T_2, D_2, w) \in [a]$, both $V(T_1)$ and $V(T_2)$ can receive message independently. Note that $r_1(= r)$ is in $D = D_1 \cup D_2$. Thus, $[a] \circ [a] \subseteq [a]$.

(ii) If $(T_1, D_1, w) \in [a]$ and $(T_2, D_2, w) \in [b]_l (0 \leq l \leq k)$, both $V(T_1)$ and $V(T_2)$ can also receive message independently. Note that $r_1(= r)$ is in $D = D_1 \cup D_2$. Thus, $[a] \circ (\cup_{l=0}^{k}[b]_l) \subseteq [a]$.

(iii) If $(T_1, D_1, w) \in [a]$ and $(T_2, D_2, w) \in [c]_h (0 \leq h \leq k)$, according to OR 1, r_2 receives the message from r_1 initially since $r_1 \in D_1$ and $r_1 r_2 \in E(T)$. Then according to the definition of $[c]_q$, we know that all vertices of T_2 can be observed at last. As $r(= r_1)$ is in D, we have $[a] \circ (\cup_{h=0}^{k}[c]_h) \subseteq [a]$.

Now we prove that $[a] \subseteq [a] \circ [a] \cup [a] \circ (\cup_{l=0}^{k}[b]_l) \cup [a] \circ (\cup_{h=0}^{k}[c]_h)$. Since $(T, D, w) \in [a]$, we know that $r(= r_1) \in D$ and hence $r \in D_1$. Since there is only one edge $r_1 r_2$ connecting the tree T_1 and the tree T_2 and $r_1 \in D_1 = D \cap V(T_1)$, T_1 does not need to obtain message from the T_2. Hence, D_1 is a $kPDS$ of T_1, i.e. $(T_1, D_1, w) \in [a]$. Consider $D_2 = D \cap V(T_2)$ below. When $r_2 \in D$, it is obvious that $(T_2, D_2, w) \in [a]$. When $r_2 \notin D$, we discuss the following two cases.

case(I) r_2 can be observed by some child in T_2.

It means that there is no message exchange between T_1 and T_2. Since D is a $kPDS$ of T and $r_2 \notin D_2$, D_2 must be a $kPDS$ of T_2^p for some $p \in \{0, 1, \cdots, k\}$, i.e., $(T_2, D_2, w) \in [b]_p$ satisfies $0 \leq p \leq k$.

case(II) r_2 cannot be observed by its children in T_2.

Since r_2 cannot be observed by its children in T_2, D_2 is not a $kPDS$ of T_2. But D is a $kPDS$ of T and by OR 1, r_2 is observed by r initially. It means that $(T_2, D_2, w) \in [c]_q$ for some $q \in \{0, 1, \cdots, k\}$.

In summary, $[a] = [a] \circ [a] \cup [a] \circ (\cup_{l=0}^{k}[b]_l) \cup [a] \circ (\cup_{h=0}^{k}[c]_h)$. □

Lemma 3. *Let p be an integer satisfying $0 \le p \le k$.*

$$[b]_p = [c]_{k-p} \circ [a] \cup [c]_{k-p} \circ (\cup_{l=1}^{k}[b]_l) \cup [b]_p \circ [a] \cup [b]_p \circ (\cup_{t=0}^{k}[b]_t) \cup [b]_{p+1} \circ (\cup_{h=0}^{k}[c]_h).$$

Proof. Let $(T, D) = (T_1, D_1) \circ (T_2, D_2)$, where $D_1 \subseteq V(T_1), D_2 \subseteq V(T_2), D_1 \cup D_2 = D$. Suppose that $r_1(= r)$ is the root of T_1 and r_2 is the root of T_2 and $r_1 r_2 \in E(T)$. We first show that $[c]_{k-p} \circ [a] \cup [c]_{k-p} \circ (\cup_{l=1}^{k}[b]_l) \cup [b]_p \circ [a] \cup [b]_p \circ (\cup_{t=1}^{k}[b]_t) \cup [b]_{p+1} \circ (\cup_{h=0}^{k}[c]_h) \subseteq [b]_p$.

(i) If $(T_1, D_1, w) \in [c]_{k-p}$ and $(T_2, D_2, w) \in [a]$. according to OR 1, r_1 receives the message from r_2 initially since $r_2 \in D_2$ and $r_1 r_2 \in E(T)$. Then according to the definition of $[c]_{k-p}$, there are exactly $k - p$ children observed by $r_1(= r)$, and all vertices of T_1 can be observed at last. Since $(T_2, D_2, w) \in [a]$, all vertices of T_2 are observed by D_2. There are exactly p extended children of r which can be observed by $r(= r_1)$. So D is a $kPDS$ of T^p but not of T^{p+1} for some $p \in \{0, 1, \cdots, k\}$, i.e., $(T, D, w) \in [b]_p$ with $0 \le p \le k$ (Definition 3). That is $[c]_{k-p} \circ [a] \subseteq [b]_p$.

(ii) If $(T_1, D_1, w) \in [c]_{k-p}$ and $(T_2, D_2, w) \in [b]_l (1 \le l \le k)$. For $1 \le l \le k$, according to OR 1 and definition of $[b]_l$, D_2 is a $kPDS$ of $T_2^l (1 \le l \le k)$, so r_1 receives the message from $V(T_2)$. And then also according to the definition of $[c]_{k-p}$, there are exactly $k - p$ children observed by $r_1(= r)$, all vertices of T_1 and T can be observed at last. There are exactly p extended children of r which can be observed by $r(= r_1)$, so D is a $kPDS$ of T^p, but not of T^{p+1} for $p \in \{0, 1, \cdots, k\}$, i.e., $(T, D, w) \in [b]_p$ satisfies $0 \le p \le k$ (Definition 3). That is $[c]_{k-p} \circ (\cup_{l=1}^{k}[b]_l) \subseteq [b]_p$.

(iii) If $(T_1, D_1, w) \in [b]_p$ and $(T_2, D_2, w) \in [a]$, both $V(T_1)$ and $V(T_2)$ can receive message independently. So D is a $kPDS$ of T, and there are exactly p extended children of r which can be observed by $r(= r_1)$, D is a $kPDS$ of T^p but not of T^{p+1} for some $p \in \{0, 1, \cdots, k\}$, i.e., $(T, D, w) \in [b]_p$ with $0 \le p \le k$ (Definition 3). Thus, $[b]_p \circ [a] \subseteq [b]_p$.

(iv) If $(T_1, D_1, w) \in [b]_p$ and $(T_2, D_2, w) \in [b]_t (0 \le t \le k)$. both $V(T_1)$ and $V(T_2)$ can receive message independently. Also there are exactly p extended children of r can be observed by $r(= r_1)$, D is a $kPDS$ of T^p but not of T^{p+1} for some $p \in \{0, 1, \cdots, k\}$, i.e., $(T, D, w) \in [b]_p$ with $0 \le p \le k$ (Definition 3). Thus, $[b]_p \circ (\cup_{t=1}^{k}[b]_t) \subseteq [b]_p$.

(v) If $(T_1, D_1, w) \in [b]_{p+1}$ and $(T_2, D_2, w) \in [c]_h (0 \le h \le k)$, according to OR 1 and definition of $[b]_{p+1}$, D_1 is a $kPDS$ of T_1^{p+1}, so r_2 receives the message from r_1. There are exactly p extended children of r which can be observed by $r(= r_1)$, D is a $kPDS$ of T^p but not of T^{p+1} for some $p \in \{0, 1, \cdots, k\}$, i.e., $(T, D, w) \in [b]_p$ satisfies $0 \le p \le k$ (Definition 3). Thus, $[b]_{p+1} \circ (\cup_{h=0}^{k}[c]_h) \subseteq [b]_p$.

Now we show that $[b]_p \subseteq [c]_{k-p} \circ [a] \cup [c]_{k-p} \circ (\cup_{l=1}^{k}[b]_l) \cup [b]_p \circ [a] \cup [b]_p \circ (\cup_{t=1}^{k}[b]_t) \cup [b]_{p+1} \circ (\cup_{h=0}^{k}[c]_h)$. Keep in mind, $(T, D) = (T_1, D_1) \circ (T_2, D_2)$, which has $D_1 \subseteq V(T_1), D_2 \subseteq V(T_2), D_1 \cup D_2 = D, r(= r_1) \notin D$. We divide two cases to discuss D_2 as follows.

case(I) D_2 is a $kPDS$ of T_2.

Suppose $(T_2, D_2, w) \in [a]$. Then the root $r = r_1$ of T has two ways to get message. One way is that r is observed by r_2, the other way is that it is observed by its children in T_1. If r cannot get message from its children in T_1, since D is a $kPDS$ of T^p, but not of T^{p+1}, there are exactly $k - p$ children of r_1 observed by $r(= r_1)$. It means that $(T_1, D_1, w) \in [c]_{k-p}$. Suppose that r obtains message from its children in T_1. Note that $r_2 \in D_2$ and hence r_1 does not need to pass message to r_2. Since D is a $kPDS$ of T^p, but not of T^{p+1} and $r(= r_1) \notin D$, it is known that D_1 is a $kPDS$ of T_1^p, but not of T_1^{p+1}. Note that $= r_1 \notin D_1$. Hence, $(T_1, D_1, w) \in [b]_p$.

Suppose $(T_2, D_2, w) \in [b]_l (1 \leq l \leq k)$. Similarly, r has two ways to get message and we have the conclusion $(T_1, D_1, w) \in [c]_{k-p}$ or $(T_1, D_1, w) \in [b]_p$. The proof is similar to the case of $(T_2, D_2, w) \in [a]$. The detail is left to readers.

Suppose $(T_2, D_2, w) \in [b]_0$. Both $V(T_1)$ and $V(T_2)$ can receive message independently. Therefore, $r(= r_1)$ is only observed by its children in T_1. So, $(T_1, D_1, w) \in [b]_p$.

case(II) D_2 is not a $kPDS$ of T_2.

We know that $r(= r_1) \notin D_1$ and D_2 is not a $kPDS$ of T_2. If r_2 can get message from $r(= r_1)$, then all vertices of T_2 can be observed at last. Otherwise it is a contradiction that D is not a $kPDS$ of T. So it means if a message is given to r_2 in advance, then all vertices of T_2 will get message. That is, $(T_2, D_2, w) \in [c]_h (0 \leq h \leq k)$. on the other hand, since $r(= r_1) \notin D_1$, we know that $(T_1, D_1, w) \notin [a]$, considering that $r(= r_1)$ still needs to send message to r_2, so there must be $(T_1, D_1, w) \in [b]_{p+1}$.

In summary, $[b]_p = [c]_{k-p} \circ [a] \cup [c]_{k-p} \circ (\cup_{l=1}^k [b]_l) \cup [b]_p \circ [a] \cup [b]_p \circ (\cup_{t=1}^k [b]_t) \cup [b]_{p+1} \circ (\cup_{h=0}^k [c]_h)$. $\square$

Lemma 4. *Let q be an integer satisfying $0 \leq q \leq k$.*

$$[c]_q = [c]_{q-1} \circ (\cup_{h=0}^k [c]_h) \cup [c]_q \circ [b]_0.$$

Proof. Let $(T, D) = (T_1, D_1) \circ (T_2, D_2)$, where $D_1 \subseteq V(T_1), D_2 \subseteq V(T_2), D_1 \cup D_2 = D$. We first show that $[c]_{q-1} \circ (\cup_{h=0}^k [c]_h) \cup [c]_q \circ [b]_0 \subseteq [c]_q$.

(i) If $(T_1, D_1, w) \in [c]_{q-1}$ and $(T_2, D_2, w) \in [c]_h (0 \leq h \leq k)$, according to the definition of $[c]_{q-1}$, there are exactly $q - 1$ children observed by $r_1(= r)$ in T_1 if a message is given to r in advance. As $r_1 r_2 \in E(T)$, there are exactly q children observed by $r_1(= r)$ in T. That is $[c]_{q-1} \circ (\cup_{h=0}^k [c]_h) \subseteq [c]_q$.

(ii) If $(T_1, D_1, w) \in [c]_q$ and $(T_2, D_2, w) \in [b]_0$, according to the definition of $[c]_q$, there are exactly q children observed by $r_1(= r)$ in T_1 if a message is given to r in advance. By the definition of $[b]_0$, r_2 can be observed by its children in $T_@$. So there are exactly q children observed by $r_1(=r)$ in T if a message is given to r in advance. That is, $[c]_q \circ [b]_0 \subseteq [c]_q$.

Now we show that $[c]_q \subseteq [c]_{q-1} \circ (\cup_{h=0}^{k}[c]_h) \cup [c]_q \circ [b]_0$. For $(T, D, w) \in [c]_q$, according to the definition of $[c]_q$, r is necessary to send message to q children who have not been observed. Note that $(T, D) = (T_1, D_1) \circ (T_2, D_2)$, $D_1 \subseteq V(T_1)$, $D_2 \subseteq V(T_2)$, $D_1 \cup D_2 = D$.

case(I) If r_2 is one of the q children of r who have not been observed.

Since $(T, D, w) \in [c]_q$, all vertices of T_2 can be observed if r_2 is given a message in advance. So there must be $(T_2, D_2, w) \in [c]_h (1 \le h \le k)$. Except that r_2, r still needs to send message to the $q - 1$ children in T_1 if r_1 is given a message in advance, so $(T_1, D_1, w) \in [c]_{q-1}$.

case(II) If r_2 is not one of the q children of r who have not been observed.

Then, D_2 is $kPDS$ of T_2 and r_2 cannot send message to r_1. So, $(T_2, D_2, w) \in [b]_0$. $r(= r_1)$ must send message to its q children in T_1 if r has message in advance. Then, $(T_1, D_1, w) \in [c]_q$.

In summary, $[c]_q = [c]_{q-1} \circ (\cup_{h=0}^{k}[c]_h) \cup [c]_q \circ [b]_0$.   □

Therefore, according to Lemmas 2–4, we conclude that

Theorem 5. *For a weighted tree T, we have that:*
$[a] = [a] \circ [a] \cup [a] \circ (\cup_{l=0}^{k}[b]_l) \cup [a] \circ (\cup_{h=0}^{k}[c]_h);$
$[b]_p = [c]_{k-p} \circ [a] \cup [c]_{k-p} \circ (\cup_{l=1}^{k}[b]_l) \cup [b]_p \circ [a] \cup [b]_p \circ (\cup_{t=0}^{k}[b]_t) \cup [b]_{p+1} \circ (\cup_{h=0}^{k}[c]_h).(0 \le p \le k);$
$[c]_q = [c]_{q-1} \circ (\cup_{h=0}^{k}[c]_h) \cup [c]_q \circ [b]_0.(0 \le q \le k).$

Based on the conclusion of Theorem 5, an effective algorithm for solving the k-power domination problem in weighted tree $T = (V, E, w)$ is given below.

For any $v_i \in V$, define the dynamic programming vector as follows: $(r^a(v_i), \overline{r^b(v_i)}, \overline{r^c(v_i)})$, where $\overline{r^b(v_i)}, \overline{r^c(v_i)}$ respectively correspond to a $(k + 1)$-dimensional vector. $r^{b_t}(v_i), r^{c_t}(v_i)$ represents the value of each t-th dimension.

The algorithm input includes a known tree sequence $\{v_1, v_2, \cdots, v_n\}$ and the algorithm starts from an independent set $\{v_1, v_2, \cdots, v_n\}$. The initial algorithm vector is: $\{w(v_i), \overline{\infty}, \overline{0, \infty}\}$, $\overline{\infty}$ indicates that each dimension has the value ∞, $\overline{0, \infty}$ means the value of $r^{c_0}(v_i)$ is 0, and other dimensions has the value ∞.

Now, we are ready to present the algorithm.

Algorithm 1. WKPDT(Weighted k-Power Domination in Trees)

Input: A weighted tree $T = (V, E, w)$ with a tree ordering $v_1, v_2, \cdots, v_n$.
Output: $\gamma_{kp}^w(T)$.

1   **for** $i := 1$ **to** n **do**
2      $r^a(v_i) \leftarrow w(v_i)$;
3      $\overline{r^b(v_i)} \leftarrow \overline{\infty}$;
4      $\overline{r^c(v_i)} \leftarrow \overline{\infty}$;
5   **end**
6   **for** $j := 1$ **to** $n - 1$ **do**
7      $Min_{bp} = \min(\overline{r^b(v_i)})$;
8      $Min_{cp} = \min(\overline{r^c(v_i)})$;
9      $v_i = F(v_j)$;
10     $r^a(v_i) = r^a(v_i) + \min\{r^a(v_j), Min_{bp}, Min_{cp}\}$;
11     $r^{c0}(v_i) = r^{c0}(v_i) + r^{b0}(v_i)$;
12     $r^{bk}(v_i) = \min\{r^{c0}(v_i) + r^a(v_j), r^{c0}(v_i) + Min_{bp}, r^{bk}(v_i) + r^a(v_j), r^{bs}(v_i) + Min_{cp}\}$;
13   **for** $t := 0$ **to** k **do**
14   **if** $t < k$ **then**
15          $r^{bt}(v_i) = \min\{r^{ck-t}(v_i) + r^a(v_j), r^{ck-t}(v_i) + Min_{bp}, r^{bt}(v_i) + r^a(v_j), r^{bt}(v_i) + Min_{bp}, r^{bt+1}(v_i) + Min_{bp}\}$;
16   **end**
17   **if** $t > 0$ **then**
18          $r^{ct}(v_i) = \min\{r^{ct-1}(v_i) + Min_{bp}, r^{ct}(v_i) + r^{b0}(v_i)\}$;
19   **end**
20   **end**
21   **end**
22   **return** $r_{kp}^w(T) = \min\{r^a(v_n), r^{b0}(v_n), r^{b1}(v_n), \cdots, r^{bk}(v_n)\}$;

From the above argument, we can obtain the following theorem.

Theorem 6. *Algorithm WKPDT can output the weighted k-power domination number of any weighted tree $T = (V, E, w)$ in linear time $O(m + n)$, where $n = |V|$ and $m = |E|$.*

References

1. Aazami, A.: Domination in graphs with bounded propagation: algorithms, formulations and hardness result. J. Comb. Optim. **19**(4), 429–456 (2010)
2. Aazami, A., Stilp, K.: Approximation algorithms and hardness for domination with propagation. SIAM J. Discret. Math. **23**, 1382–1399 (2009)
3. Baldwin, T.L., Mili, L., Boisen Jr., M.B., Adapa, R.: Power system observability with minimal phasor measurement placement. IEEE Trans. Power Systems. **8**, 707–715 (1993)
4. Brandstadt, A., Kratsch, D.: On domination problems on permutation and other graphs. Theoret. Comput. Sci. **54**(2–3), 181–198 (1987)
5. Chang, G.J.: Algorithmic aspects of domination in graphs. In: Handbook of Optimization, vol. 3, pp. 339–405 (1998)
6. Chang, G.J., Dorbec, P., Montassier, M., Raspaud, A.: Generalized power domination of graphs. Discret. Appl. Math. **160**, 1691–1698 (2012)

7. Chang, M.S., Liu, Y.C.: Polynomial algorithms for the weighted perfect domination problems on chordal graphs and split graphs. Inf. Process. Lett. **48**(4), 205–210 (1993)
8. Chang, M.S., Liu, Y.C.: Polynomial algorithms for weighted perfect domination problems on interval and circular-arc graphs. J. Inf. Sci. Eng. **11**, 549–568 (1994)
9. Dorbec, P., Mollard, M., Klavžar, S., Špacapan, S.: Power domination in product graphs. SIAM J. Discret. Math. **22**, 554–567 (2008)
10. Dorfling, M., Henning, M.A.: A note on power domination in grid graphs. Discret. Appl. Math. **154**, 1023–1027 (2006)
11. Guo, J., Niedermeier, R., Raible, D.: Improved algorithms and complexity results for power domination in graphs. Algorithmica **52**, 177–202 (2008)
12. Haynes, T.W., Hedetniemi, S.M., Hedetniemi, S.T., Henning, M.A.: Domination in graphs applied to electric power networks. SIAM J. Discret. Math. **15**, 519–529 (2002)
13. Liao, C.-S., Lee, D.-T.: Power domination problem in graphs. In: Wang, L. (ed.) COCOON 2005. LNCS, vol. 3595, pp. 818–828. Springer, Heidelberg (2005). https://doi.org/10.1007/11533719_83
14. Lu, C., Mao, R., Wang, B.: Power domination in regular claw-free graphs. arXiv preprint arXiv:1808.02613 (2018)
15. Kratsch, D., Stewart, L.: Domination on cocomparability graphs. Discret. Appl. Math. **63**(3), 215–222 (1995)
16. Natarajan, K.S., White, L.J.: Optimum domination in weighted trees. Inf. Process. Lett. **7**(6), 261–265 (1978)
17. Novak, T., Zerovnik, J.: Weighted domination number of cactus graphs. Int. J. Appl. Math. **29**(4), 401–423 (2016)
18. Wimer, T.V.: Linear algorithms on k-terminal graphs. Ph.D. Thesis, Clemson University (1987)
19. Yen, C.C., Lee, R.: The weighted perfect domination problem. Inf. Process. Lett. **35**(6), 295–299 (1990)
20. Yeh, H.G., Chang, G.J.: Weighted connected domination and Steiner trees in distance-hereditary graphs. Discret. Appl. Math. **87**(1–3), 245–253 (1998)

General Rumor Blocking: An Efficient Random Algorithm with Martingale Approach

Qizhi Fang[1], Xin Chen[1(✉)], Qingqin Nong[1], Zongchao Zhang[2],
Yongchang Cao[1], Yan Feng[1], Tao Sun[1], Suning Gong[1], and Ding-Zhu Du[1,3]

[1] School of Mathematical Sciences, Ocean University of China, Qingdao 266100,
Shandong Province, People's Republic of China
cxin0307@163.com
[2] Guangzhou VIP Information Technology Co, Ltd., Guangzhou, China
[3] Department of Computer Science, University of Texas,
Richardson, TX 75080, USA

Abstract. Rumor Blocking, an important optimization problem in social network, has been extensively studied in the literature. Given social network $G = (V, E)$ and rumor seed set A, the goal is asking for k protector seeds that protect the largest expected number of social individuals by truth. However, the source of rumor is always uncertain, rather than being predicted or being known in advance in the real situations, while rumor spreads like wildfire on the Internet.

This paper presents General Rumor Blocking with unpredicted rumor seed set (randomized A) and various personal profits while being protected (weights of nodes in V). We first show that the objective function of this problem is non-decreasing and submodular, and thus a $(1 - 1/e)$ approximate solution can be returned by greedy approach. We then propose an efficient random algorithm R-GRB which returns a $(1 - 1/e - \varepsilon)$ approximate solution with at least $1 - n^{-\ell}$ probability. We show that it runs in $O\left(m(n - r)(k \log(n - r) + \ell \log n)/\varepsilon^2\right)$ expected time, where $m = |E|$, $n = |V|$, $r = |A|$ and k is the number of protector seeds.

Keywords: Rumor Blocking · Random algorithm
Greedy approach · Martingale

1 Introduction

In the last few decades, the tremendous development of the Internet creates large-scale communication platforms, such as WeChat and Twitter. Each platform can be modeled as a social network which is a graph of relationships and interactions within a group of individuals. People communicate their information, ideas with each other on the social networks. The networks make communications more

This work is supported by NSFC (No. 11271341 and No. 11201439).

© Springer Nature Switzerland AG 2018
S. Tang et al. (Eds.): AAIM 2018, LNCS 11343, pp. 161–176, 2018.
https://doi.org/10.1007/978-3-030-04618-7_14

easier and bring great benefit to the society. But in some case they also bring harm. For instance, in July 2017, a rumor that Yao Ming was arrested for drug abuse circulated widely in Twitter and Weibo. Although it has been clarified, this rumor brings serious impacts on Yao Ming's personal life and even Chinese Basketball Association's reputation. In fact, similar rumors appear on social networks from time to time and they may make significant inroads into the population. Therefore, it is necessary to research effective strategies to block the diffusion of a rumor.

Most of Rumor Blocking problems are considered in the independent cascade (IC) model, which is formulated by Kempe et al. [8]. The (IC) model captures the intuition that influence can spread stochastically through a network. Specifically, this model can be described as a directed edge-weighted graph $G = (V, E)$ with n nodes and m edges. There is a weight p_{uv} on each edge e_{uv}, representing the probability that the process spreads along this edge from u to v. Influence spreads via a random process that begins at a set of seed nodes. Each node, once activated, has a chance of subsequently activating its neighbors. Its information diffusion process looks like cascades starting with a set of seed users. Note that when two opposing information (truth and rumor) diffuse in a network, a user is likely to believe and spread the information arriving first. One of effective strategies for rumor containment is to select a certain number of protector seeds to compete against the rumor by spreading truth. Due to limits of budget, the number of protector seeds cannot be very large. A natural problem is how to select k seeds to maximize the number of users protected by the truth.

1.1 Related Works

The work related to the Rumor Blocking problems is the study of competitive influence diffusion. Some researchers concentrate on the case that there is only one decision-maker, such as [2,13]. Budak et al. [3] firstly study influence limitation problem under competing campaigns and they show that the problem is NP-hard. He et al. [7] study information blocking maximization (IBM) under competitive linear threshold (CLT) diffusion model. Fan et al. [5] and [6] proposed two new models of competitive influence diffusion and study rumor containment maximization problem. The approach used most often is to prove the monotone increasing property and submodularity of an objective function and then apply the classic greedy algorithm to obtain a $1 - 1/e$-approximation solution [10]. But due to the massive size and the randomness of a probabilistic diffusion model, the calculation of its objective function often very complicated. Thus running greedy algorithm on it is computationally expensive. Inspired by the reverse algorithm ideas proposed in [1], and [14], Tong et al. [16] present an efficient randomized algorithm for the rumor blocking problem which runs in $O(\frac{km \ln n}{\delta^2})$ expected time and provides a $(1 - 1/e - \delta)$-approximation with a high probability. Competitive influence diffusion is also studied in game environments where multiple decision-makers try to maximize their own objectives, like [12].

The paper is organized as follows. Section 2 presents General Rumor Blocking problem and depicts a greedy algorithm which returns a $(1 - 1/e)$-approximate solution. In Sect. 3, we propose an efficient random algorithm R-GRB for General Rumor Blocking problem. By using the tools of reverse methods and martingale's approaches, we analyze the approximate ratio and time complexity.

2   General Rumor Blocking Problem

2.1   Problem Definition

Let $G = (V, E)$ be a directed social network, which contains n nodes and m directed edges. Let r and k be two parameters. For each node $v \in V$, we denote $w_v \in [\frac{1}{n-r}, 1]$ as individual v's personal interest for the rumor information, as well as the personal profit (while being protected). For each edge $e_{uv} \in E$, denote $p_{uv} \in [0, 1]$ as an influence (protected or infected) probability.

Suppose that the rumor seed set is unpredicted and uncertain in advance. Let $\mathcal{A} = \{A | A \subseteq V, |A| = r\}$ be the collection of random rumor seed sets. Without loss of generality, assume that there is a probability distribution over $\mathcal{A}$.

Our goal is asking for a protector seed set B such that $|B| \leq k$ and it maximizes the expected the social profit $\sigma(B)$. To be specific, we firstly denote $I_{A,g}(B)$ as the set of nodes protected by B in an outcome g with a rumor seed set A, i.e.,

$$I_{A,g}(B) = \{v \mid t_{pu}^g(B) < t_{ru}^g(A) < +\infty\}.$$

Then the objective function $\sigma(B)$ is defined as

$$\sigma(B) = \sum_{A \in \mathcal{A}} \Pr[A] \sum_g \Pr[g] \sum_{v \in I_{A,g}(B)} w_v.$$

From the above definition, $\sigma(B)$ tends to increase with the expansion of protector seed set B. Besides, the marginal utility of $\sigma(B)$ is decreasing. Thus we obtain the following observation.

Obsersation 1. *The objective function $\sigma(\cdot)$ of the novel rumor blocking problem is non-decreasing and submodular.*

2.2   Greedy Algorithm

Since the original Rumor Blocking problem [3] is NP-hard, it is clear that General Rumor Blocking problem is also NP-hard. However, we observe the submodularity of the objective function $\sigma(B)$, thus we can use greedy approach [10] to obtain a $(1 - 1/e)$-approximate solution.

Lemma 1. *The General Rumor Blocking problem can obtain a $(1-1/e)$ approximate solution by the greedy algorithm.*

Algorithm 1. Greedy Algorithm for General Rumor Blocking

Input: a directed graph $G = (V, E)$, a parameter k, a probability distribution over $\mathcal{A}$.

Output: a protector seed set B.

1: Initialize: $B = \emptyset$, $\sigma(\emptyset) = 0$;
2: **for** $i = 1$ to k **do**
3:     $B = B \cup \arg\max_{u \in V \setminus A} (\sigma(B \cup \{u\}) - \sigma(B))$;
4: **return** B.

---

3   Random Algorithm and Theoretical Analysis

This section provides an efficient random algorithm R-GRB for General Rumor Blocking. At the beginning, we introduce several key definitions.

3.1   Definitions and Properties

Definition 1. (Random Reverse Protected Set)

   A random Reverse Protected (RP) set R is generated by executing the following three steps:

1. randomly sample a size-r rumor seed set A from $\mathcal{A}$, and then sample an outcome g by removing each edge e_{uv} in G with $1 - p_{uv}$ probability;
2. select a target $v \in V \setminus A_r$ with $\Pr[v] = \frac{w_v}{\sum_{u \in V \setminus A} w_u}$ probability;
3. take the node set $R_{A,g}(v) = \{u \mid t_{uv}^g < t_{rv}^g(A) < +\infty\}$.

   Intuitively, $R_{A,g}(v)$ consists of nodes that is capable of protecting the target v while the rumor seed set is A and the outcome is g. For any protector seed set B, denote $x(B, R_{A,g}(v))$ and $h_{A,g}(B, v)$ as follows:

$$x(B, R_{A,g}(v)) = \begin{cases} 1, B \cap R_{A,g}(v) \neq \emptyset; \\ 0, \text{otherwise}. \end{cases} \quad h_{A,g}(B, v) = \begin{cases} 1, v \in I_{A,g}(B); \\ 0, \text{otherwise}. \end{cases}$$

Note that $x(B, R_{A,g}(v)) = 1$ if there exists a protector seed $u_0 \in B$ and u_0 is capable of protecting the target v while the rumor seed set is A and the outcome is g. That is, $x(B, R_{A,g}(v)) = 1$ if the target v can be protected by B, which implies $h_{A,g}(B, v) = 1$, vice versa.

Obsravation 2. *The random variable* $x(B, R_{A,g}(v)) = 1$ *iff* $h_{A,g}(B, v) = 1$.

   Now consider a protector seed set B and a collection $\mathcal{R} = \{R_1, \dots, R_\rho\}$ which consists of ρ random RP sets. Note that each R_i corresponds to an independent rumor seed set A_i and outcome g_i. For any A_i, denote $w(\overline{A_i}) = \sum_{u \in V \setminus A_i} w_u$ and define $f_{\mathcal{R}}(B)$ with respect to $\mathcal{R}$ and B as follows:

$$f_{\mathcal{R}}(B) = \frac{1}{\rho} \sum_{R_i \in \mathcal{R}} w(\overline{A_i}) \cdot x(B, R_i).$$

The following lemma, with the proof presented in the Appendix, illustrates that $f_{\mathcal{R}}(B)$ is non-decreasing and submodular.

Lemma 2. *For any collection $\mathcal{R}$ of random RP sets, the function $f_{\mathcal{R}}(\cdot)$ is non-decreasing and submodular.*

Consider the relations between $\mathbb{E}[f_{\mathcal{R}}(B)]$ and $\sigma(B)$, we have

Lemma 3. *For any protector seed set B,*

$$\mathbb{E}[f_{\mathcal{R}}(B)] = \sigma(B),$$

where the expectation of $f_{\mathcal{R}}(B)$ is taken over the randomness in $\mathcal{R}$.

Proof. Based on the linearity of expectation, we have

$$\mathbb{E}[f_{\mathcal{R}}(B)] = \frac{1}{\rho} \sum_{R_i \in \mathcal{R}} \mathbb{E}[w(\overline{A_i}) \cdot x(B, R_i)] = \mathbb{E}[w(\overline{A_i}) \cdot x(B, R_i)],$$

According to the process of generating a random RP set in Definition 1, we obtain that the probability of sampling a RP set R_i is

$$\Pr[R_i] = \Pr[A_i] \cdot \Pr[g_i] \cdot \Pr[v_i] = \Pr[A_i] \cdot \Pr[g_i] \cdot \frac{w_{v_i}}{\sum_{u \in V \setminus A_i} w_u}.$$

Let $\mathcal{P}$ be a set consisting of all possible RP sets. We have

$$\mathbb{E}[w(\overline{A_i}) \cdot x(B, R_i)] = \sum_{A_i} \Pr[A_i] \sum_{g_i} \Pr[g_i] \sum_{v_i \in V \setminus A_i} x(B, R_i) \cdot w_{v_i}.$$

Due to Observation 1, that is $x(B, R_i) = 1$ if and only if $h_{A_i, g_i}(B, v_i) = 1$. We have

$$\mathbb{E}[f_{\mathcal{R}}(B)] = \sum_{A_i} \Pr[A_i] \sum_{g_i} \Pr[g_i] \sum_{v_i \in V \setminus A_i} h_{A_i, g_i}(B, v_i) \cdot w_{v_i}$$

$$= \sum_{A} \Pr[A] \sum_{g} \Pr[g] \sum_{v \in I_{A,g}(B)} w_v = \sigma(B),$$

implying the lemma.   □

Accordingly, the random variable $f_{\mathcal{R}}(B)$ over $\mathcal{R}$ can be used as an estimator of $\sigma(B)$. In order to ensure the accuracy of this estimation, we introduce *martingales* [4] and their related properties [4].

Definition 2 (Martingale). *A martingale is a sequence of random variables $Y_1, Y_2, Y_3, \cdots$ such that $\mathbb{E}[|Y_i|] < +\infty$ and $\mathbb{E}[Y_i \mid Y_1, \ldots, Y_{i-1}] = Y_{i-1}$ for any i.*

Property 1 (Martingale's Property). *Let $Y_1, Y_2, Y_3, \ldots$ be a martingale, such that $|Y_1| \leq a, |Y_j| - |Y_{j-1}| \leq a$ for each $j \in \{2, \ldots, i\}$, and*

$$\mathrm{Var}[Y_1] + \sum_{j=2}^{i} \mathrm{Var}[Y_j \mid Y_1, \ldots, Y_{j-1}] \leq b.$$

Then for any $\gamma > 0$,

$$\Pr[Y_i - \mathbb{E}[Y_i] \geq \gamma] \leq \exp\left(-\frac{\gamma^2}{\frac{2}{3}a\gamma + 2b}\right); \quad \Pr[Y_i - \mathbb{E}[Y_i] \leq -\gamma] \leq \exp\left(-\frac{\gamma^2}{2b}\right).$$

Algorithm 2. Node Selection (the 2nd stage of R-GRB)

Input: a sampling result $\mathcal{R} = \{R_1, \cdots, R_\rho\}$, a parameter k.
Output: a protector seed set $\tilde{B}$.
1: Initialize: $\tilde{B} = \emptyset$
2: **for** $i = 1$ to k **do**
3:     $\tilde{B} = \tilde{B} \cup \arg\max_{u \in V \setminus \tilde{B}} f_\mathcal{R}(\tilde{B} \cup \{u\}) - f_\mathcal{R}(\tilde{B})$
4: **return** $\tilde{B}$.

Consider the collection of random RP sets $\mathcal{R} = \{R_1, \cdots, R_\rho\}$. Let B be any protector seed set and let $q = \mathbb{E}[w(\overline{A_i}) \cdot x(B, R_i)]$. From the proof of Lemma 3, $q = \sigma(B)$. We construct a sequence of random variables

$$Z_i = \sum_{j=1}^{i} (w(\overline{A_j}) \cdot x(B, R_j) - q), i = 1, 2, \ldots, \rho.$$

By the properties of random RP sets, we can verify the following lemma.

Lemma 4. *The sequence of random variables $Z_1, \cdots, Z_\rho$ is a martingale and for any $\varepsilon > 0$, we have*

$$\Pr\left[\sum_{i=1}^{\rho}(w(\overline{A_i}) \cdot x(B, R_i) - \rho q \geq \varepsilon \cdot \rho q\right] \leq \exp\left(-\frac{\varepsilon^2}{(2 + \frac{2}{3}\varepsilon)(n - r)} \cdot \rho q\right); \quad (1)$$

$$\Pr\left[\sum_{i=1}^{\rho}(w(\overline{A_i}) \cdot x(B, R_i)) - \rho q \leq -\varepsilon \cdot \rho q\right] \leq \exp\left(-\frac{\varepsilon^2}{2(n - r)} \cdot \rho q\right). \quad (2)$$

The proof of Lemma 4 is presented in the Appendix.

3.2   Random Algorithm for General Rumor Blocking

This section presents R-GRB algorithm that is inspired by ideas of IMM algorithm [15]. Basically, the R-GRB contains two stages:

- **Sampling of random RP sets.** This stage iteratively generates random RP sets until reaches a certain stopping instruction. Then collect these RP sets into a set $\mathcal{R}$.
- **Node Selection.** This stage adopts greedy algorithm for maximizing weighted coverage and it derives a size-k node set $\tilde{B}$ that covers a considerable large weight of RP-sets in $\mathcal{R}$.

Node Selection. Algorithm 2 provides the pseudo-code of the second stage of R-GRB algorithm. Assume that $\mathcal{R} = \{R_1, \cdots, R_\rho\}$ is the collection of random RP sets generated by the sampling stage. Based on Lemma 2, $f_\mathcal{R}(B)$ is non-decreasing and submodular. One can see that the greed algorithm is a $(1 - 1/e)$-approximate algorithm to find a size-k node set maximizing $f_\mathcal{R}(\cdot)$.

The remainer of this section will confirm the minimum number of RP sets in $\mathcal{R}$ and show that the output of R-GRB $\tilde{B}$ guarantees a $(1-1/e-\varepsilon)$-approximate solution with high probability, while the number $\rho = |\mathcal{R}|$ is greater than some certain number.

Consider the General Rumor Blocking on $G = (V, E)$. We denote B^* as the optimal solution and denote by OPT$= \sigma(B^*)$ the optimal value. Let ℓ be a probability indicator. For any $\varepsilon_1, \varepsilon_2 \in (0, 1)$ and any $\delta_1, \delta_2 \in (0, 1)$, we define

$$\rho_1 = \frac{2(n-r)\log(1/\delta_1)}{\varepsilon_1^2 \cdot \text{OPT}}, \quad \rho_2 = \frac{(2 + \frac{2}{3}\varepsilon_2)(n-r)\log\left(\binom{n}{k}/\delta_2\right)}{\varepsilon_2^2 \cdot \text{OPT}}.$$

Let $\mathcal{R}$ be a size-ρ set of random RP sets. Since $f_{\mathcal{R}}(B) = \frac{1}{\rho}\sum_{R_i \in \mathcal{R}} \left(w(\overline{A_i}) \cdot x(B, R_i)\right)$ and $\sigma(B) = \mathbb{E}[f_{\mathcal{R}}(B)]$. That is, the random variable $\frac{1}{\rho}\sum_{R_i \in \mathcal{R}} \left(w(\overline{A_i}) \cdot (B, R_i)\right)$ over $\mathcal{R}$ is an estimator of $\sigma(B)$. To ensure the accuracies of this estimation and our algorithm, ρ needs to be sufficiently large. The following lemma and theorem consider this issue.

Lemma 5. *If the number of random RP sets in the sampling result $\mathcal{R}$ of R-GRB algorithm suffices that $\rho \geq \rho_1$,*

$$f_{\mathcal{R}}(\tilde{B}) \geq (1 - 1/e)(1 - \varepsilon_1)\text{OPT}$$

holds with at least $1 - \delta_1$ probability.

The proof of Lemma 5 is presented in the Appendix.

Theorem 1. *If the number of random RP sets in the sampling result $\mathcal{R}$ suffices that $\rho \geq \max\{\rho_1, \rho_2\}$, Algorithm 2, the second stage of R-NRB, returns a $(1 - 1/e-\varepsilon)$ approximate solution of the Novel Rumor Blocking problem with at least $1 - n^{-\ell}$ probability.*

Proof. Let $\varepsilon_2, \delta_2 \in (0, 1)$, $\varepsilon_2 = \varepsilon - (1 - 1/e)\varepsilon_1$, $\delta_2 + \delta_1 \leq n^{-\ell}$. From Lemma 5, we know that if $\rho \geq \rho_1$,

$$f_{\mathcal{R}}(\tilde{B}) \geq (1 - 1/e)(1 - \varepsilon_1)\text{OPT}$$

holds with at least $(1-\delta_1)$ probability. Note that $(1-\delta_1)(1-\delta_2) > 1-(\delta_1+\delta_2) \geq 1 - n^{-\ell}$. The theorem holds if we can show that when $\rho \geq \rho_2$,

$$\Pr[\sigma(\tilde{B}) - f_{\mathcal{R}}(\tilde{B}) \geq -\varepsilon_2 \cdot \text{OPT}] \geq 1 - \delta_2. \tag{3}$$

To prove the inequality 3, it is sufficient to show that if $\rho \geq \rho_2$,

$$\Pr[f_{\mathcal{R}}(\tilde{B}) - \sigma(\tilde{B}) \geq \varepsilon_2 \cdot \text{OPT}] \leq \delta_2.$$

Based on Lemmas 3 and 4, we can verify that if $\rho \geq \rho_2$, for any size-k protector seed set B,

$$\Pr[f_{\mathcal{R}}(B) - \sigma(B) \geq \varepsilon_2 \cdot \text{OPT}] \leq \delta_2 \Big/ \binom{n}{k}.$$

Thus the theorem is proved.   □

Algorithm 3. Sampling of Random RP Sets (the first stage of R-NRB)

Input: a directed network $G = (V, E)$, parameters k, r, ε and ℓ.
Output: a set $\mathcal{R}$ of random RP sets.
1: Initialize: $\mathcal{R} = \emptyset$, NPT$= 1$, $\varepsilon_0 = \sqrt{2}\varepsilon$
2: $\mu_0 = (2 + \frac{2}{3}\varepsilon_0) \cdot (n - r) \cdot \left(\log \binom{n}{k} - \log \delta_0\right) \varepsilon_0^{-2}$
3: $\tilde{\mu} = (4 - 2/e) \cdot (2 - 1/e + \frac{1}{3}\varepsilon) \cdot (n - r) \cdot \left(\log \binom{n}{k} + \log(n^\ell) + \log 2\right) \varepsilon^{-2}$
4: **for** $i = 1$ to $\log_2(n - r) - 1$ **do**
5:     $\xi_i = (n - r) \cdot 2^{-i}$, $\rho_i = \mu_0/\xi_i$
6:     **while** $|\mathcal{R}| \leq \rho_i$ **do**
7:         generate a random RP set and put it into $\mathcal{R}$;
8:     $B_i = $ Algorithm 2 $(\mathcal{R}, k)$
9:     **if** $f_\mathcal{R}(B_i) \geq (1 + \varepsilon_0) \cdot \xi_i$ **then**
10:         NPT $= f_\mathcal{R}(B_i)/(1 + \varepsilon_0)$
11:         **break**;
12:     $\rho = \tilde{\mu}/$NPT;
13:     **while** $|\mathcal{R}| \leq \rho$ **do**
14:         generate a random RP set and put it into $\mathcal{R}$;
15: **return** $\mathcal{R}$.

According to Theorem 1, Algorithm 2 can obtain a $(1 - 1/e - \varepsilon)$ approximate solution for novel rumor blocking problem with high probability when the number of random RP sets in the sampling result $\mathcal{R}$ s.t. $\rho \geq \max\{\rho_1, \rho_2\}$. To minimize ρ as far as possible, let $\delta_1 = \delta_2 = 1/(2n^\ell)$ and $\varepsilon_1 = \varepsilon_2 = \varepsilon/(2 - 1/e)$. Denote

$$\tilde{\rho} = \frac{(4 - 2/e)(2 - 1/e + \varepsilon/3)(n - r)\left(\log \binom{n}{k} + \log(2n^\ell)\right)}{\varepsilon^2 \cdot \text{OPT}},$$

and $\tilde{\mu} = \tilde{\rho} \cdot \text{OPT}$. It can be verified that $\tilde{\rho} \geq \max\{\rho_1, \rho_2\}$.

Unfortunately, it is NP-hard to obtain the OPT of the General Rumor Blocking problem. In the next section, we find a lower bound of the optimal solution called NPT, instead of real OPT and use $\rho = \tilde{\mu}/$NPT $\geq \max\{\rho_1, \rho_2\}$ to be the number of random RP sets in the sampling result $\mathcal{R}$.

In the end of this section, we show the time complexity of the second stage of R-GRB.

Lemma 6. *Algorithm 2 runs in $O(\sum_{R \in \mathcal{R}} |R|)$ time.*

The proof of Lemma 6 is presented in the Appendix.

Sampling of Random RP Sets. Let $\varepsilon_0 \in (0, 1)$ and $\delta_0 = n^{-\ell}/\log_2(n - r)$. Algorithm 3 (Sampling phase), the first stage of R-GRB, aims to generate a set $\mathcal{R}$ with a sufficiently large number of random RP sets, $|\mathcal{R}| \geq \tilde{\rho} = \tilde{\mu}/\text{OPT}$. Since it is hard to obtain the real OPT, the algorithm searches various lower bounds of OPT by constantly generating random RP sets and repeatedly calling Algorithm 2 (Node Selection). When the lower bound is sufficiently precise, it stops generating and calling. The final $|\mathcal{R}|$ can be identified by iterations of the for-loop, the lower bound NPT and parameters $\tilde{\mu}, \mu_0$.

As for the generation of a random RP set R, we adopt reverse Breath First Search (BFS) algorithm [9]. First we randomly select a rumor seed set A and select a target node v with $\Pr[v] = w_v/(\sum_{u \in V \backslash A} w_u)$ probability. Then we use reverse BFS algorithm to search nodes until one rumor seed is visited, then put these nodes into R.

Note that the random RP sets generated by Algorithm 3 are not independent. It is because the generation of the i_{th} random RP set is dependent on whether the first $i-1$ random RP sets can estimate a satisfactory lower bound of OPT. As a result, we adopt martingale method on the theoretical analysis of R-GRB algorithm. The rest of this section will show the theoretical rationality and the time complexity of Algorithm 3.

Theorem 2. *Algorithm 3 returns a sampling result $\mathcal{R}$ with $|\mathcal{R}| \geq \tilde{\mu}/OPT$ with at least $1 - n^{-\ell}$ probability.*

Proof. Consider the i_{th} round of the for-loop in Algorithm 3. Observe that in i_{th} round, $\mathcal{R}$ is a collection of random RP sets with

$$|\mathcal{R}| \geq \rho_i = \frac{\mu_0}{\xi_i} = \frac{(2 + \frac{2}{3}\varepsilon_0)(n - r)\left(\log\binom{n}{k} - \log(\delta_0)\right)}{\varepsilon_0^2 \cdot \xi_i}.$$

B_i is the output of calling Algorithm 2 on $\mathcal{R}$.

Claim 1. *If $OPT < \xi_i$, $\Pr[f_{\mathcal{R}}(B_i) \geq (1 + \varepsilon_0) \cdot \xi_i] \leq \delta_0$.*

The proof of Claim 1 is presented in the Appendix.

Claim 2. *If $OPT \geq \xi_i$, $\Pr[OPT \geq f_{\mathcal{R}}(B_i)/(1 + \varepsilon_0)] \geq 1 - \delta_0$.*

The proof of Claim 2 is presented in the Appendix.

These two claims can imply the correctness of the theorem. To see this, let $i^* = \lceil \log_2((n - r)/OPT) \rceil$, then $\xi_j > OPT$ for each $j \in \{1, \ldots, i^* - 1\}$ and $\xi_j \leq OPT$ for $j \geq i^*$. First, Claim 1 implies that Algorithm 3 terminates its for-loop before the i^*_{th} iteration with at most $(i^* - 1) \cdot \delta_0$ probability. Note that when the algorithm enters into the j_{th} for loop with $j \geq i^*$, we have $\xi_j \leq OPT$. Then by Claim 2, NPT $= f_{\mathcal{R}}(\tilde{B})/(1 + \varepsilon_0) \geq OPT$ with at most $\log_2(n - r) - i^* + 1$ probability. Overall, NPT can be derived from the teminate condition (line 11-13 of the Algorithm 3) to be a lower bound of OPT with at least $1 - \log_2(n - r) \cdot \delta_0$ probability. Thus Algorithm 3 can return a sampling result $\mathcal{R}$ with $|\mathcal{R}| \geq \tilde{\mu}/OPT$ probability and the theorem holds. $\square$

In the rest of this section we analyse the time complexity of Algorithm 3 (Sampling phase). The analysis will be in progress from the time of generating a random RP set, the expected time of generating the sampling result $\mathcal{R}$ to the expected time of calling Algorithm 2.

At the beginning, we focus on the time of generating one random RP set. Recall that we use reverse BFS to search each incoming neighbor and related edge constantly until one rumor seed being visited. Let $G[R]$ be the subgraph induced by the nodes in R. Denote by $d(R)$ the number of edges in the induce subgraph $G[R]$. Note that $G[R]$ is a connected graph and $|R| \le d(R)+1$. Thus we need to visit $d(R)$ edges to generate RP set R, that is, we need $\mathbb{E}[d(R)]$ expected time to generate a random RP set. Then the upper bound of $\mathbb{E}[d(R)]$ can be verified by the following lemma with the proof presented in the Appendix.

Lemma 7
$$\mathbb{E}[d(R)] \le m \cdot \text{OPT}.$$

In the following, we discuss the expected time of generating random RP sets and calling Algorithm 2 in Algorithm 3.

For generating random RP sets, by Lemma 7, it requires $\mathbb{E}[d(R)] \le m \cdot \text{OPT}$ time to obtain one random RP set. As for the expected number of RP sets in the final sampling result $\mathcal{R}$, we observe that $|\mathcal{R}| = \max\{\mu_0/\xi_{i_t}, \tilde{\mu}/\text{NPT}\}$, where i_t is the stopping iteration of the for-loop and $\xi_{i_t} \le \text{NPT} \le \text{OPT}$. Thus the expected number

$$\mathbb{E}[|\mathcal{R}|] = O\left(\frac{\max\{\mu_0, \tilde{\mu}\}}{\text{OPT}}\right) = O\left(\frac{(n-r)(k\log(n-r) + \ell\log n)/\varepsilon^2}{\text{OPT}}\right), \quad (4)$$

where the last equation holds by magnifying values of μ_0 and $\tilde{\mu}$. Recall that generate one random RP set requires $\mathbb{E}[d(R)]$ time. Thus the expected time of generating the sampling result $\mathcal{R}$ is $\mathbb{E}[\sum_{R \in \mathcal{R}} d(R)]$. However the number of random RP sets in $\mathcal{R}$ are dependent on terminate condition of Algorithm 3, it is not trivial to obtain that

$$\mathbb{E}[\sum_{R \in \mathcal{R}} d(R)] = \mathbb{E}[|\mathcal{R}|] \cdot \mathbb{E}[d(R)]. \quad (5)$$

We can verify the equality 5 by introducing another property of martingale [18].

Property 2. Let $Y_1, Y_2, \ldots$ be a martingale and $\tau < +\infty$ be a random variable such that the event $\tau = i$ is independent of $Y_1, \ldots Y_{i-1}$. Then we have

$$\mathbb{E}[Y_\tau] = \mathbb{E}[Y_1].$$

By the inequality 4 and Lemma 7, the expected time of generating $\mathcal{R}$ is

$$\mathbb{E}[\sum_{R \in \mathcal{R}} d(R)] = \mathbb{E}[|\mathcal{R}|] \cdot \mathbb{E}[d(R)] = O\left(m(n-r)(k\log(n-r) + \ell\log n)/\varepsilon^2\right).$$

Then for calling Algorithm 2, based on Lemma 6, and for any RP set R, the induced subgraph $G[R]$ is connected, implying $|R| \le d(R) + 1$. Thus

$$O\left(\mathbb{E}[\sum_{R \in \mathcal{R}} |R|]\right) = O\left(\mathbb{E}[\sum_{R \in \mathcal{R}} d(R)]\right) = O\left(m(n-r)(k\log(n-r) + \ell\log n)/\varepsilon^2\right).$$

In summary, the following theorem holds.

Theorem 3. *Algorithm 3 runs in $O\left(m(n-r)(k\log(n-r) + \ell\log n)/\varepsilon^2\right)$ expected time.*

Summary of R-GRB. From the preceding two sections, we have known that R-NRB firstly executes Algorithm 3 to obtain a sampling $\mathcal{R}$ of random RP sets, then inserts the sampling result $\mathcal{R}$ into Algorithm 2 to obtain an approximate protector seed set $\tilde{B}$. Based on Theorems 1, 2 and 3, we obtain the final theorem.

Theorem 4. *R-GRB algorithm returns a $(1 - 1/e - \varepsilon)$ approximate solution with at least $1 - n^{-\ell}$ probability and its expected running time is*

$$O\left(m(n-r)(k \log(n-r) + \ell \log n)/\varepsilon^2\right).$$

4 Contributions

In this paper we present General Rumor Blocking problem with an unpredicted rumor seed set (randomized A) and various personal profits while being protected (weights of nodes in V), where A is the rumor seed set, V is the node set of the social network $G = (V, E)$. We first show that the objective function is non-decreasing and submodular, and thus a $(1 - 1/e)$ approximate solution can be returned by a greedy approach. Inspired by the reverse algorithm ideas in [15], we then propose an efficient random algorithm R-GRB which returns a $(1 - 1/e - \varepsilon)$ approximate solution with at least $1 - n^{-\ell}$ probability. Futher, we show that it runs in $O\left(m(n-r)(k \log(n-r) + \ell \log n)/\varepsilon^2\right)$ expected time, where $r = |A|$ and k is the number of protector seeds.

Appendix

Proof of Lemma 2. First, we show the monotonicity. For any node set B and any protector seed $u \notin B$, we have

$$f_{\mathcal{R}}(B \cup \{u\}) - f_{\mathcal{R}}(B) = \frac{1}{\rho} \sum_{R_i \in \mathcal{R}} w(\overline{A_i}) \cdot (x(B \cup \{u\}, R_i) - x(B, R_i)).$$

Since $x(B, R_i) = 1$ leads to $x(B \cup \{u\}) = 1$, $f_{\mathcal{R}}(B \cup \{u\}) - f_{\mathcal{R}}(B) \geq 0$. It implies the monotonicity.

Then, we show the submodularity. For any pair of B_1, B_2 with $B_1 \subseteq B_2$ and $u \notin B_2$, we have

$$(f_{\mathcal{R}}(B_1 \cup \{u\}) - f_{\mathcal{R}}(B_1)) - (f_{\mathcal{R}}(B_2 \cup \{u\}) - f_{\mathcal{R}}(B_2))$$
$$= \frac{1}{\rho} \sum_{R_i \in \mathcal{R}} w(\overline{A_i})[(x(B_1 \cup \{u\}, R_i) - x(B_1, R_i)) - (x(B_2 \cup \{u\}, R_i) - x(B_2, R_i))]$$

Observe that if $x(B_2 \cup \{u\}, R_i) - x(B_2, R_i) = 1$, $B_2 \cap R_i = \emptyset$ and $u \in R_i$. Then $B_1 \cap R_i = \emptyset$ and $u \in R_i$, implying that $x(B_1 \cup \{u\}) - x(B_1, R_i) = 1$. Thus,

$$(f_{\mathcal{R}}(B_1 \cup \{u\}) - f_{\mathcal{R}}(B_1)) - (f_{\mathcal{R}}(B_2 \cup \{u\}) - f_{\mathcal{R}}(B_2)) \geq 0.$$

The submodularity follows.   □

Proof of Lemma 4. We first show that the sequence $Z_1, \ldots, Z_\rho$ is a martingale. Since $Z_i = \sum_{j=1}^{i}(w(\overline{A_j}) \cdot x(B, R_j)) - q)$, we have $\mathbb{E}[Z_i] = 0$ and $\mathbb{E}[|Z_i|] < +\infty$. Based on the process of generating random RP sets, we can observe that the value of $x(B, R_i)$ is independent of $x(B, R_1), \ldots, x(B, R_{i-1})$. Therefore,

$$\mathbb{E}[Z_i \mid Z_1, \cdots, Z_{i-1}] = \mathbb{E}[Z_{i-1} + w(\overline{A_i}) \cdot x(B, R_i) - q \mid Z_1, \cdots, Z_{i-1}]$$
$$= Z_{i-1} + \mathbb{E}[w(\overline{A_i}) \cdot x(B, R_i)] - q = Z_{i-1},$$

implying that $Z_1, \cdots, Z_\rho$ is a martingale.

Then we find the value of a and b in the conditions of Martingale's Property respect with $Z_1, \cdots, Z_\rho$. Recall that $w(\overline{A_i}) = \sum_{u \in V \setminus A_i} w_u$ and $w_u \in [\frac{1}{n-r}, 1]$ for any u. Since

$$|Z_1| = |w(\overline{A_1}) \cdot x(B, R_1) - q| \leq n-r \quad \text{and} \quad |Z_j - Z_{j-1}| = |w(\overline{A_j}) \cdot x(B, R_j) - q| \leq n-r$$

for each $j \in \{2, \ldots, i\}$, we can set $a = n-r$. Based on the properties of variance and $Z_\rho = \sum_{j=1}^{\rho}(w(\overline{A_j}) \cdot x(B, R_j) - q)$, we can set $b = (n-r) \cdot \rho q$. It is because

$$\text{Var}[Z_1] + \sum_{j=2}^{\rho} \text{Var}[Z_j \mid Z_1, \cdots, Z_{j-1}] = \sum_{j=1}^{\rho} \text{Var}[w(\overline{A_j}) \cdot x(B, R_j)]$$

$$= \sum_{j=1}^{\rho} \{\mathbb{E}[(w(\overline{A_j}) \cdot x(B, R_j))^2] - (\mathbb{E}[w(\overline{A_j}) \cdot x(B, R_j)])^2\}$$

$$\leq \sum_{j=1}^{\rho} \{\mathbb{E}[(n-r) \cdot w(\overline{A_j}) \cdot x(B, R_j)] - 0\} \leq (n-r) \cdot \rho q,$$

where the second inequality from the end holds from the facts that $w(\overline{A_i}) \leq n-r$ and $x^2(B, R_j) = x(B, R_j)$. Applying Martingale's Property, one can see that inequalities (1) and (2) hold.   □

Proof of Lemma 5. Let $\tilde{B}$ be the solution of Algorithm 2 (Node Selection). Let B^* be the optimal solution. Based on the greedy approach in Algorithm 2, we have

$$f_{\mathcal{R}}(\tilde{B}) \geq (1 - 1/e) \cdot f_{\mathcal{R}}(B^*).$$

In the sequel we show that if $\rho \geq \rho_1$, $\Pr[f_{\mathcal{R}}(B^*) \leq (1 - \varepsilon_1) \cdot \text{OPT}] \leq \delta_1$.

The result means that $\Pr[f_{\mathcal{R}}(B^*) \geq (1 - \varepsilon_1) \cdot \text{OPT}] \geq 1 - \delta_1$, and thus $f_{\mathcal{R}}(\tilde{B}) \geq (1 - 1/e)(1 - \varepsilon_1) \cdot \text{OPT}$ holds with at least $1 - \delta_1$ probability when $\rho \geq \rho_1$, implying the lemma.

By Lemmas 3 and 4, we can verify that if $\rho > \rho_1$, the following inequality holds.

$$\Pr[f_{\mathcal{R}}(B^*) \leq (1 - \varepsilon_1) \cdot \text{OPT}] \leq \delta_1. \tag{6}$$

Consider the sampling result $\mathcal{R}$ of R-NRB algorithm. First, for any $R_i \in \mathcal{R}$, we denote by $x(B^*, R_i)$ a random variable such that $x(B^*, R_i) = 1$ if

$B^* \cap R_i \neq \emptyset$ and $x(B^*, R_i) = 0$ otherwise. Based on Lemma 4, the sequence of $Z_i = \sum_{j=1}^{i} \left(w(\overline{A_i}) \cdot x(B^*, R_i) - q^* \right)$, $i \in \{1, \ldots, \rho\}$ is a martingale. Let $q^* = \mathbb{E}[f_\mathcal{R}(B^*)]$. By Lemma 3, $q^* = \sigma(B^*) = \text{OPT}$. We have

$$\Pr\left[f_\mathcal{R}(B^*) \leq (1 - \varepsilon_1) \cdot \text{OPT} \right] = \Pr\left[\frac{1}{\rho} \sum_{R_i \in \mathcal{R}} w(\overline{A_i}) \cdot x(B^*, R_i) \leq (1 - \varepsilon_1) \cdot q^* \right]$$

$$= \Pr\left[\sum_{R_i \in \mathcal{R}} w(\overline{A_i}) \cdot x(B^*, R_i) - \rho q^* \leq -\varepsilon_1 \cdot \rho q^* \right].$$

By the inequality (2) of Lemma 4 and $\rho \geq \rho_1$,

$$\Pr\left[\sum_{R_i \in \mathcal{R}} w(\overline{A_i}) \cdot x(B^*, R_i) - \rho q^* \leq -\varepsilon_1 \cdot \rho q^* \right]$$

$$\leq \exp\left(-\frac{\varepsilon_1^2}{2(n-r)} \cdot \rho q^* \right) \leq \exp\left(-\frac{\varepsilon_1^2}{2(n-r)} \cdot \rho_1 q^* \right) = \delta_1.$$

Thus the inequality 6 holds. This completes the proof of the lemma.   □

Proof of Lemma 6

Proof. Let $\mathcal{R} = \{R_1, \ldots, R_\rho\}$ be an input of Algorithm 2 and let $|R|$ be the number of nodes in a random RP set R. Recall that Algorithm 2 is a greedy process which returns a protector seed set $\tilde{B}$ by maximizing the marginal utility of $f_\mathcal{R}(\cdot)$. Due to

$$f_\mathcal{R}(B) = \frac{1}{\rho} \sum_{R_i \in \mathcal{R}} (w(\overline{A_i}) \cdot x(B, R_i)),$$

it is clear that Algorithm 2 is equivalent to the greedy approach for a *maximum weighted coverage* problem. We have known that the time complexity of greedy approach for the maximum weighted coverage is $O(\sum_{R \in \mathcal{R}} |R|)$ in [17]. Thus, Algorithm 2 runs in $O(\sum_{R \in \mathcal{R}} |R|)$ time.   □

Proof of Claim 1. We prove Claim 1 by showing that if $\text{OPT} < \xi_i$,

$$\Pr[f_\mathcal{R}(B) \geq (1 + \varepsilon_0) \cdot \xi_i] \leq \delta_0 \Big/ \binom{n}{k},$$

for any size-k set B.

Based on Lemma 4, the sequence $Z_d = \sum_{j=1}^{d} \left(w(\overline{A_j}) \cdot x(B, R_j) \right)$, $d \in \{1, \ldots, \rho_i\}$ is a martingale. Denote $q = \mathbb{E}[f_\mathcal{R}(B)]$. It is clear that

$$\Pr[f_\mathcal{R}(B) \geq (1 + \varepsilon_0) \cdot \xi_i]$$

$$= \Pr\left[\rho_i \cdot f_\mathcal{R}(B) - \rho_i q \geq ((1 + \varepsilon_0) \cdot \xi_i / q) - 1) \cdot \rho_i q \right]$$

$$\leq \exp\left(-\frac{\zeta^2}{(2 + \frac{2}{3}\zeta)(n-r)} \cdot \rho_i q \right),$$

where $\zeta = (1 + \varepsilon_0) \cdot \xi_i/q - 1$ and the last inequality holds from the inequality (1) of Lemma 4. By Lemma 3, $q = \sigma(B) \leq \mathrm{OPT} < \xi_i$, we have $\zeta = (1+\varepsilon_0) \cdot \xi_i/q - 1 > \varepsilon_0 \cdot \xi_i/q > \varepsilon_0$. Then the right side of above inequality

$$= \exp\left(-\frac{\zeta}{(2/\zeta + \frac{2}{3})(n-r)} \cdot \rho_i q\right) < \exp\left(-\frac{\varepsilon_0 \cdot \xi_i/q}{(2/\varepsilon_0 + \frac{2}{3})(n-r)} \cdot \rho_i q\right).$$

Since $\rho \geq \mu_0/\xi_i$, the right side of above inequality

$$\leq \exp\left(-\frac{\varepsilon_0}{(2/\varepsilon_0 + \frac{2}{3})(n-r)} \cdot \mu_0\right) = \delta_0 \Big/ \binom{n}{k}.$$

Thus the Claim 1 is proved.   □

Proof of Claim 2. We prove Claim 2 by showing that if $\mathrm{OPT} \geq \xi_i$,

$$\Pr[\mathrm{OPT} < f_\mathcal{R}(B)/(1 + \varepsilon_0)] \leq \delta_0 \Big/ \binom{n}{k}.$$

for any size-k set B. Based on Lemma 4, the sequence $Z_d = \sum_{j=1}^{d}(w(\overline{A_j}) \cdot x(B, R_j))$, $d \in \{1,\ldots,\rho\}$ is a martingale. Recall that $f_\mathcal{R}(B) = \frac{1}{\rho}\sum_{R_i \in \mathcal{R}}(w(\overline{A_i}) \cdot x(B, R_i))$ and denote $q = \mathbb{E}[f_\mathcal{R}(B)] < \mathrm{OPT}$. It is clear that

$$\Pr[\mathrm{OPT} < f_\mathcal{R}(B)/(1 + \varepsilon_0)] = \Pr[\rho \cdot f_\mathcal{R}(B) - \rho \cdot \mathrm{OPT} > \varepsilon_0 \cdot \rho \cdot \mathrm{OPT}]$$
$$\leq \Pr\left[\rho \cdot f_\mathcal{R}(B) - \rho q \geq (\varepsilon_0 \cdot \mathrm{OPT}/q) \cdot \rho q\right].$$

By inequality (1) of Lemma 4 and $q < \mathrm{OPT}$,

$$\Pr\left[\rho \cdot f_\mathcal{R}(B) - \rho q \geq (\varepsilon_0 \cdot \mathrm{OPT}/q) \cdot \rho q\right] \leq \exp\left(-\frac{(\varepsilon_0 \cdot \mathrm{OPT}/q)^2}{(2 + \frac{2}{3}(\varepsilon_0 \cdot \mathrm{OPT}/q))(n-r)} \cdot \rho q\right)$$
$$= \exp\left(-\frac{\varepsilon_0^2 \cdot \mathrm{OPT}^2}{(2q + \frac{2}{3}\varepsilon_0 \cdot \mathrm{OPT})(n-r)} \cdot \rho\right) < \exp\left(-\frac{\varepsilon_0^2 \cdot \mathrm{OPT}}{(2 + \frac{2}{3}\varepsilon_0)(n-r)} \cdot \rho\right).$$

By $\rho \geq \mu_0/\xi_i$ and $\mathrm{OPT} \geq \xi_i$, the right side of above inequality

$$\exp\left(-\frac{\varepsilon_0^2 \cdot \mathrm{OPT}}{(2 + \frac{2}{3}\varepsilon_0)(n-r)} \cdot \rho\right)$$
$$\leq \exp\left(-\frac{\varepsilon_0^2 \cdot \mathrm{OPT} \cdot (2 + \frac{2}{3}\varepsilon_0) \cdot (n-r) \cdot \log\left(\binom{n}{k}/\delta_0\right)}{(2 + \frac{2}{3}\varepsilon_0) \cdot (n-r) \cdot \varepsilon_0^2 \cdot \xi_i}\right) \leq \delta_0 \Big/ \binom{n}{k}.$$

Thus, the Claim 2 holds.   □

Proof of Lemma 7. Denote by $\hat{v}$ a random node is subjected to some probability distribution $\mathcal{V}$ with $\Pr^*[\hat{v}]$. Let $\Pr^*[\hat{v}] = \frac{d(\hat{v})}{2m}$, where $d(\hat{v})$ is the in-degree of $\hat{v}$ in G and m is the number of edges in G. For any random node $\hat{v}$, denote by $p_{\hat{v}}$ the probability that $\hat{v}$ is covered by a random RP set. Let $\varphi(\hat{v}, R)$ be a function as follows:

$$\varphi(\hat{v}, R) = \begin{cases} 1, & \hat{v} \in R; \\ 0, & otherwise. \end{cases}$$

Recall that for any protector seed set B, $x(B, R) = 1$ if $B \cap R \neq \emptyset$ and $x(B, R) = 0$ otherwise. $\mathcal{P}$ is a collection consisting of all possible RP sets. Then we obtain that

$$\mathbb{E}[d(R)] \leq \sum_{R \in \mathcal{P}} \Pr[R] \sum_{\hat{v} \in R} d(\hat{v})$$

$$= \sum_{R \in \mathcal{P}} \Pr[R] \sum_{\hat{v} \in V} d(\hat{v}) \cdot \varphi(\hat{v}, R)$$

$$= \sum_{\hat{v} \in V} d(\hat{v}) \sum_{R \in \mathcal{P}} \Pr[R] \cdot x(\{\hat{v}\}, R),$$

where the last inequality holds from the fact that $\varphi(\hat{v}, R) = 1$ if and only if $x(\hat{v}, R) = 1$. Recall the Definition 1 (Random RP Set), assume that the sampled rumor seed set is A, then the target v is selected with $\Pr[v] = \frac{w_v}{w(A)}$. Since $w(\overline{A_i}) = \sum_{u \in V \setminus A_i} w_u \geq (n - r) \cdot \frac{1}{n-r} = 1$, the right side of above inequality

$$\leq \sum_{\hat{v} \in V} d(\hat{v}) \sum_{A, g, \hat{v} \in V \setminus A} \Pr[A] \Pr[g] \Pr[\hat{v}] \cdot w(\overline{A}) \cdot x(\{\hat{v}\}, R)$$

$$= \sum_{\hat{v} \in V} d(\hat{v}) \sum_A \Pr[A] \sum_g \Pr[g] \sum_{u \in I_{A,g}(\{\hat{v}\})} w_u$$

$$= \sum_{\hat{v} \in V} d(\hat{v}) \cdot \sigma(\{\hat{v}\}) \leq m \cdot \mathrm{OPT},$$

where the first and second equalities can be derived from the proof of Lemma 3. Then the last inequality holds by $\sum_{\hat{v} \in V} d(\hat{v}) = m$ and $\sigma(\{\hat{v}\}) \leq \mathrm{OPT}$, Therefore the lemma holds. $\qquad\square$

References

1. Borgs, C., Brautbar, M., Chayes, J., Lucier, B.: Maximizing social influence in nearly optimal time. In: The Proceeding of the 25th SODA, SIAM, pp. 946–957 (2014)
2. Borodin, A., Filmus, Y., Oren, J.: Threshold models for competitive influence in social networks. In: Saberi, A. (ed.) WINE 2010. LNCS, vol. 6484, pp. 539–550. Springer, Heidelberg (2010). https://doi.org/10.1007/978-3-642-17572-5_48
3. Budak, C., Agrawal, D., El Abbadi, A.: Limiting the spread of misinformation in social networks. In: The Proceeding of the 20th WWW, pp. 665–674. ACM (2011)
4. Chung, F.R.K., Lu, L.: Concentration inequalities and martingale inequalities: a survey. Internet Math. **3**(1), 79–127 (2006)
5. Fan, L., Wu, W., Zhai, X., Xing, K., Lee, W., Du, D.Z.: Maximizing rumor containment in social networks with constrained time. Soc. Netw. Anal. Min. **4**(1), 214 (2014)
6. Fan, L., Wu, W., Xing, K., Lee, W.: Precautionary rumor containment via trustworthy people in social networks. Discret. Math. Algorithms Appl. **8**(01), 1650004 (2016)

7. He, X., Song, G., Chen, W., Jiang, Q.: Influence blocking maximization in social networks under the competitive linear threshold model. In: The Proceeding of the 2012 SIAM International Conference on Data Mining, Society for Industrial and Applied Mathematics, pp. 463–474 (2012)

8. Kempe, D., Kleinberg, J., Tardos, E.: Maximizing the spread of influence through a social network. In: The Proceeding of the 9th SIGKDD, pp. 137–146. ACM (2003)

9. Moore, E.F.: The shortest path through a maze. In: The Proceeding of International Symposium Switching Theory, pp. 285–292 (1959)

10. Nemhauser, L.G., Wolsey, A.L., Fisher, L.M.: An analysis of approximations for maximizing submodular set functional. Math. Program. **14**(1), 265–294 (1978)

11. Nguyen, H.T., Thai, M.T., Dinh, T.N.: Stop-and-stare: optimal sampling algorithms for viral marketing in billion-scale networks. In: The Proceedings of the 2016 International Conference on Management of Data, pp. 695–710. ACM (2016)

12. Nguyen, H.T., Tsai, J., Jiang, A., Bowring, E., Maheswaran, R., Tambe, M.: Security games on social networks. In: The Proceeding of the 2012 AAAI fall symposium series (2012)

13. Pathak, N., Banerjee, A., Srivastava, J.: A generalized linear threshold model for multiple cascades. In: The Proceeding of the ICDM, pp. 965–970 (2010)

14. Song, C., Hsu, W., Lee, M.L.: Targeted influence maximization in social networks. In: The Proceeding of the 25th ACM International on Conference on Information and Knowledge Management, pp. 1683–1692 (2016)

15. Tang, Y., Shi, Y., Xiao, X.: Influence maximization in near-linear time: a martingale approach. In: The Proceeding of the 2015 ACM SIGMOD International Conference on Management of Data, pp. 1539–1554 (2015)

16. Tong, G., et al.: An efficient randomized algorithm for rumor blocking in online social networks. arXiv preprint arXiv:1701.02368 (2017)

17. Vaziran, V.V.: Approximation Algorithms. Springer, Heidelberg (2002). https://doi.org/10.1007/978-3-662-04565-7

18. Williams, D.: Probability with Martingales. Cambridge University Press, Cambridge (1991)

19. Zhang, H., Zhang, H., Li, X., Thai, M.T.: Limiting the spread of misinformation while effectively raising awareness in social networks. In: Thai, M.T., Nguyen, N.P., Shen, H. (eds.) CSoNet 2015. LNCS, vol. 9197, pp. 35–47. Springer, Cham (2015). https://doi.org/10.1007/978-3-319-21786-4_4

A Robust Power Optimization Algorithm to Balance Base Stations' Load in LTE-A Network

Jihong Gui, Wenguo Yang, Suixiang Gao, and Zhipeng Jiang[✉]

University of Chinese Academy of Sciences, No. 19(A) Yuquan Road,
Shijingshan District, Beijing 100049, People's Republic of China
jiangzhipeng@ucas.ac.cn

Abstract. The explosive growth of communication device and user data has stressed the dense Long Term Evolution Advanced (LTE-A) network. In order to relieve communication congestion in high-load base stations (BSs) in the downlink network, it is necessary for network operators to balance these loads meanwhile guarantee the quality of service (QoS). In this work, a robust $Min - Max$ generalized linear fractional programming (GLFP) model about power optimization under QoS constraints is established for load balancing, where signal coverage and user access are mathematically described by $sigmod$ function and $softmax$ function, respectively. Since GLFP is a well-known NP-hard problem, a heuristic algorithm named *generalized bisection method* (GBM) is proposed and its time complexity is at most $O(MN^2 \log \frac{W}{\epsilon})$. Simulation results demonstrate the effectiveness and rapidity of the proposed algorithm.

Keywords: Power optimization · Load balancing
LTE-A network · Robust optimization

1 Introduction

With the rapid and remarkable growth of user equipment (UE), the explosive data traffic driven by various applications such as voice and video has stressed the enormous and dense Long Term Evolution Advanced (LTE-A) network in recent years [1]. In order to ensure the quality of service (QoS) of users, the conventional design principle of the downlink network is to guarantee the signal coverage of base station (BS) and reduce the interference of different signals.

Nevertheless, engineering approaches such as increasing bandwidth to expand system capacity or enhancing transmit power of some BSs with poor measured performance will cause serious load unbalanced and traffic congestion, which will ultimately deteriorate the system performance and reduce global network quality inevitably due to unbalanced distribution of user position and number. Additionally, the prevalent user access rule in LTE-A network is to associate with the specific BS providing the maximum received signal strength (Max-RSS) [2]. As a consequence, transmit power of total BSs should be considered

© Springer Nature Switzerland AG 2018
S. Tang et al. (Eds.): AAIM 2018, LNCS 11343, pp. 177–189, 2018.
https://doi.org/10.1007/978-3-030-04618-7_15

comprehensively to approach a trade-off between the load balancing and QoS of users.

Present works shown below have provided numbers of excellent contributions to load balancing. Based on the signal-to-interference-plus-noise ratio (SINR) prediction and UE measurement, an engineering method for load estimation and balancing after hand over (HO) is presented by [3]. Moreover, various emerging access technologies such as Wi-Fi, Wireless Local Area Network (WLAN) and millimeter wave (mmWave) are also proved as effective ways to balance load [4–6]. Meanwhile, there are also numbers of researches to balance loads by optimized method. A cell offset optimization framework for load balancing in heterogeneous networks is presented in [7]. Achieving load balancing by optimal user association scheme in heterogeneous cellular networks is presented by [8] and the logarithmic utility formulation is adopted by researchers as objective function to approach the fairness of loads.

Different from previous work neglecting the relationship between user association scheme and received signal strength in LTE-A network, we originally make mathematical descriptions for signal coverage probability and user access probability based on the actual scene. In addition, the fairness of loads adopts a robust $Min - Max$ optimization rather than a log function, where relieving load of the most clogged BS benefits to reduce traffic congestion. Then for load balancing, we establish a $Min - Max$ generalized linear fractional programming (GLFP) model about power optimization under the constraints of users' QoS including signal coverage quality and interference performance. The GLFP model is well-known NP-hard for summation terms more than three [9]. Therefore, a *generalized bisection method* (GBM) is proposed to approach a good feasible solution. Specifically, its basic idea is to decline the upper bound by solving a tighter problem and enhance the lower bound by solving a more relaxed problem. These two problems are both linear feasibility problem and easily solved. Moreover, the time complexity of GBM is $O(MN^2 \log \frac{W}{\epsilon})$ in the worst case.

The rest of this paper is organized as follows. The robust optimization model for load balancing is established in Sect. 2. Algorithm GBM is designed and the time complexity of it is provided in Sect. 3. Ultimately, the simulation results and numerical analysis are presented in Sect. 4.

2   Problem Formulation

Considering the downlink network in LTE-A system illustrated in Fig. 1, there are N base stations (BSs) and M hot spots (HSs) in the region, where the set of BSs and HSs are denoted by $\mathcal{B} = \{1, 2, 3, \cdots, N\}$ and $\mathcal{S} = \{1, 2, 3, \cdots, M\}$, respectively. Let S_i and B_j denote the position of HS i and BS j respectively, then the distance between them is $d_{ij} = \|S_i - B_j\|_2^2$. Then transmit power (TP) of BS B_j and received power (RP) of S_i from B_j are respectively denoted by P_j and P_{ij} in $Watt$ dimension, where the power P in $Watt$ is converted by power P^{dB} in dB via $P = 10^{\frac{P^{dB}}{10}}$. In addition, the relation between transmit power

P_j and receive power P_{ij} is formulated as $P_{ij} = P_j h_{ij}$, where h_{ij} is the channel gain between HS i and BS j and is easily calculated by numbers of mature radio propagation model, such as *Okumura Model* and *Hata Model* [10–12].

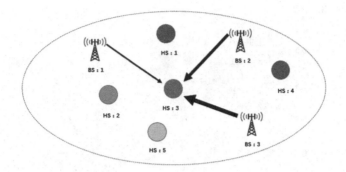

Fig. 1. N BSs and M HSs in the LTE-A downlink network

The calculation of traffic loads of these N BSs should consider the user amount of each HS S_i and the detail association between users and BSs. Firstly, let w_i denote the user amount of S_i since the distribution of user number is not uniform in practice. Moreover, it's assumed that users in each S_i will be associated to the same BS and the detail access scheme of each user conventionally follows the conventional criterion of Max-RSS in LTE-A network. Namely, the users at HS i will associate to the strongest signal from BS j if the RP P_{ij}^{dB} is the maximum strength among these BSs. Then the access phenomenon is mathematically expressed by the indicator function about P_{ij}^{dB} as follows.

$$\chi_{\{P_{ij}^{dB}=\max\limits_{k} P_{ik}^{dB}\}} = \begin{cases} 1, & P_{ij}^{dB} \geq P_{ik}^{dB}, \quad \forall k \\ 0, & \text{Otherwise} \end{cases} \tag{1}$$

Since the indicator function 1 of user access is discrete and the *maximum* function is inconvenient to optimize, a *softmax* function about P_{ij} formed as 2 is introduced to approach it. The 2 relaxes the binary range of $\chi_{\{P_{ij}^{dB}=\max\limits_{k} P_{ik}^{dB}\}}$ to the interval from 0 to 1, which meets the value range of probability. Moreover, q_{ij} is a monotonically increasing function of P_{ij}^{dB}, which meets that the stronger RP, the higher user access probability. The equation $\sum\limits_{k=1}^{N} q_{ik} = 1, \forall i$ is equivalent that users in each HS can all definitely associate to only one BS. Then q_{ij} is regarded as the probability of users in HS i accessing to BS j.

$$q_{ij} = \frac{e^{\lambda_q P_{ij}^{dB}}}{\sum\limits_{k=1}^{N} e^{\lambda_q P_{ik}^{dB}}} \tag{2}$$

The positive factor $\lambda_q > 0$ in 2 doesn't affect the order of P_{ij}^{dB}. In detail, the relationship between access probability q_{ij} and P_{ij}^{dB} is illustrated in Fig. 2. Clearly, the larger λ_q, the bigger gaps between the maximum and minimum of P_{ij}^{dB}. In order to uniform the unit of power, q_{ij} is converted to the function of P_{ij}^{dB} formed as Eq. 3.

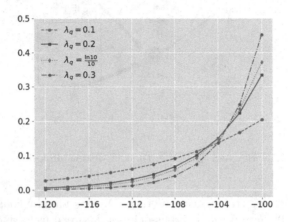

Fig. 2. The relationship between $softmax$ function q_{ij} and P_{ij}^{dB} with different λ_p

$$q_{ij} = \frac{P_{ij}^{\frac{10}{\ln 10}\lambda_q}}{\sum\limits_{k=1}^{N} P_{ik}^{\frac{10}{\ln 10}\lambda_q}} = \frac{P_{ij}}{\sum\limits_{k=1}^{N} P_{ik}} \tag{3}$$

The default value of λ_q in 3 is set to $\frac{\ln 10}{10}$ for the convenience of model analysis.

Based on the definition of user amount and user access probability, the load expectation of BS j is calculated by 4.

$$y_j = \sum_{i=1}^{M} w_i q_{ij} = \sum_{i=1}^{M} \frac{w_i P_{ij}}{\sum\limits_{k=1}^{N} P_{ik}} = \sum_{i=1}^{M} \frac{w_i h_{ij} P_j}{\sum\limits_{k=1}^{N} h_{ik} P_k} \tag{4}$$

Generally, the high load at some BSs may increase the energy consumption of them and reduce resource utilization in the downlink network. To balance these loads, minimizing the maximum load y_j is imperative and the $Min-Max$ optimization is a robust scheme. Therefore, the objective function about tramit power P_j is $\min\limits_{P_1,P_2,\cdots,P_N} \max\limits_{j} \sum\limits_{i=1}^{M} \frac{w_i h_{ij} P_j}{\sum\limits_{k=1}^{N} h_{ik} P_k}$.

However, the power optimization for load balancing should consider the demand of users' QoS including the improvement of signal coverage and the decline of signal interference at the same time.

On the one hand, the signal coverage mainly depends on the strength of RP and a given empirical threshold P_0^{dB}. Then the coverage quality of user i covered by BS j is indicated as the indicator function of $P_{ij}^{dB} - P_0^{dB} \geq 0$ as 5.

$$\chi_{\{P_{ij}^{dB} - P_0^{dB} \geq 0\}} = \begin{cases} 1, & P_{ij}^{dB} - P_0^{dB} \geq 0 \\ 0, & \text{Otherwise} \end{cases} \tag{5}$$

which implies that users in HS i are covered by BS j if RP P_{ij}^{dB} should be greater than the threshold P_0^{dB}. Similar to the approximation of 1, the discrete function $\chi_{\{P_{ij}^{dB} - P_0^{dB} \geq 0\}}$ is approximated by the *sigmod* function of P_{ij}^{dB} formed as 6

$$p_{ij} = \frac{1}{1 + e^{-\lambda_p(P_{ij}^{dB} - P_0^{dB})}} \tag{6}$$

where $\lambda_p > 0$ is the gradient control factor at origin because of

$$\frac{\partial p_{ij}}{\partial P_{ij}^{dB}} \mid_{P_{ij}^{dB} = 0} = \frac{\lambda_p}{4} \tag{7}$$

The growth trend of p_{ij} from zero to one with a larger λ_p is faster, illustrated in Fig. 3.

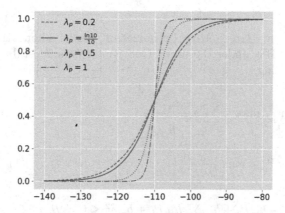

Fig. 3. The *sigmod* function of $P_{ij}^{dB} - P_0^{dB}$ with different λ_p

Therefore, the *sigmod* function with big enough λ_p solves the discontinuity of the indicator function well and it's also very relevant to the actual situation. Moreover, p_{ij} is a monotonically increasing function of P_{ij}^{dB}, which implies that the stronger RP, the higher probability of being covered. Meanwhile, $p_{ij} = \frac{1}{2}$ if $P_{ij}^{dB} = P_0^{dB}$, indicating that the threshold P_0^{dB} is the critical power value for judging signal coverage. In order to enhance the coverage probability, 8 should be guaranteed and a_{ij} is the threshold of network coverage performance.

$$p_{ij} = \frac{h_{ij}P_j}{P_0 + h_{ij}P_j} \geq a_{ij}, \forall i, j \iff P_j \geq \max_i \frac{a_{ij}P_0}{h_{ij}(1 - a_{ij})}, \forall j \tag{8}$$

On the other hand, the signal interference is traditionally measured by the signal-to-interference-plus-noise ratio (SINR) in wireless network. The interference strength of signal from BS j to HS i will be weaker if the SINR between them is bigger. Then the SINR between HS i and BS j is denoted by r_{ij} and formulated as 9.

$$r_{ij} = \frac{\overbrace{P_j h_{ij}}^{\text{Received singal}}}{\underbrace{\sum_{k \neq j}^{N} P_k h_{ik}}_{\text{interference from other cells}} + \underbrace{\delta^2}_{\text{system noise}}} \tag{9}$$

The numerator in 9 is received signal P_{ij} in $Watt$ and the denominator is interference from other cells and additive white Gaussian noise (AWGN), respectively. In order to reduce the interference, 10 should be also satisfied and b_{ij} in 10 is the threshold of SINR performance between BS j to HS i.

$$r_{ij} = \frac{h_{ij} P_j}{\sum_{k \neq j}^{N} h_{ik} P_k + \delta^2} \geq b_{ij}, \forall i,j \iff -h_{ij} P_j + b_{ij} \sum_{k \neq j}^{N} h_{ik} P_k + b_{ij} \delta^2 \leq 0, \forall i,j \tag{10}$$

Consequently, a $Min - Max$ optimization model about transmit power $P_1, P_2, \cdots, P_N$ of BSs under the users' QoS constraints is established as model (I) for load balancing.

$$\min_{P_1, P_2, \cdots, P_N} \max_j \sum_{i=1}^{M} \frac{w_i h_{ij} P_j}{\sum_{k=1}^{N} h_{ik} P_k}$$

$$s.t. \quad P_j \geq \max_i \frac{a_{ij} P_0}{h_{ij}(1-a_{ij})}, \qquad \forall j \quad (1)$$

$$-h_{ij} P_j + b_{ij} \sum_{k \neq j}^{N} h_{ik} P_k + b_{ij} \delta^2 \leq 0, \quad \forall i,j \quad (2) \quad (I)$$

$$0 \leq P_j \leq P_{max}, \qquad \forall j$$

The objective function in model (I) is to balance the loads of BSs and the linear constraints in model (I) is to guarantee the users' QoS. In detail, the constraint (1) in model (I) is to guarantee the signal coverage quality corresponding to 8 and the constraint (2) is to reduce the signal interference strength corresponding to 10. Moreover, the lower bound of each decision variable P_j is zero and its upper bound is P_{max}, which is the maximum transmit power of BS.

Obviously, the constraint (1) in model (I) is integrated into the lower bound of decision variables P_j and the model (I) is simplified as model (II).

$$\min_{P_1,P_2,\cdots,P_N} \max_j \sum_{i=1}^{M} \frac{w_i h_{ij} P_j}{\sum_{k=1}^{N} h_{ik} P_k}$$

$$s.t. \quad -h_{ij}P_j + b_{ij}\sum_{\substack{k\neq j}}^{N} h_{ik}P_k + b_{ij}\delta^2 \leq 0, \quad \forall i,j \quad (II)$$

$$\max_i \frac{a_{ij}P_0}{h_{ij}(1-a_{ij})} \leq P_j \leq P_{max}, \qquad \forall j$$

The model (II) is a $Min - Max$ generalized linear fractional programming (GLFP) model and it's NP-hard when $M \geq 3$. However, larger M is more significant in actual network.

3   Algorithm

In this section, a heuristic algorithm named *Generalized Bisection Method* (GBM) is designed for model (II). The basic idea and time complexity of it are presented as follows.

The model (II) is hard to solve mainly because of the $Min - Max$ optimization and the nonconvex objective function $\sum_{i=1}^{M} \frac{w_i h_{ij} P_j}{\sum_{k=1}^{N} h_{ik} P_k}$. Inspired by the ordinary *Bisection Method* proposed in [13], a scalar variable t is introduced to eliminate the effect of $Min - Max$ optimization model. Let $t = \max_j \sum_{i=1}^{M} \frac{w_i h_{ij} P_j}{\sum_{k=1}^{N} h_{ik} P_k}$, then

$$\sum_{i=1}^{M} \frac{w_i h_{ij} P_j}{\sum_{k=1}^{N} h_{ik} P_k} - t \leq 0, \forall j \tag{11}$$

The model (II) is transformed as

$$\min \quad t$$

$$s.t. \quad \sum_{i=1}^{M} \frac{w_i h_{ij} P_j}{\sum_{k=1}^{N} h_{ik} P_k} \leq t, \qquad \forall j \quad (1)$$

$$-h_{ij}P_j + b_{ij}\sum_{\substack{k\neq j}}^{N} h_{ik}P_k + b_{ij}\delta^2 \leq 0, \quad \forall i,j \quad (2) \quad (III)$$

$$\max_i \frac{a_{ij}P_0}{h_{ij}(1-a_{ij})} \leq P_j \leq P_{max}, \qquad \forall j$$

where the constraint (1) in model (III) is too complex to solve and the ordinary *Bisection Method* doesn't work now.

Since it's hard to solve model (III) directly, GBM is designed by solving two linear optimization models to approach the optimal solution of model (III). Specifically, the constraint (1) in model (III) is tighted as

$$\frac{w_i h_{ij} P_j}{\sum\limits_{k=1}^{N} h_{ik} P_k} \le \frac{t}{M} \Longleftrightarrow w_i h_{ij} P_j - \frac{t}{M} \sum_{k=1}^{N} h_{ik} P_k \le 0, \forall i, j \qquad (12)$$

because the constraint (1) in model (III) will be satisfied if linear inequalities 12 have feasible solutions. Therefore, the feasible solution of model (III) can be found if the tighted model (IV) is feasible.

$$
\begin{aligned}
\min \quad & t \\
s.t. \quad & w_i h_{ij} P_j - \frac{t}{M} \sum_{k=1}^{N} h_{ik} P_k \le 0, & \forall i, j \quad (1) \\
& -h_{ij} P_j + b_{ij} \sum_{k \neq j}^{N} h_{ik} P_k + b_{ij}\delta^2 \le 0, & \forall i, j \quad (2) \quad (IV) \\
& \max_i \frac{a_{ij} P_0}{h_{ij}(1 - a_{ij})} \le P_j \le P_{max}, & \forall j
\end{aligned}
$$

On the contrary, the constraint (1) in model (III) is relaxed as

$$\frac{w_i h_{ij} P_j}{\sum\limits_{k=1}^{N} h_{ik} P_k} \le t \Longleftrightarrow w_i h_{ij} P_j - t \sum_{k=1}^{N} h_{ik} P_k \le 0, \forall i, j \qquad (13)$$

because $\frac{w_i h_{ij} P_j}{\sum\limits_{k=1}^{N} h_{ik} P_k} \ge 0, \forall i, j$. The model (III) will be infeasible if the relaxed model (V) formed as (V) has no feasible solutions.

$$
\begin{aligned}
\min \quad & t \\
s.t. \quad & w_i h_{ij} P_j - t \sum_{k=1}^{N} h_{ik} P_k \le 0, & \forall i, j \quad (1) \\
& -h_{ij} P_j + b_{ij} \sum_{k \neq j}^{N} h_{ik} P_k + b_{ij}\delta^2 \le 0, & \forall i, j \quad (2) \quad (V) \\
& \max_i \frac{a_{ij} P_0}{h_{ij}(1 - a_{ij})} \le P_j \le P_{max}, & \forall j
\end{aligned}
$$

Totally, model (IV) and (V) are both linear feasibility problems which can be easily solved with a fixed t. Additionally, the initial upper and lower bound of t in model (III) are $W = \sum\limits_{i}^{M} w_i$ and 0, respectively. Then let t be the midpoint of the upper bound and lower bound and solving model (IV) and (V) in each iteration.

There are four possible combinations for the two solutions. Firstly, model (IV) and (V) are both infeasible, then t is two small to find anyone feasible solution.

Hence, the lower bound of t should be enhanced because original model (III) is also infeasible in this case. Secondly, the model (IV) and (V) are both feasible, then the optimal solution of model (III) may be found with a smaller t and the upper bound of t should be declined in this case. Thirdly, model (IV) is infeasible while model (V) is feasible, then the constraint (2) in model (III) is also satisfied with the feasible solution $\boldsymbol{P^*}$ in model (V). However, the objective value in model (III) calculated by $\boldsymbol{P^*}$ may not lower than the current t in this iteration, therefore, the upper bound of t should be declined to a smaller value between the calculated value with $\boldsymbol{P^*}$ and the current t. Lastly, model (IV) is feasible while model (V) is infeasible. This situation can't appear because the solution space of model (IV) is a subset of the solution space of model (V).

Therefore, the upper bound of t will be declined and the lower bound of t will be enhanced after multiple iterations. A good feasible solution of model (III) will be eventually obtained when the upper and lower bounds are relatively closed. Conclusively, the pseudo code of GBM is shown as follows. where $\epsilon > 0$ is a control factor of solution precision.

Algorithm 1. GBM

Input: $\underline{t} = 0, \bar{t} = \sum_{i}^{M} w_i = W$, tolerance $\epsilon > 0$.

Output: The optimal solution $\boldsymbol{P^*}$ and optimal value f^*.

1: **while** $\bar{t} - \underline{t} > \epsilon$ **do**

2:     $t := \frac{\bar{t} + \underline{t}}{2}$.

3:     Solve the linear feasibility problem (IV) with the fixed t.

4:     **if** (III) is feasible and $\boldsymbol{P^*}$ is the feasible solution of it **then**

5:         $\bar{t} := \max_{j} \sum_{i=1}^{M} \frac{w_i h_{ij} P_j^*}{\sum_{k=1}^{N} h_{ik} P_k^*}$.

6:     **else**

7:         Solve the linear feasibility problem (V) with the fixed t.

8:         **if** (IV) is feasible and $\boldsymbol{P^*}$ is the feasible solution of it **then**

9:             $\bar{t} := \min\{\max_{j} \sum_{i=1}^{M} \frac{w_i h_{ij} P_j^*}{\sum_{k=1}^{N} h_{ik} P_k^*}, t\}$.

10:         **else**

11:             $\underline{t} := t$

12: **return** $\boldsymbol{P^*}$ and $f^* = \max_{j} \sum_{i=1}^{M} \frac{w_i h_{ij} P_j^*}{\sum_{k=1}^{N} h_{ik} P_k^*}$

The time complexity of GBM depends on the amount of iteration which is at most $\lceil \log_2(\frac{W}{\epsilon}) \rceil$, where $W = \sum_{i}^{M} w_i$. In each iteration, judging the feasibility of model (IV) and model (V) cost totally at most $O(MN^2)$ time complexity because the amount of inequalities is MN meanwhile the feasible judgment of an inequality requires $O(N)$ basic operations. Therefore, the total time complexity of GBM is $O(MN^2 * \lceil \log_2(\frac{W}{\epsilon}) \rceil) = O(MN^2 \log \frac{W}{\epsilon})$.

4   Performance Simulation

To verify the validity and rapidity of GBM, numerical simulation experiments implemented by MATLAB r2015b with 2.4 GHz CPU and 32 core processors are presented in this section. The first subsection is the system parameters of these simulation experiments including the parameters of wireless network topology and ratio propagation model. The second subsection is to analyse the validity of GBM from the perspective of the change of BS amount N and total user amout W, respectively. Moreover, a detail load distribution example will be also presented in this subsection as a supplementary explanation. The last part is to verify the $O(MN^2 \log \frac{W}{\epsilon})$ time complexity of GBM.

4.1   Simulation Parameters

The simulation parameters of these random experiments are listed in Table 1. The simulation wireless network is made up by some HSs, BSs and users. And HS amount M, BS amount N and total user amount W are a random integer from 10 to 50, 10 to 100 and 100 to 1000, respectively. Moreover, the maximum transmit power $P_{max} = 15\ dB$ and the adopted ratio propagation model refers to the distribution of frequency bands and empirical parameters in actual network. Additionally, the tolerance $\epsilon = 10^{-8}$.

Table 1. Simulation parameters

Parameters	Value
HS amount (M)	10–50
BS amount (N)	10–100
Total user amount (W)	100–1000
Maximum transmit power (dB)	15
Frequency (f:Hz)	1800–2600
Distance between BSs and HSs (d:km)	0.0833–1.4142
Attenuation coefficient in free space (a)	20–30
Building blockage factor (b)	20
Propagation model	$32.45 + a * \lg d + b * \lg(f)$

4.2 The Calculation Validity Analysis

Figure 4 illustrates the final load distribution based on algorithm GBM with $M = 20, N = 100, W = 1000$. The blue line in this figure represents the absolute equilibrium value (AEV) because $\frac{W}{N} = \frac{1000}{100} = 100$. Since the problem needs to satisfy user's QoS, AEV is almost impossible to achieve. Therefore, the load of some BSs will be higher than AEV, as red bars shown in this figure. On the contrary, green bars represent the lower load of BSs. It can be clearly seen that under the action of GBM, there is no excessive load on one BS and other BSs are idle.

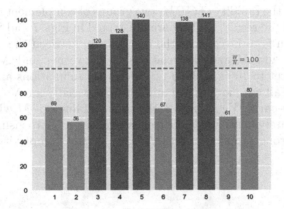

Fig. 4. The load distribution based on GBM with $M = 20, N = 100, W = 1000$ (Color figure online)

Furthermore, in order to observe the difference between the optimal value of GBM and AEV, Figs. 5 and 6 show the effects of parameters N and W on the performance of GBM, respectively. In Fig. 5, AEV is inversely proportional

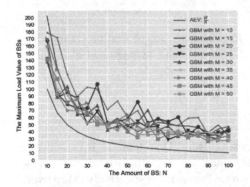

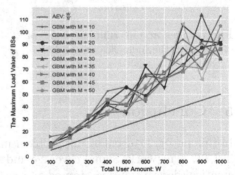

Fig. 5. The comparison between GBM and AEV as N increases with $W = 1000$

Fig. 6. The comparison between GBM and AEV as W increases with $N = 20$

to N because W is fixed to 1000 in these groups of experiments. Then $AEV = \frac{W}{N} = \frac{1000}{N}$ becomes smaller as N increases. Although the optimal values of GBM with different M are all bigger than AEV, their trends are roughly the same as AEV. In Fig. 6, AEV is proportional to W because N is fixed to 20 in these groups of experiments. $AEV = \frac{W}{N} = \frac{W}{20}$ becomes bigger as W increases. There is still a certain gap between GBM and AEV in this figure. At the same time, the target value after GBM optimization is very far from the worst case when all loads are concentrated on only one BS.

4.3   The Calculation Rapidity Analysis

In order to verify the rapidity of GBM, two groups of complementary simulations are executed and the results of them are illustrated in Figs. 7 and 8, respectively. The aims of them is to investigate the relationship between the actual running time t of GBM and M and the relationship between t and N, respectively. Moreover, W is set to 1000 in these two groups of simulations and their actual running time are all not more than 1 min (60 s), especially for $N = 100, M = 50$. Additionally, comparing Figs. 7 and 8, the trends of lines $t - M$ with different N are all approximately linear while the trends of lines $t - N$ are visually quadratic, which provide strong support for $O(MN^2 \log \frac{W}{\epsilon})$ time complexity of GBM.

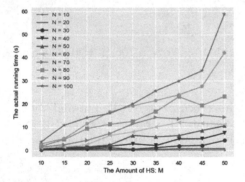

Fig. 7. The actual running time of GBM with the increasement of M

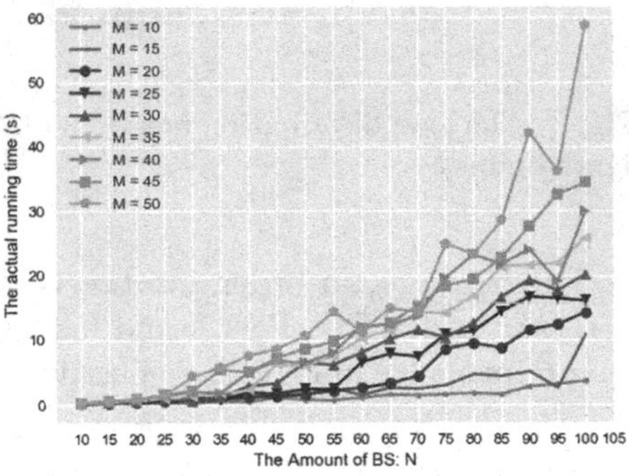

Fig. 8. The actual running time of GBM with the increasement of N

5   Conclusion

In this paper, load balancing problem meeting users' Qos demand is considered in LTE-A network. Signal coverage and user access are both formed as a mathematical description via *sigmod* and *softmax* function, respectively. Moreover, a $Min - Max\ GLFP$ model with the constraints of users' QoS is established to balance the traffic load of BS in LTE-A network. However, $GLFP$ is a well-known NP-hard problem with more than three summation items. Therefore,

GBM algorithm is designed by constantly solving a tightproblem and a relaxed problem. These two problems are linear feasibility problems and their solutions can be easily approached. Moreover, the time complexity of GBM is proved as $O(MN^2 \log \frac{W}{\epsilon})$ where ϵ is a given solution accuracy. Ultimately, simulation results verify the effectiveness and rapidity of GBM.

Acknowledgments. Thanks to China Mobile Group Beijing Company Limited (CMBJ) and Datang Telecom Technology and Industry Group (DTmobile).

References

1. Li, S., Da Xu, L., Zhao, S.: 5G internet of things: a survey. J. Ind. Inf. Integr. **10**, 1–9 (2018)
2. Liu, D., et al.: User association in 5G networks: a survey and an outlook. IEEE Commun. Surv. Tutorials **18**(2), 1018–1044 (2015)
3. Lobinger, A., Stefanski, S., Jansen, T., Balan, I.: Load balancing in downlink LTE self-optimizing networks. In: 2010 IEEE 71st Vehicular Technology Conference, pp. 1–5, May 2010
4. Zhou, F., Feng, L., Yu, P., Li, W.: Energy-efficiency driven load balancing strategy in LTE-WiFi interworking heterogeneous networks. In: 2015 IEEE Wireless Communications and Networking Conference Workshops (WCNCW), pp. 276–281, March 2015
5. Sun, L., Wang, L., Qin, Z., Yuan, Z., Chen, Y.: A novel on-line association algorithm for supporting load balancing in multiple-ap wireless lan. Mobile Netw. Appl. **23**(3), 395–406 (2018)
6. Goyal, S., Mezzavilla, M., Rangan, S., Panwar, S., Zorzi, M.: User association in 5G mmWave networks. In: Wireless Communications and Networking Conference, pp. 1–6 (2017)
7. Siomina, I., Yuan, D.: Load balancing in heterogeneous LTE: range optimization via cell offset and load-coupling characterization. In: 2012 IEEE International Conference on Communications (ICC), pp. 1357–1361, June 2012
8. Ye, Q., Rong, B., Chen, Y., Al-Shalash, M., Caramanis, C., Andrews, J.G.: User association for load balancing in heterogeneous cellular networks. IEEE Trans. Wirel. Commun. **12**(6), 2706–2716 (2013)
9. Bitran, G.R., Novaes, A.G.: Linear programming with a fractional objective function. Oper. Res. **21**(1), 22–29 (1973)
10. Ersoy, T., Yalizay, B., Cilesiz, I., Akturk, S.: Propagation of diffraction-free and accelerating laser beams in turbid media. Atti della Accademia Peloritana dei Pericolanti - Classe di Scienze Fisiche, Matematiche e Naturali **89**(31), 5 (2011). https://doi.org/10.1478/C1V89S1P031. ISSN: 1825-1242
11. Hauri, D.D., et al.: Exposure to radio-frequency electromagnetic fields from broadcast transmitters and risk of childhood cancer: a census-based cohort study. Am. J. Epidemiol. **179**(7), 843 (2014)
12. Okumura, Y., Ohmori, E., Kawano, T., Fukuda, K.: Field strength and its variability in VHF and UHF land-mobile radio service. Rev. Electr. Commun. Lab. **16**(9–10), 825–873 (1968)
13. Boyd, V., Faybusovich, L.: Convex optimization. IEEE Trans. Autom. Control **51**(11), 1859 (2006)

Faster Compression of Patterns
to Rectangle Rule Lists

Ian Albuquerque Raymundo Da Silva[1], Gruia Calinescu[2(✉)],
and Nathan De Graaf[3]

[1] Pontifical Catholic University of Rio de Janeiro, Rio de Janeiro, Brazil
`ian.albuquerque.silva@gmail.com`
[2] Illinois Institute of Technology, Chicago, IL, USA
`calinescu@iit.edu`
[3] Iowa State University, Ames, IA, USA
`nathandegraaf@gmail.com`

Abstract. Access Control Lists (ACLs) are an essential security component in network routers. ACLs can be geometrically modeled as a two-dimensional black and white grid; our interest is in the most efficient way to represent such a grid. The more general problem is that of Rectangle Rule Lists (RRLs), which is finding the least number of rectangles needed to generate a given pattern. The scope of this paper focuses on a restricted version of RRLs in which only rectangles that span the length or width of the grid are considered. Applegate et al.'s paper "Compressing Rectilinear Pictures and Minimizing Access Control Lists" gives an algorithm for finding an optimal solutions for strip-rule RRLs in $O(n^3)$ time, where n is the total number of rows and columns in the grid. Following the structure of Applegate et al.'s algorithm, we simplify the solution, remove redundancies in data structures, and exploit overlapping sub-problems in order to achieve an optimal solution for strip-rule RRLs in $O(n^2 \log n)$ time.

1 Introduction

We consider the following problem of generating patterns by drawing rectangles. A target pattern is given as a grid of black and white cells. We begin with a white grid and place solid black or white rectangles, each rectangle covering any previously placed rectangles. Placing arbitrary sized rectangles makes the problem NP-hard, so we consider only placing rectangles that extend either the full height or width of the grid. Our goal is to find the smallest number of rectangles required to create the target pattern.

1.1 Problem Definition

Define a pattern P to be an n_R by n_C grid of black and white squares. Let $n = n_R + n_C$. A rectangle strip-rule on P is either a black or white rectangle that

© Springer Nature Switzerland AG 2018
S. Tang et al. (Eds.): AAIM 2018, LNCS 11343, pp. 190–208, 2018.
https://doi.org/10.1007/978-3-030-04618-7_16

extends from one side of the pattern to the opposite side. Precisely, a rectangle in P is either a set of contiguous rows of P or a set of contiguous columns of P.

A rectangle strip-rule list (a RSRL) that generates P is an ordered list of rectangle strip-rules, that when applied in order to a blank (white) grid the size of P creates the target pattern. See Fig. 1 for an example of a pattern generated by a RSRL.

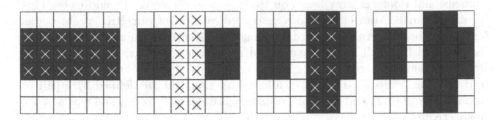

Fig. 1. A pattern generated by a RSRL of 3 elements.

We say a pattern P is a strip-rule pattern if there is a RSRL that generates P. Note that not every pattern P is a strip-rule pattern. See Fig. 2 for an example of a pattern that is not a strip-rule pattern.

Fig. 2. The 2×2 checkerboard: an example of a pattern that is not a strip-rule pattern.

An RSRL is considered optimal if it has the minimum number of rectangle strip-rules of any RSRL that generates P.

1.2   Problem Background

Rectilinear Pictures and Access Control Lists. The method of stacking rectangles to create patterns has applications in both graphics and network routers. A common method for drawing graphics is to allow the user to repeatedly apply a rectangle tool to a blank canvas (see Xfig or PowerPoint). Each rectangle is of a solid color and covers everything in a defined rectangular region. This sequence of drawn rectangles can be represented by a rectangle rule list (a RRL). The problem of finding the minimum length RRL needed to create a given pattern is a generalization of the RSRL problem we explore.

Alternatively, instead of being given an n_R by n_C grid as input, one can start with the numbers n_R and n_C, and a list of m rules. The goal is to find a shortest list that gives the same pattern as the input list. As shown in an extended version of Applegate et al. [1] available on some of the paper authors' homepages, it is possible to construct the n_R by n_C grid in time $O(n_R \cdot n_C + m^2)$.

One important application of RRLs is in access routers. An internet service provider might use access control lists (ACLs) on network router line cards in order to choose whether to forward or drop a packet based on the sending or receiving IP address. This decision could be answered by checking a two-dimensional Boolean array based on the sender and receiver IP addresses. This problem can be translated into a restricted version of our RRL problem (as explained below). An in-depth discussion of this translation and its properties is provided in detail in Applegate et al.'s paper [1]. Briefly, the idea in ACL minimization is to construct a given grid of size 2^w by 2^w, indexed by binary strings from 0^w to 1^w. Rectangles allowed on this grid are defined by a pair of binary strings (y, z) and cover any squares whose indices have y and z as a prefix respectively.

Related Work. Unfortunately, the general problem of finding minimum length RRLs has been shown to be NP-hard by Applegate et al. [1]. Instead, we work on a restricted version of the problem in which any rectangle rules applied must extend either the height or width of the original pattern and only black or white rectangles are allowed. This 2-color strip-rule problem was originally posed in [1], where an optimal solution is given in $O(n^3)$-time. Applegate et al. [1] obtains an optimal solution to the strip-rule version of the ACL problem with a similar (but more complicated) $O(wn^3)$-time algorithm. A 1.5 ratio approximation algorithm is given for the problem in $O(n^2)$-time. While there exist numerous results related to various other restricted problems regarding RRLs, we know of no other work related to this 2-color strip-rule problem.

ACL can be used in firewalls [7]. Kang et al. [7] introduces axioms with the goal of creating and analyzing algorithms for optimizing the rewriting of rules in Software Defined Networks. Liu et al. [9] considers classifiers in dimensions higher than two, which reduce to either ACL or RRL lists in dimension two. They propose and experimentally evaluate heuristics without performance guarantees, as well as relate these classifiers to the Firewall Decision Diagrams of Gouda and Liu [5]. Pao and Lu [12], and Comerford et al. [3] also consider higher dimensional classification based on rectangle rules. Kang et al. [6] and Zhang et al. [15] introduce more general rule-minimization problems. Sun and Kim [13] uses rule minimization as a start for a solution to an extended problem. Efficiently removing redundant rules has been proposed in Sun and Kim [14].

The complexity of finding minimum length ACLs in two dimensions is still unknown, but with an arbitrary number of dimensions, this problem is NP-hard according to Kang et al. [8]. Applegate et al. [1] gives a $O(\min(m^{1/3}, OPT^{1/2}))$-approximation algorithm for finding minimum length RRLs, where m is the length of the input RRL. This is still the best published ratio.

Daly et al. [4] provides heuristics for higher-dimensional ACL and RRL minimization (reference [4] calls the ACLs minimization prefix-ACL, while range-ACL is their terminology for finding minimum length RRLs), on the way improving the approximation ratios provided by Applegate et al. [1] for ACLs minimization. The approximation algorithms of [4] and [1] use as a subroutine the strip-rule version that we study.

Our Results. Using the structure defined in Applegate et al.'s [1] $O(n^3)$ exact algorithm for 2-color RSRLs, we give an improved $O(n^2 \log n)$ exact algorithm for the RSRL problem. Given the similar structures of the RRL and ACL optimal solutions given by [1], we expect our solution can be extended to improve the running time of an exact algorithm for the strip-rule ACL problem from $O(wn^3)$ to $O(wn^2 \log n)$. As strip-rule ACLs occur in a high percentage of ACL minimization cases [1], this is one reason to study RSRL. Another two reasons are: we believe RSRL is a natural problem, and RSRL is used in the approximation algorithms for RRL minimization.

Our result is obtained by digging deeper into the structure of the dynamic programming of Applegate et al. [1] combined with the use of geometric data structures to speed up the process. We use fast two-dimensional orthogonal range queries, using existing data structures (a time bound of $O(\log n)$ time per query being textbook material [2]).

2 Preliminaries

We begin by exploring the strategies and tools for finding an optimal RSRL. The Pick-Up-Sticks algorithm will detail the basic structure used to find an RSRL, but not necessarily an optimal RSRL. Then we will explain the concept of equivalence classes, which can be grouped into "segments". All of these concepts will be put together in our algorithm for finding an optimal solution.

2.1 The Maximum Pick-Up-Sticks (MPUS) Algorithm

As proposed by Applegate et al.'s paper, we can build an algorithm for finding whether a pattern is a strip-rule pattern and, if so, finding an RSRL that generates it.

The Pick-Up-Sticks (PUS) Algorithm. The Pick-Up-Sticks algorithm of Applegate et al. [1] is an algorithm for generating a RSRL of a pattern P if such a list exists. The idea is to pick up monochromatic rows and columns in order to build the RSRL backwards. Every time we pick a row or column, we color it gray.

Let P be a black and white pattern. We define a column or row in P as being pseudo-monochromatic if it is composed of only gray and white cells or black and gray cells. Note that monochromatic columns and rows are also pseudo-monochromatic. The Pick-Up-Sticks algorithm builds an RSRL as follows:

While there are still black cells, choose a pseudo-monochromatic column or row and color all its cells gray. After a row or column is colored gray, add a rectangle that corresponds to covering that row or column with whatever non-gray color was left in that row or column to the beginning of our RSRL. Note that, in our problem definition, we begin applying RSRLs with a white grid, so we can stop picking up sticks when no black cells exist. If, on the other hand, we modify the problem to begin with a grid of some fourth color, then we would only stop picking up sticks when all cells are gray. There may be a difference of one rule between optimum solutions to these two problems (white initial grid or some fourth color initial grid) with the same target pattern. For the sake of symmetry, from now on we use this *modified version* of the problem. The method works with minor modifications for the original version.

The algorithm may not succeed, as it may not find a pseudo-monochromatic column or row. If the algorithm does end in an all-gray grid, then the algorithm has "picked up" the rectangles of the RSRL in reverse order. Figure 3 gives a representation of one possible execution of the Pick-Up-Sticks algorithm and Fig. 4 shows the generation of the original pattern from the resultant RSRL.

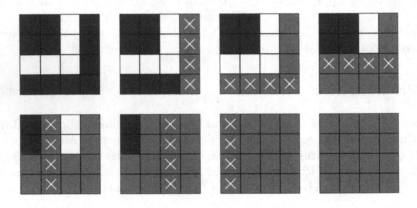

Fig. 3. The execution of the Pick-Up-Sticks algorithm on a strip-rule pattern.

The Pick-Up-Sticks algorithm is guaranteed to generate an RSRL if one exists (see below), but it is not necessarily an optimal RSRL of minimum length. At any stage of the algorithm there could be more than one option on which row or column to pick up. Different choices can lead to different sizes of RSRLs. In the next sections we will discuss on how Applegate et al.'s paper narrows down the number of choices for each stage in order to find an optimal RSRL.

Theorem 1 (part of Theorem 3.1 of Applegate et al. [1]). *A black and white pattern P is a strip-rule pattern if and only if the Pick-Up-Sticks algorithm results in a grid with all of its cells gray. In that case, the reverse list of picked up rows and columns is a RSRL for P.*

Improving by Picking Maximal Sticks. One important observation by Applegate et al.'s paper on the Pick-Up-Sticks (PUS) algorithm is that there can be no harm in always using maximal strip-rules. A maximal strip-rule is a strip-rule that picks up a maximal contiguous sequences of pseudo-monochromatic

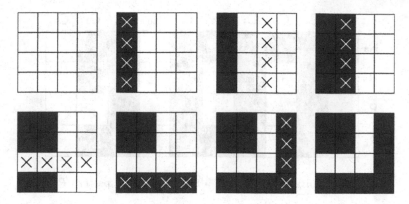

Fig. 4. The RSRL generated by the execution of the MPUS algorithm on the pattern of the Fig. 3.

rows or columns. Following Applegate et al. [1], we call such a contiguous sequence a *block*.

This is possible because replacing a non-maximal strip-rule by the maximal one that contains it does not affect the pseudo-monochromaticity of any later rule. Hence, we may always use the Maximal Pick-Up-Sticks (MPUS) algorithm instead of the Pick-Up-Sticks (PUS) algorithm. In this new algorithm, once we choose the pseudo-monochromatic column (or row) to pick, we pick the maximal contiguous set of pseudo-monochromatic columns (or rows, respectively) that contains it.

2.2   Equivalence Classes

We now introduce the concept of equivalence classes of rows, as used by Applegate et al.'s paper. This definition will allow us to give structure on the order in which columns and rows should be picked up during an execution of MPUS in order to find an optimal RSRL.

Definition and Properties. Given a pattern P, we define two rows or columns as being in the same equivalence class if and only if they are both of the same type (row or column) and their cells have the exact same colors in P, in the same order; that is, they are identical.

We will denote a column equivalence class in a pattern P as being C_x, where x is the number of black cells in the original pattern. Analogously, we will denote R_y as being the row class with y black cells in the original pattern. See Fig. 5 for an example of a pattern and its equivalence classes. It follows from the monotonicity property (Theorem 2 below) that all the columns in C_x are identical in all positions, for all x, and the same property holds for all the rows of R_y, for all y.

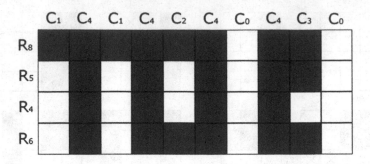

Fig. 5. Equivalence classes of a black and white pattern P.

Theorem 2 (part of Theorem 3.1 of Applegate et al. [1]). *(Monotonicity Property) Let P be a black and white pattern. P is a strip-rule pattern if and only if the two equivalent properties hold:*

(i) *For any color $c \in \{black, white\}$, the rows of P are hierarchical: given any two rows of P, the set of columns where c is present in the first row either contains or is contained in the set of columns where c is present in the second.*

(ii) *The same property of item (i) holds with the roles of row and column switched.*

Note that every row and column belongs to exactly one equivalence class. Also, if two columns belong to an equivalence class in the beginning of the MPUS algorithm, then they will remain of the same class until one of them is picked up by the algorithm. This is since picking up other columns does not change these two columns at all, while picking up rows modify these two columns in exactly the same way. The same property holds for two rows that belong to an equivalence class.

We also define an equivalence class as being *active* during the execution of the MPUS algorithm if some member of that class is pseudo-monochromatic but not all gray. We will use the next proposition.

Proposition 1 (Applegate et al. [1], Observation 6. on page 1070). *During the execution of the MPUS algorithm on a black and white strip-rule pattern, there are always exactly two active classes at any given time. The two active classes are either a row and a column class of the same color or both classes of the same kind (rows or columns), being one of each color.*

The intuition on why this happens is that in order for a new class to become active, all the members of an old active class need to be picked up.

Embedded Rows and Columns. We can start improving the MPUS algorithm by introducing the concept described by Applegate et al.'s paper as embedded rows and columns. This will allow us to pick up more rows and columns using the same number of rectangles.

First, we will define the concept of an embedded column or row at a given stage of the MPUS algorithm. Given a point during the execution of the MPUS algorithm, let b be a column of an active equivalence class B. Let a_1 be the first column to the left of b that does not belong to B and has not yet been picked up. Similarly, let a_2 be the first column to the right of b that does not belong to B and has not yet been picked up. We say that b is *embedded* in equivalence class A if both a_1 and a_2 belong to that class A. The definition is analogous for rows. Note that columns can be embedded only in column classes and rows only in row classes. See Fig. 6 for an example.

Fig. 6. Example of a pattern P where one member of C_0, the second column, is embedded in C_2.

If during the execution of the MPUS algorithm we have the option to pick up two column-blocks that embed a set of columns of another active class, then we can pick up the two column-blocks that embed the third one with two rectangles and then the embedded one, totaling three rectangles. However, it is better to pick the embedded block first with one rectangle, and then pick the two blocks that were previously embedding it with only one rectangle, thus using a total of two rectangles for the same set of columns. As an example, in Fig. 6, one cannot benefit by picking up the first and the third column, when one can pick up first the second column, followed by picking up the block of the first three columns.

Equivalence Classes Ordering. As we pick up all the members of a column class, we change the rows of the pattern. Since the other columns remain intact, the new class to become active must be a row class. By using the same argument for row classes, we see that whenever we pick up all the members of a class, the next class to become active is of the opposite type. This is formally stated below.

Proposition 2 (Implicit in Applegate et al. [1]). *For a pattern P of two colors, if one picks up all columns of an active class and does not make all the grid gray, then a row class of the opposite color becomes active. If one picks up all rows of an active class and does not make all the grid gray, then a column class of the opposite color becomes active.*

We can order the equivalence classes of both rows and columns in a given pattern P by the number of black cells on it. In that ordering, we can see from the proposition below that all the rows that two consecutive column classes differ form exactly an equivalence class. This holds analogously for columns.

Proposition 3 (Implicit in Applegate et al. [1]). *Let i be such that $1 \leq i < N_C$ and j be such that $1 \leq j < N_R$.*

(a) *Let S_i be the set of rows that have white cells on the intersections of the columns of C_{x_i} and black cells in the intersection of the columns of $C_{x_{i+1}}$. Then S_i is a row class ($S_i \in R$).*

(b) *Let S_j be the set of columns that have white cells on the intersections of the rows of R_{y_j} and black cells in the intersection of the columns of $R_{y_{j+1}}$. Then S_j is a column class ($S_j \in C$).*

This can be used to prove:

Proposition 4. *The number of row classes and the number of column classes differ by at most one; that is $|N_R - N_C| \leq 1$.*

Using that information, we can set up the following ordering for the equivalence classes:

1. Start with the column class with only white cells. If this class does not exist, then start with the row class with only black cells (the existence of this row class follows immediately from the monotonicity property – Theorem 2 in the Appendix).
2. Alternate columns and rows, putting the columns in ascending order of black cells and the rows in descending order of black cells.

Let $C_{x_0}, C_{x_1}, \ldots, C_{x_{(N_C-1)}}, C_{x_{N_C}}$ be the ascending ordering of column classes by the number of black cells. Let $R_{y_0}, R_{y_1}, \ldots, R_{y_{(N_R-1)}}, R_{y_{N_R}}$ be the ascending ordering of row classes by the number of black cells. Using the rules above, we should get an ordering such as $C_{x_0}, R_{y_{N_R}}, C_{x_1}, R_{y_{(N_R-1)}}, \ldots, R_{y_1}, C_{x_{(N_C-1)}}, R_{y_0}, C_{x_{N_C}}$ (alternating row and column classes). This array may start and/or end with row equivalence classes, instead of columns as above.

Proposition 5. *There is an array $E_1, E_2, \ldots, E_N$ of equivalence classes that respect the following property: Given a and b such that $1 \leq a < b \leq N$, if E_a and E_b are the active classes, then for every c such that $1 \leq c < a$ or $b < c \leq N$, E_c has already been picked up. Also, if $b = a + 1$, then either $E_a \in R$ and $E_b \in C$ or $E_a \in C$ and $E_b \in R$. Moreover, none of the columns/rows of a class E_c for $a < c < b$ are pseudo-monochromatic.*

We omit the proof for lack of space. We will call this ordering the *hierarchical array* of the pattern P. The hierarchical array is a useful extension of Observation 5 on page 1070 of Applegate et al. [1].

We will also refer to this ordering of equivalence classes as $E_1, E_2, \ldots, E_N$. That is, if the hierarchical array starts with a column, then $E_1 = C_{x_0}$, $E_2 = R_{y_{N_R}}$, $E_3 = C_{x_1}$, and so on, and if the hierarchical array starts with a row, then $E_1 = R_{y_{N_R}}$, $E_2 = C_{x_0}$, $E_3 = R_{y_{(N_R-1)}}$, and so on. Note that N_C, the number of column classes, is at most n_C, the number of columns, and the same property holds for rows.

2.3   Segments

During MPUS, there are always two active classes. A "segment" succinctly represents the blocks that are available for pick up while a certain pair of classes are active. These segments can be translated into nodes of a graph (the *segment graph* that is discussed in detail later), that can be used for finding an optimal solution, as described below. Note that our representation of segments is a slightly condensed version of the one described in the original Applegate et al.'s paper [1]. The difference is that the original definition is a five tuple with two redundant elements.

Definition and Properties. Whenever a new class becomes active during the execution of the MPUS algorithm on a pattern P, we will define a *segment* S as being a tuple of three elements (E_x, E_y, U), where E_x and E_y are the active equivalence classes and U is the subset of members (rows or columns) left unpicked of E_y $(U \subseteq E_y)$.

Proposition 6. *Let E_1, E_2, ..., E_N be the hierarchical array of a pattern. Given a and b such that $1 \leq a < b \leq N$ it is possible to know what colors correspond to each active class of (E_a, E_b).*

We omit the proof. It is not here where the running time is improved.

We can represent the sequence of actions of the MPUS execution as a sequence of segments, as described in this paragraph. Given the pick-up order obtained through the execution of the MPUS, let's create a segment (E_0, E_N, E_N) for the starting state of pattern P and a segment (E_x, E_y, U), $U \subseteq E_y$, for every time a new class E_x becomes active; we say that the RSRL *includes* the segment whenever such a segment becomes active during the MPUS execution corresponding to the RSRL.

Since E_x is the new class, all of the members of E_x are still in the pattern. E_y, in the other hand, is a class that was already active. It may have some of its members already picked up. We store that information in U by maintaining the members that have not been picked up yet. Note also that any classes that are between E_x and E_y in the hierarchical array of P have not become active yet, resulting in the fact that all of their members are still there. At the same time, members that come before and after the interval delimited by E_x and E_y have all been picked up. Hence, for each one of those segments that are created whenever a class becomes active it is possible to reconstruct the pattern at that moment of the execution of the MPUS algorithm.

Segment Branching. Applegate et al.'s paper [1] shows that if we are looking for an optimal solution, then we can narrow down the number of segments significantly by reducing the number of options into which a segment can branch to two.

Suppose a minimum-length RSRL for a strip-rule pattern P includes the segment $S = (E_x, E_y, U)$ and that to get to the next segment, if it exists, we pick up all members of E_x (possibly picking up some members of $U \subseteq E_y$) while at least one member of U remains active.

- If $U \subseteq E_y$ has a member that is not embedded in E_x, then there exists a minimum-length RSRL that includes the segment $S = (E_x, E_y, U)$ and that to get to the next segment you pick up all the members of $U \subseteq E_y$ that are embedded in E_x, followed by picking all members of E_x.
- If all members of $U \subseteq E_y$ are embedded in E_x, then there exists another minimum-length RSRL that includes the segment $S = (E_x, E_y, U)$ and that to get to the next segment you pick up all the members of $U \subseteq E_y$ (thus, finishing picking up every member of E_y in the pattern). Note that in this case, one can get to the next segment by picking up all the members of $U \subseteq E_y$ first.

The same property holds with the roles of E_x and U switched. In this case, we assume the next segment from S, if it exists, is originally reached by picking up all members of $U \subseteq E_y$. Intuitively, this happens because we can always pick embedded members at no extra cost, if one is to pick all the blocks of the embedding class.

Proposition 7 (statement 2(a) of Lemma 4.2 of Applegate et al. [1]).
Suppose we have two sets U_x and U_y of active blocks of the two active classes E_x and E_y. Suppose there is an optimal solution that completely picks up E_x before completely picking up E_y. Then there also exists an optimal solution that picks up all embedded members of E_y before picking up E_x.

Moreover, suppose we have two sets U_x and U_y of active blocks of the two active classes E_x and E_y. To finish picking up blocks (except at the very end), we must reach the situation that either E_x or E_y are not active anymore; say E_x becomes not active, while E_y is still active. Picking up any block from the class E_y that is not embedded in E_x can be done, without loss of optimality, later. So we can assume that only the members of U_y that are embedded in E_x are picked up while E_x is active.

Note also that for every segment we can pick the latest class that became active or the oldest, a total of two options. We can then say that every segment will branch into two other segments. In both cases, we can determine the next segment reached by eliminating every embedded member of the class not being picked up, followed by eliminating every member of the class being picked up. The only exception to this occurs when we pick every embedded member of the class not being picked up we end up picking all of its members, thus already reaching a new segment. In that case, the original segment branches into only one segment, instead of two.

Note that there could be a segment that does not have any next segment. In this case, the segment is a segment at the very end of MPUS execution. If we pick up any of either class, then we will obtain an all-gray grid and the algorithm will be over. Also, note that for this to happen we must have that E_x and E_y are next to each other in the hierarchy array, and picking-up one of them will also result in picking-up the other one.

Segment Graph. We build a "segment graph" using the segments as nodes, as described by the following structure proposed by Applegate et al. [1]:

- Start with the segment (E_0, E_N, E_N), that represents the grid as it is in the very beginning of the MPUS algorithm. This will be the starting node.
- For every node (segment) in the graph, generate the two segments (possibly one, as described in the previous subsection) that it branches into and add a directed edge from it to the new generated nodes. There will be one segment for each of the two options of picking up the newest or the oldest of the two classes in the segment. In both cases, we can determine the next segment reached by eliminating every embedded member of the class not being picked up, followed by eliminating every member of the class being picked up.
- Define the cost for each edge as the number of rules in the MPUS algorithm used to go from one segment to another.
- Create an artificial node corresponds to an all-gray grid. This will be the end node of the graph. Every time a segment has its two classes adjacent to each other in the hierarchy array, create an edge from it to this all-gray grid node. This edge should have cost one (since we solve the modified version); if we were to solve the original version, then this edge should have cost zero if the members of the classes of the segment were white and one otherwise.

Because of what was discussed in the Segments Branching section, at least one of the paths from the starting node to the end node of the graph will be a minimum-length RSRL. Note that having (E_N, E_0, E_0) as the starting node will lead to a graph that represents the same possible steps for the MPUS because it also represents the same starting pattern. Note also that some nodes may be reached by more than one node. Every node will only branch into nodes that correspond to patterns with a smaller number of total of columns and rows. This implies that there are no cycles in this graph.

We say that one segment *reaches* another if there is a path in this graph from that segment to the other one. The distance from one segment to another is the sum of the costs of the edges that compose the path between them, if such a path exists. One very important property of this graph is that if two segments (E_x, E_y, U) and (E_x, E_y, U') are reachable from the starting node of the graph, then either $U \subseteq U'$ or $U' \subseteq U$.

Lemma 1 (Containment Lemma – Lemma 4.3 of Applegate et al. [1]).
Let S and S' be two segments on the same equivalence classes (in the same order). Then either $U \subseteq U'$ or $U' \subseteq U$.

Intuitively, this follows from the fact that we have a strict ordering on the equivalence classes that could embed columns in U. See Applegate et al.'s paper [1] for a rigorous inductive proof of this lemma. We will see that this implies that the number of nodes in this graph is $O(n^2)$, where n is the number of rows and columns in the original pattern.

3   Algorithm

The flow of our algorithm closely follows that of the $O(n^3)$ algorithm given by Applegate et al. We similarly create a segment graph of the reachable states in a

MPUS of the pattern. However, we create this graph faster by grouping similar segments into S-groups, which can be processed together.

3.1   Setup

We begin by reading the input pattern into an equivalence class list and creating the first node in our segment graph.

Equivalence Class List. We are given as input an $n_R \times n_C$ grid of black and white cells. Assign each of the rows an index from 1 to n_R from top to bottom, similarly each column, an index from 1 to n_C from left to right. Group rows and columns into equivalence classes and order the classes by number of black cells as previously described in the Subsect. 2.2. Precisely, construct the hierarchical array:

$$C_{x_0}, R_{y_{N_R}}, C_{x_1}, R_{y_{(N_R-1)}}, \ldots, R_{y_1}, C_{x_{(N_C-1)}}, R_{y_0}, C_{x_{N_C}}$$

(this array may start and/or end with row equivalence classes, instead of columns as above).

We will need to determine from a range of column indexes which columns are also in a range of equivalence classes, and the similar question with rows instead of columns. A data structure such as an orthogonal range tree [2] accomplishes these queries in $O(\log n)$ time with $O(n \log n)$ space and setup time.

S-Group. Define an S-group (E_1, E_2) (for equivalence classes E_1 and E_2, where the order of E_1 and E_2 in the tuple above matters) to be a collection of all segments $\{C_1, C_2, U'\}$ such that $C_1 = E_1$ and $C_2 = E_2$. To completely represent each of these segments, an S-group (E_1, E_2) maintains a list of segments S and a master list U. This list U will be used to keep track of each of the segment's individual list (the list U' of this segment), as described below. This allows each segment to be stored in constant space, while the S-group gets stored in $O(|E_1| + |E_2|)$ space.

The list U contains the index of every member of E_2 ordered such that for every segment $\{E_1, E_2, U'\}$ in the S-group, the first $|U'|$ members of U comprise the same set as U'. Such an ordering is guaranteed to exist by the Containment Lemma 1. We generate this ordering by noting that indexes appearing in more segments appear earlier in the list U.

The list S of segments maintained by the S-group, keeps these segments ordered by the size of their set U'. Each segment s_i will have three values, $u(s_i)$, $d(s_i)$, and $p(s_i)$. Set $u = |U'|$. The field d gives the minimum number of sticks required to reach the state of segment starting from the original pattern. We want to maintain the shortest path to each node, so we use the value p to store a reference to the segment right before this one on this path.

Origin S-Group. We begin with the S-group (E_1, E_2) with $U = \overline{E_2}$ and S comprised of a single segment $(u, d, p) = (|\overline{E_2}|, 0, null)$, where $\overline{E_2}$ represents the set of all the members of E_2. The classes E_1 and E_2 are the first and last members of the hierarchical array. We will explore outwards from this S-group in a breadth-first fashion, enumerating all reachable segments. The graph on the

S-groups is guaranteed to be acyclic because the number of picked up classes is strictly increasing along each edge. This graph with S-groups implicitly stores all the useful information of the segment graph.

3.2   Finding the Next S-Group

Suppose we have some S-group (E_1, E_2) with associated lists U and S. Without loss of generality we will assume E_1 to be a column class. We wish to generate all S-groups reachable from this S-group. The two S-groups immediately reachable from this S-group correspond to completely removing either class E_1 or E_2.

To determine the cost of removing a class E_i from a segment s_j, we count how many members of E_{3-i} can be picked up for free, as described in detail later. If two members of E_i may be picked up with the same stick while E_1 and E_2 are the active classes, then we will call the two members *contiguous*. It follows from Proposition 5 that a pair of columns is contiguous in an S-group (E_1, E_2) if every column between them is either already picked up or of type E_1 or E_2.

Before the class E_i can be removed from a segment (E_1, E_2, U') (all the members of the class being picked up), we pick up every embedded member of E_{3-i}. To determine the cost of moving to the next segment, we must count how many of the embedded columns are also contiguous. Then any embedded columns must be removed from either the set U' (if E_1 is removed) or E_1 (if U' is removed) in the next segment and, by doing this for all segments, in the next S-group. In fact, we process one S-group at a time, and obtain segments of another at most two S-groups, as explained below.

Because E_1 and E_2 are not symmetric (E_2 comes with its set U), we will describe the process of removing classes E_1 and E_2 separately.

Remove E_1. Let E_3 be the equivalence class adjacent to E_1 in the equivalence class list, which has not yet been removed. The S-group corresponding to the result of removing E_1 will be (E_3, E_2) with lists $\hat{U}$ and $\hat{S}$.

Count Contiguous Consecutive Pairs of Columns of E_1. To determine the distances to the segments of the next S-group, we must count the number of contiguous E_1 ranges.

To do so, we must determine the range of equivalence classes which have not yet been picked up. Let E_1 be the column class C_a. If E_2 is also a column class, then call it C_b. If E_2 is a row class, then C_b is the column class adjacent to E_2 in the equivalence class list that has already been picked up. We assume $a < b$, with the other case being symmetric. By Proposition 5, a column class C_i has not been picked up if and only if $i \in (a, b)$.

For each pair of adjacent members of E_1 we check to see if they are contiguous. Precisely, for each member $e_i \in E_1$, we query our column range tree for members in the range $[a + 1, b - 1]$ with index in the range $[e_i, e_{i+1}]$, where e_{i+1} is the column of E_i that has the smallest index among those with index higher than e_i (e_{i+1} is the next column of E_i after e_i). Let c be the number of pairs of consecutive members of E_1 which are also contiguous. The value $(|E_1| - c)$

corresponds to the number of sticks required to completely remove E_1 after any embedded columns of E_2 are picked up.

Get Embedded E_2 Blocks. If E_2 is also a column class, then we must count and remove any members of E_2 embedded in E_1. Similar to the way we checked that columns of E_1 were contiguous, we will check if adjacent columns of E_1 are both contiguous and contain columns of E_2. In order to accurately keep track of the cost to remove these embedded columns we must give each embedded column a tag based on which two columns surrounded the embedded column. Two members of E_2 can be picked up with the same stick if and only if they have the same tag.

When an embedded column is found, tag it with the index of the E_1 column to its left. Then add this column to a list, B, sorted by the column's index. After we have found all of the embedded columns, we are ready to generate $\hat{U}$ and $\hat{S}$.

Build the Next S-Group. Begin with $\hat{U}$ as an empty list. We also need a set T to keep track of which tags have been accounted for. Iterate through U, searching for each member m of U in the set of embedded columns, B. If $m \notin B$, then append m to the end of $\hat{U}$. Otherwise add m's tag value to the set T if it is not already there. Once we have checked all of the members U' in a segment s_i (U' being the first $u(s_i)$ elements of U), add a new segment $\hat{s}_i$ to $\hat{S}$ with the following values: $u(\hat{s}_i) = |\hat{U}|$, $d(\hat{s}_i) = d(s_i) + |T| + |E_1| - c$, and $p(\hat{s}_i) = s_i$ (note that $|\hat{U}|$ and $|T|$ are computed for the sets $\hat{U}$ and T exactly when having finished processing the last element of U' while iterating through U; $\hat{U}$ and T can change later on). If $\hat{S}$ already contains a segment $\hat{s}_j$ with $u(\hat{s}_j) = u(\hat{s}_i)$, then keep only the segment with shorter distance d in the set $\hat{S}$, and remove the other one. As an aside, one can see that if we do not remove duplicates, if two segments from the same S-group s_i and s_j have $u(s_i) > u(s_j)$, then $u(\hat{s}_i) \geq u(\hat{s}_j)$, which is used in the proof of the Containment Lemma 1.

Remove E_2. Let E_3 be the equivalence class adjacent to E_2 in the equivalence class list, which has not yet been removed. The S-group corresponding to the result of removing E_2 will be (E_3, E_1) with lists $\hat{U}$ (whose elements are members of the class E_1) and $\hat{S}$.

Count Contiguous Consecutive Pairs of Rows/Columns of E_2. To count contiguous ranges of E_2, we must count the contiguous ranges within each segment separately. We use a counter c, initially set to 0. Fortunately, if a pair of rows/columns is contiguous in one segment then it is also contiguous in all larger (with bigger value of u) segments. For each index $u_j \in U$, insert u_j into a sorted list of indexes, I. Get the predecessor and successor of u_j in I and check if these ranges from u_j to its neighbors are contiguous. If either of these ranges exists and is contiguous, then this column can be picked up for free. If not, we increment our cost counter c. Once we have iterated over the first $u(s_i)$ blocks, we save the state of our counter in a value $c_i = c$. This value corresponds to the cost to completely pick up the rows/columns in the segment s_i after any embedded members of E_1 have been picked up. Proceed (with c possibly increasing) until we finish the list U.

Get Embedded E_1 Blocks. If E_2 is also a column class, then we must count and remove the embedded columns of E_1. Each segment could have a unique number of embedded columns, where the larger the segment, the more columns can be embedded and the smaller the resulting segment will be in the next S-group. As an aside, this is an argument used in the proof of the Containment Lemma 1. To generate the ordering of $\hat{U}$ and the values in $\hat{S}$ and we must carefully count the embedded columns.

We iterate through the columns of U, keeping track of which columns are embedded and will get picked up. To help us with this task, we start with a sorted list V of the indexes of E_1, an empty list for $\hat{U}$, and a counter d, initialized to $d = 0$, to measure the number of required sticks. Also start with B, a set of columns of E_1, initialized as the empty set.

Similar to the way we counted contiguous blocks, for each column $u_i \in U$ we insert u_i into a sorted list of indexes I. Get the predecessor and successor of u_i in I, called u_j and u_k respectively, if they exist (in which case $j < i$ and $k < i$). Use range search to check if u_j and u_i are contiguous, and if u_i and u_k are contiguous; if a pair does not exist, then treat it as not being contiguous. If neither of these two pairs is contiguous, then do nothing. If exactly one of these pairs is contiguous, then use binary search in V to obtain the set of columns of E_1 embedded between the pair, add this set to B, and increment d. If both of these pairs are contiguous, then use binary search to determine if there are elements of $\overline{E_1}$ (defined earlier as all the columns of class E_1) between u_j and u_i and between u_i and u_k; in which case we increment d. After we have iterated through $u(s_i)$ columns of U, it is time to add to $\hat{U}$ and create a new segment $\hat{s}_i \in \hat{S}$ with the following values: $u(\hat{s}_i) = |V| - |B|$, $d(\hat{s}_i) = d(s_i) + d + c_i$, and $p(\hat{s}_i) = s_i$. Remove all of the members of B from V, then add all of these members to the beginning of $\hat{U}$. Reset B to be the empty set. Continue iterating through U.

Once we have finished iterating through U, add the remaining elements in V to the beginning of $\hat{U}$.

Build the Next S-Group. If E_2 is a row class, then $\hat{U} = \overline{E_1}$ and $\hat{S}$ contains one segment $\hat{s}$. To find $\hat{s}$, we iterate over each segment $s_i \in S$ looking for the segment s_i with minimum value $d(s_i) + c_i$. $\hat{s}$ has the following values: $u(\hat{s}) = |\overline{E_1}|$, $d(\hat{s}) = d(s_i) + c_i$, and $p(\hat{s}) = s_i$.

If E_2 is a column class, $\hat{U}$ and $\hat{S}$ were created while we found embedded columns.

3.3   Merging Identical S-Groups

Once we have generated a new S-group (E_1, E_2) we add it to a two-dimensional table, where the row is determined by E_1's index in the original list of equivalence classes, and the column similarly determined from E_2's index.

If an S-group (E_1, E_2) already exists, then we must merge the two S-groups. Given two (E_1, E_2) S-groups, $G_1 = \{U_1, S_1\}$ and $G_2 = \{U_2, S_2\}$, we will compute a new S-group, $G_3 = \{U_3, S_3\}$, that encompasses both of these S-groups which will then get stored in our table. We iterate through both G_1 and G_2 concurrently in order to create G_3, as described below. We start with U_3 and S_3 being empty.

Merge S_1 and S_2, maintaining the ordering based on u. Iterate through each s_i in this merged list. Let S_j be the list which contains s_i. Remove elements from the start of U_j, adding them to the end of U_3 until $|U_3| = u(s_i)$. (Do not add a value to U_3 if it is already in U_3.) This works because of the Containment Lemma 1. (As an aside, the proof in Applegate et al. [1] of the Containment Lemma relies on proving the fact that all the values of $u(s)$ with $s \in S_1$ are at most the minimum of the values of $u(s)$ with $s \in S_2$, or vice versa.) Add s_i to S_3. If G_1 and G_2 both have segments such that $u(s_a) = u(s_b)$, then choose the segment of smaller distance to add to S_3.

3.4   Finding an Optimum RSRL

While we build the graph on S-groups, we keep track of the segment which is a valid endpoint of the smallest distance. We define a *valid endpoint* to be a segment which is completely gray (this is since we solve the modified version of the problem; for the original problem, we would have a segment which is completely white and gray). A segment (E_1, E_2) is an endpoint if $E_1 = E_2$. (In the original version, we also have the case where E_1 and E_2 are adjacent in the equivalence class list and the cell where E_1 and E_2 intersect is white in the original pattern.) Then once the graph on S-groups is finished, we build a path to the valid endpoint using the values stored in p. This path will give an optimal order to remove equivalence classes, which can then be translated into an optimal list of rectangle strip-rules.

This algorithms returns an optimum solution, as follows from all the discussion above.

4   Time and Space Complexity

In this section we will show the time and space complexity of the previously defined algorithm is $O(nN \log n)$ and $O(nN)$ respectively – where N is the number of equivalence classes and n is the number of rows and columns in the original pattern. The ideas of our proof are partially taken from a more complete version of Applegate et al. [1], which showed the number of reachable segments to be in $O(n^2)$. Indeed, our contribution is a faster way of processing a segment, cutting down this processing time down from $O(n)$ to $O(\log n)$.

Following Applegate et al. [1], we will show that if the complexity for an S-group (E_a, E_b) is in $O(|E_a| + |E_b|)$, then the overall complexity of all S-groups is $O(nN)$. For each S-group, add the first term of its complexity, $O(|E_a|)$, to one two-dimensional array, and its second, $O(|E_b|)$, to a second array – each at location (E_a, E_b).

$$\begin{bmatrix} E_1 & E_1 & \dots & E_1 \\ E_2 & E_2 & \dots & E_2 \\ \vdots & \vdots & \ddots & \vdots \\ E_N & E_N & \dots & E_N \end{bmatrix} \begin{bmatrix} E_1 & E_2 & \dots & E_N \\ E_1 & E_2 & \dots & E_N \\ \vdots & \vdots & \ddots & \vdots \\ E_1 & E_2 & \dots & E_N \end{bmatrix}$$

By noting that the sum of all members of all equivalence classes equals the total number of rows and columns, we get $\sum_{i=0}^{N} |E_i| = n$. The sum of the columns in the first matrix and the sum of the rows in the second matrix both equal n. The sum of all terms in both matrices is $2nN$, so the complexity of all S-groups is in $O(nN)$.

The space complexity of the algorithm is determined by the sizes of the lists storing the S-groups. Each S-group maintains a list U, which holds members of E_b, and is therefore in $O(|E_b|)$. The list S contains all segments. Each segment is guaranteed to have a unique size u, and the values of u are positive values at most $|E_b|$. The complexity of each segment is constant, so we again have $O(|E_b|)$. As we have previously shown, since each segment is in $O(|E_a| + |E_b|)$, the overall space complexity is $O(nN)$.

Each operation we perform on an S-group (E_a, E_b) happens in time either $O(|E_a| \log n)$ or $O(|E_b| \log n)$, as indeed every "check" from the algorithm's description takes time $O(\log n)$, after $O(n \log n)$ initialization of the range search data structure or the ordered list represented by a balanced binary tree. So the runtime for processing the equivalence class list is $O(nN \log n)$.

Since reading the input takes $O(n^2)$ time and N is in $O(n)$ (this follows immediately from Theorem 2), we relax our bounds and say our space complexity is $O(n^2)$ and the overall runtime is $O(n^2 \log n)$.

5 Conclusions

As noted in Norige et al. [10, 11], these solutions do not generalize to dimensions higher than two. This is the most interesting open question, in our opinion.

We believe that range trees can be replaced by ad-hoc methods to obtain a $O(n^2)$ algorithm for exact RRL strip-rule minimization. The savings of $O(\log n)$ in running time comes at the expense of a more complicated algorithm which we decided not to present.

Acknowledgments. Ian's research was done while at Illinois Institute of Technology, and supported by the Brazil Scientific Mobility Program. Gruia's and Nathan's work was done while at Illinois Institute of Technology, and supported by the National Science Foundation under awards NSF-1461260 (REU).

References

1. Applegate, D., Călinescu, G., Johnson, D.S., Karloff, H.J., Ligett, K., Wang, J.: Compressing rectilinear pictures and minimizing access control lists. In: Bansal, N., Pruhs, K., Stein, C. (eds.) Proceedings of the Eighteenth Annual ACM-SIAM Symposium on Discrete Algorithms, SODA 2007, New Orleans, Louisiana, USA, 7–9 January 2007, pp. 1066–1075. SIAM (2007)
2. de Berg, M., van Kreveld, M., Overmars, M., Schwarzkopf, O.: Computational Geometry, 2nd edn. Springer, Heidelberg (2000). https://doi.org/10.1007/978-3-662-04245-8

3. Comerford, P., Davies, J.N., Grout, V.: Reducing packet delay through filter merging. In: Proceedings of the 9th International Conference on Utility and Cloud Computing, UCC 2016, pp. 358–363. ACM, New York (2016). https://doi.org/10.1145/2996890.3007854

4. Daly, J., Liu, A.X., Torng, E.: A difference resolution approach to compressing access control lists. IEEE/ACM Trans. Netw. **24**(1), 610–623 (2016). https://doi.org/10.1109/TNET.2015.2397393

5. Gouda, M.G., Liu, A.X.: Structured firewall design. Comput. Netw. **51**(4), 1106–1120 (2007). https://doi.org/10.1016/j.comnet.2006.06.015

6. Kang, N., Liu, Z., Rexford, J., Walker, D.: Optimizing the "One Big Switch" abstraction in software-defined networks. In: Proceedings of the Ninth ACM Conference on Emerging Networking Experiments and Technologies, CoNEXT 2013, pp. 13–24. ACM, New York (2013). https://doi.org/10.1145/2535372.2535373

7. Kang, N., Reich, J., Rexford, J., Walker, D.: Policy transformation in software defined networks. In: Proceedings of the ACM SIGCOMM 2012 Conference on Applications, Technologies, Architectures, and Protocols for Computer Communication, SIGCOMM 2012, pp. 309–310. ACM, New York (2012). https://doi.org/10.1145/2342356.2342424

8. Kogan, K., Nikolenko, S., Culhane, W., Eugster, P., Ruan, E.: Towards efficient implementation of packet classifiers in SDN/OpenFlow. In: Proceedings of the Second ACM SIGCOMM Workshop on Hot Topics in Software Defined Networking, HotSDN 2013, pp. 153–154. ACM, New York (2013). https://doi.org/10.1145/2491185.2491219

9. Liu, A.X., Meiners, C.R., Torng, E.: TCAM Razor: a systematic approach towards minimizing packet classifiers in TCAMs. IEEE/ACM Trans. Netw. **18**(2), 490–500 (2010). https://doi.org/10.1109/TNET.2009.2030188

10. Norige, E., Liu, A.X., Torng, E.: A ternary unification framework for optimizing TCAM-based packet classification systems. In: Proceedings of the Ninth ACM/IEEE Symposium on Architectures for Networking and Communications Systems, ANCS 2013, pp. 95–104. IEEE Press, Piscataway (2013)

11. Norige, E., Liu, A.X., Torng, E.: A ternary unification framework for optimizing TCAM-based packet classification systems. IEEE/ACM Trans. Netw. **26**(2), 657–670 (2018). https://doi.org/10.1109/TNET.2018.2809583

12. Pao, D., Lu, Z.: A multi-pipeline architecture for high-speed packet classification. Comput. Commun. **54**(C), 84–96 (2014). https://doi.org/10.1016/j.comcom.2014.08.004

13. Sun, Y., Kim, M.S.: Bidirectional range extension for TCAM-based packet classification. In: Crovella, M., Feeney, L.M., Rubenstein, D., Raghavan, S.V. (eds.) NETWORKING 2010. LNCS, vol. 6091, pp. 351–361. Springer, Heidelberg (2010). https://doi.org/10.1007/978-3-642-12963-6_28

14. Sun, Y., Kim, M.S.: Tree-based minimization of TCAM entries for packet classification. In: Proceedings of the 7th IEEE Conference on Consumer Communications and Networking Conference, CCNC 2010, pp. 827–831. IEEE Press, Piscataway (2010)

15. Zhang, Y., Natarajan, S., Huang, X., Beheshti, N., Manghirmalani, R.: A compressive method for maintaining forwarding states in SDN controller. In: Proceedings of the Third Workshop on Hot Topics in Software Defined Networking, HotSDN 2014, pp. 139–144. ACM, New York (2014). https://doi.org/10.1145/2620728.2620759

Algorithm Designs for Dynamic Ridesharing System

Chaoli Zhang, Jiapeng Xie, Fan Wu$^{(\boxtimes)}$, Xiaofeng Gao, and Guihai Chen

Shanghai Jiao Tong University, Shanghai, China
chaoli_zhang@sjtu.edu.cn, {fwu,gao-xf,gchen}@cs.sjtu.edu.cn

Abstract. A ridesharing system mitigates traffic congestion and car pollution by allowing passengers to share their travel cost with others. Nowadays, with the development of the smartphone technology, dynamic ridesharing systems enable passengers request a car anytime and anywhere. This paper mainly considers the problems of how to allocate passengers to drivers, how to charge the passengers and how to design feasible schedules for the driver in such online environment. The allocation problem is modeled as an online weighted matching problem with the graph changing over time. Firstly, we give a fair pricing method which is easy to be understood and accepted by the passengers. We develop a greedy algorithm called LIQMAX_GRE for the purpose of maximizing liquidity. The schedule problem which is similar with the hamiltonian path problem is NP-hard and we design a heuristic nearest neighbor algorithm to solve it.

1 Introduction

The widespread use of private cars has brought great convenience to our life compared with the public transport, but it also raises a lot of challenges, such as traffic congestion and environment pollution. Taxicab seems to be a good choice, but it is usually expensive. On the other hand, however, there are often many empty seats in a taxicab or a private car, resulting in a waste of seat resources. Thus, the demand for ridesharing has increased sharply in recent years [4]. A ridesharing system aims to bring together people with similar itineraries and time schedules, and offer rides for them [1,6]. It has generated more and more interest in recent years.

Researches on ridesharing systems, such as [13,16], usually considered offline scenarios, where the requests are reported to the platform in advance. However, with the development of wireless networks and the proliferation of smartphones, the online scenarios, where a passenger submits a request dynamically and the

This work was supported in part by the National Key R&D Program of China 2018YFB1004703, in part by China NSF grant 61672348, 61672353, and 61472252. The opinions, findings, conclusions, and recommendations expressed in this paper are those of the authors and do not necessarily reflect the views of the funding agencies or the government.

S. Tang et al. (Eds.): AAIM 2018, LNCS 11343, pp. 209–220, 2018.
https://doi.org/10.1007/978-3-030-04618-7_17

ridesharing system responds to the request within a limited time, are more common in practice. These new characteristics makes the system more convenient. However, it also brings more challenges to the design of an efficient ridesharing system as well.

In a ridesharing system, drivers and passengers usually have their own trip plan, that is, different origins and destinations, different departure time and arrival time limits. Different allocations result in rather different trip schedules which result in rather different detours. The ridesharing system has to take all of these into consideration and give an effective allocation design for their customers. There are different criterions for efficiency.

Liquidity is an important criterion for efficiency. It aims to satisfy as many customers as possible, which is very important to attract more people to participate in ridesharing.

In a real ridesharing situation, it is natural to consider the consumers are rational and want to be charged fairly. When different passengers are allocated to even the same driver, it results in rather different route designs. The detour problem is complex which means even only one passenger changes her origin or destination, the whole route design may change widely.

To charge the passengers fairly, we first need to figure out their influences to the route results. It is not as simple as it seems. For convenience and simplification, consider the route design in ridesharing as one kind of the travel salesman problems. Even in the classical TSP, a tiny change in travel points brings significant differences to the results.

1.1 Challenges

As have been showed above, allocation and pricing are the most important and basic parts in the ridesharing system which require special and elaborate designs.

For the allocation, we want to make the system efficient. How to measure the efficiency? How to achieve the efficiency goal? Liquidity criterion can be modeled as a kind of bipartite matching. Different from traditional bipartite matching, a driver can be allocated to several passengers as long as there are available seats in her car. It brings challenges to the design.

For the pricing, we want to charge the passengers fairly and make sure the driver be budget-balanced at the same time. It is complex to design and measure the fairness of the pricing method with desirable properties.

Besides that, when the online dynamic ridesharing system is considered, the significant challenge is that, the input data, *i.e.*, the request of the users, are revealed over time. If a new request is accepted, the system needs to allocate a driver to it. It needs to design a new schedule which satisfies not only the constraints of this new request, but also the constraints of the requests the driver is maintaining.

How to do the matching between the passengers and the drivers in the online situation? How to design a feasible schedule for the drivers in real-time? How to fairly charge each passenger with the constraint that their payment doesn't exceed the cost without ridesharing? What's more, as the passengers are usually sensitive to the price, the pricing method should be easy to understood and accepted.

1.2 Our Contributions

i. We model the matching process as an online weighted matching problem [9] with the whole graph changing over time. That is different from existing works.

ii. For pricing design, we give a sectional charging method where each active passenger shares the cost of each section of distance proportional to their requesting distances. This design is based on the feasible schedule design. It achieve fairness in some straight-forward aspects. That is, one's payment is mainly decided by their travel distance. It is easy to be understood and accepted by the passengers.

iii. For the liquidity maximization, we give a greedy algorithm for the online matching and prove that the competitive ratio is $\frac{1}{\lambda+1}$, where λ is the maximal number of passengers a car can take.

iv. For schedule design, we give a Nearest Neighbour Schedule for the schedule problem where the driver always drives to the nearest positions contained in the active requests. We also make such feasible schedule satisfy the guarantee of non-negative utility for each participant.

The rest of the paper is organized as follows. We briefly review the related work in Sect. 2. In Sect. 3, We introduce the model of the dynamic ridesharing system, and give some important definitions. In Sect. 4, we present our detailed designs, including a charging method, a matching algorithms and a schedule algorithm. Finally, we conclude this paper in Sect. 5.

2 Related Work

The problem of dynamic ridesharing has been extensively studied in recent years [15,18,19]. Riquelme *et al.* focused on the incentives of both drivers and passengers [2,8]. They build a queueing-theoretic economic model and studied optimal pricing strategies for ridesharing platform. Kleiner *et al.* designed a mechanism for dynamic ridesharing based on parallel auctions [12]. They are the first to present an auction-based solution for dynamic ridesharing system. Kamar and Horvitz present a methodology for determining ridesharing formation and a fair payment mechanism for the ridesharing platform. Zhao *et al.* studied the well known VCG mechanism and prove it results in a very high deficit [20]. They then proposed an inefficient mechanism but with deficit control and considered a VCG mechanism with two-sided reserve prices. Fang *et al.* studied the pricing and subsidies in ridesharing [5].

How to design an efficient schedule for a car also attracts a lot of attention. Santos and Xavier focused on the dynamic taxi sharing with time windows problem and showed its NP-Hardness by reducing the metric hamiltonian path problem to it [17]. They gave a greedy randomized adaptive search procedure to solve it. Huang *et al.* introduced two approaches: Branch-And-Bound and Mixed Integer Programming algorithms [7]. Then, they gave a kinetic tree structure to maintain the previous computations. However, he failed to consider the budget constraint of passengers.

In this paper, we mostly focus on the allocation of cars to passengers, which can be modeled as an online matching problem. Previously, plenty of works have been done in this real-time situation. Karp *et al.* introduced the online bipartite matching in 1990 and introduced a Ranking algorithm, which can achieve a competitive ratio of $1 - 1/e$ [11]. Kalyanasundaram and Pruhs studied an online weighted greedy matching problem which chooses the available point with the largest weight to handle the current request [9]. They proved that the Farthest Neighbour algorithm produce a matching of weight at least $1/3$ the maximum weight perfect matching and the bound is tight. Blum *et al.* studied the market clearing problem, which can be modeled as a matching problem [3]. They considered two objects: profit and liquidity. Adwords problem is also a generalization of the online bipartite matching problem where each advertiser puts in a set of bid values for keywords and a budget, representing the maximum he can afford [14]. Kalyanasundaram and Pruhs found the trade-off between the bid and unspent budget and proposed a BALANCE algorithm [10]. Except the greedy algorithm, A BALANCE algorithm is introduced, which matches a keyword with the advertiser which has spent the least fraction of its budget so far. This is an optimal algorithm for online b-MATCHING problem, a special case of the Adwords problems [10,14].

3   Preliminaries

In this section, we present the model of a dynamic ridesharing system and give some related definitions. To make our paper easy to follow, we list the frequently used notations in Table 1.

3.1   Ridesharing System Overview

We consider a ridesharing system with a central platform, a set of registered drivers $\mathbb{D} = \{D_1, D_2, ..., D_M\}$ and a set of registered passengers $\mathbb{P} = \{P_1, P_2, ..., P_N\}$. All drivers and passengers are called participants in this ridesharing system. With the registration of the participants, such as genders, ages, etc, the central platform can combine the participants from more aspects. Thus, the user experience can be improved.

In this paper, we focus on providing services for passengers. The drivers and the central platform are on the same side and the drivers only follow schedules made by the platform to serve the passengers. We divide the total time into T time slots, *i.e.*, $\mathbb{T} = \{1, 2, ..., T\}$. Requests arrive over a sequence of time slots. Let s_i and d_i indicate the origin and destination of passenger i respectively.

Table 1. Frequently used notations

Notation	Remark
$\mathbb{D}$	Set of cars
$\mathbb{P}$	Set of passengers
$\mathbb{T}$	Set of time slots
r_i	The request of passenger i
s_i	The origin of passenger i
d_i	The destination of passenger i
t_i^1	The departure time limit of passenger i
t_i^2	The arrival time limit of passenger i
$\mathbb{C}_i$	The collaborator tuple of driver i

Let t_i^1 and t_i^2 indicate the departure time and arrival time limits respectively. Then, a request r_i of passenger i is denoted as $r_i = \{s_i, d_i, t_i^1, t_i^2\}$. When a request r arrives at time $t \in \mathbb{T}$, the platform determines whether to select a car to serve her. A framework is shown in Fig. 1.

Fig. 1. A ridesharing system framework

With the help of the map application, the distance between any two locations can be calculated. Denote this distance function as $Dis(loc_1, loc_2) \in \mathbb{R}^+$ and the time cost can be estimated as $T(loc_1, loc_2) \in \mathbb{R}^+$. We make an assumption that the time cost between two locations can be calculated as well and the time map function is denoted as $T(loc_1, loc_2) \in \mathbb{R}^+$.

We assume both of the distance function and the time function are symmetric. Let a denote the unit cost in distance for a driver and b denote the payment of unit distance without ridesharing for a passenger. Thus, for a passenger P_i, her monetary utility of participating in the ridesharing is

$$U(P_i) = b * Dis(s_i, d_i) - Pay_i,$$

Pay_i is her payment to this platform. For a driver, his monetary utility is

$$U(D_i) = \sum_{i \in \{passengers\ carried\}} Pay_i - a * (driving\ distance).$$

We define the system utility as the total saving vechile miles compared with no ridesharing. An efficient ridesharing system should promise each participant a non-negative monetary utility.

For a driver, in the setting of no ridesharing, he takes passengers one by one. However, in the setting of ridesharing, he can pick up a passenger and then drivers to another starting point and picks up another passenger, rather than deliveries the first passenger immediately. Our object is to design a ridesharing system aiming at maximizing the liquidity.

3.2   Basic Definitions

Definition 1 (Active Request). *We call a request $r = \{s, d, t^1, t^2\}$ is active between the time when it is accepted or matched with a driver and the time when this passenger is travelled to her destination.*

Definition 2 (Collaborator Tuple). *Several passengers with active requests are set to be in the same collaborator tuple if they share the same car. The collaborator tuple of driver i is denoted as $\mathbb{C}_i$.*

Definition 3 (Active Driver). *We call a driver is active when his car contains empty seats. The information of an active driver is denoted as $D_i = \{loc_i, \mathbb{C}_i, schedule, \eta_i\}$, where loc_i is his current location and η_i indicates the number of empty seats. A schedule is a sorted list of locations he will travel. For example, if this driver has l active requests $r_1, r_2, ..., r_l$ now, his schedule may look like $loc, s_1, s_2, ..., s_l, d_1, d_2, ..., d_l$ where loc is the current location of the driver.*

A schedule is feasible if it satisfies all the constraints of passengers in the collaborator tuple. *i.e.*,

$$\forall P_j \in \mathbb{C}, t_0 + T(loc, s_j) \leq t_j^1$$
$$\forall P_j \in \mathbb{C}, t_0 + T(loc, d_j) \leq t_j^2$$

In addition, the starting point of each passenger should be head of his destination and each participant should earn a non-negative monetary utility. The most thorny and important part of a dynamic ridesharing system is how to match passengers with cars and how to design a best feasible schedule. We also design an elaborate charging method to ensure a fair share of traveling expenses. The details are given in the next section.

4   System Design

In this section, we propose our detailed design for matching passengers with drivers, scheduling the drivers and fairly charging the passengers.

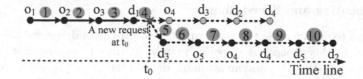

Fig. 2. Pricing example

Algorithm 1. CalPayment

Input : A schedule $(x_0(= loc), x_1, ..., x_{2l})$, Request profiles of a collaborator
tuple $\{r_1, r_2, ..., r_l\}$, Budgets of the passengers $\{b_1, b_2, b_3, ..., b_l\}$.
Output: Payment of each passenger

1 $(dis_1, dis_2, ..., dis_l)$=$(\text{Dis}(s_1, d_1), \text{Dis}(s_2, d_2), ..., \text{Dis}(s_l, d_l))$;
2 $(c_1, c_2, ..., c_l) = (0, 0, ..., 0)$;
3 **for** $i = 1$ *to* $i = 2l$ **do**
4 $\quad$ $Dis = \text{Dis}(x_{i-1}, x_i)$;
5 $\quad$ **for** *each active* r_i **do**
6 $\quad\quad$ $c_i = c_i + b_i * \dfrac{dis_i}{\sum\limits_{all\ active\ i} dis_i} \times Dis$;
7 $\quad\quad$ $b_i = b_i - c_i$;
8 $\quad$ **if** x_i *represents some* d_j **then**
9 $\quad\quad$ set r_j inactive;
10 **return** $c_1, c_2, ..., c_l$;

4.1 Pricing

In this subsection, we give a method to fairly calculate each passenger's payment to the driver. We first give an example (Fig. 2) to explain our charging method.

The driver at the location loc_1 has a collaborator tuple $\{P_1, P_2, P_3, P_4\}$ and a feasible schedule $(loc_1, s_1, s_2, s_3, d_1, s_4, d_3, d_2, d_4)$. At position loc_2, she is allocated a new request $r_5 = \{s_5, d_5, t_5^1, t_5^2\}$. Then his schedule is updated. Assume the new schedule is $(loc_2, d_3, s_5, s_4, d_4, d_5, d_2)$. We separate the driving cost among the passengers whose request is active. Each passenger pays the money proportional to his distance between starting point and destination in his request.

We give Algorithm 1 to calculate each passenger's payment. Passenger i' budget is b_i. Remaining budget is the amount a passenger can afford for his rest travel. With this charging method, a feasible schedule should not only satisfy the time constraints but also guarantee a non-negative utility for each participant. In the next section, we will give the method to allocate passengers with drivers and design feasible schedules for them.

4.2   Allocating and Scheduling

Whenever a request comes, the platform tries to provide a driver for the passenger. We model this process as an online match. At time t, assume there is a new request $r^t = \{s^t, d^t, t^1, t^2\}$ and an available drivers set $D^t = \{d^t_1, d^t_2, \ldots, d^t_n\}$. For the purpose of writing convenience, we omit t here in the case of no confusion, i.e., $r = \{s, d, t^1, t^2\}$ and $D = \{d_1, d_2, \ldots, d_n\}$. To solve this online matching problem, we should solve two sub-problems, i.e.,

- When can a passenger's request be accepted?
- How to do an optimal allocation, or an allocation with guaranteed performance in the online setting?

4.3   When to Accept a Request?

Assume at time t_0, there is a passenger and a driver, denoted as P and D. The request of passenger P is $r = \{s, d, t^1, t^2\}$ and the information of driver D is $\{loc, \mathbb{C}, schedule, \lambda\}$. Assume $\mathbb{C}$ consists of $u + v$ passengers and their remaining budgets are $b_1, b_2, \ldots, b_{u+v}$ respectively. The first u passengers have been picked up but haven't arrived and the last v passengers are still waiting for the car. Denote the starting points of waiting passengers as $s_{u+1}, s_{u+2}, \ldots, s_{u+v}$ and all the destinations as $d_1, d_2, \ldots, d_u, \ldots, d_{u+v}$.

A schedule is a permutation of these positions, including initial position loc of the driver and positions in the requests. If there exists a permutation of these positions satisfying all constraints, this passenger can be matched with this driver. Denote all positions (including the initial position of the driver and the starting points, the destinations of the drivers' requests) as

$$x_0, x_1, \ldots, x_u, x_{u+1}, \ldots, x_{u+v+1}, x_{u+v+2}, \ldots, x_{u+2v+2},$$

where x_i, $i = 1, \ldots, u + v + 1$ stands for the destination, x_i, $i = 0, u + v + 2, \ldots, u + 2v + 2$ stands for the starting point. x_{u+v+1}, x_{u+2v+2} are the starting point and destination of the new passenger's request. A schedule is equivalent to a permutation

$$<0, \sigma(1), \ldots, \sigma(u + 2v + 2)>$$

of

$$<0, 1, \ldots, u + 2v + 2>.$$

Under the assumption of knowing the travel time relationship of the positions in a map, the time series of this schedule is

$$<t_{\sigma(0)}, t_{\sigma(1)}, \ldots, t_{\sigma(u+2v+2)}>$$

correspondingly, where we set $\sigma(0) = 0$.

We show this problem formally,

$$\min_{\sigma} t_{\sigma(u+2v+2)}$$

subject to

$$t_{\sigma(i)} = \sum_{j=0}^{i-1} T(x_{\sigma(j)}, x_{\sigma(j+1)}) \qquad \forall i \tag{1}$$

$$t_{u+i} > t_{u+i+v+1}, \qquad \forall i = 1, 2, ..., v+1 \tag{2}$$

$$t_i < t_i^2, \qquad\qquad \forall i = 1, 2, ..., u+v+1 \tag{3}$$

$$t_{u+2+i} < t_{u+2+i}^1, \qquad \forall i = v, ..., 2v \tag{4}$$

$$CalPayment(schedule) \leq (b_1, ..., b_{u+v+1}), \tag{5}$$

where function T defines the travel time relationship corresponding to positions. Constraint (2) restricts the order of the starting point and the destination. Constraint (3)(4) meet the time constraints in the requests. Constraint (5) guarantees a non-negative utility for each passenger.

This problem can be reduced to the well-known hamiltonian path problem [17] which an NP-Hard problem. Here we introduce a NEAREST NEIGHBOUR heuristic algorithm Mechanism 2, which picks the next vertex closest to current vertex and does not violate the precedence constraint. If it can't meet the constraint (5), the procedure terminates and returns false.

In each iteration we choose a nearest position from the current position and calculate the corresponding cost, allocate it to all passengers. If it exhausts a passenger's budget, the procedure terminates and returns false. After the iteration, it will check other constraints.

Whether a passenger can be allocated has been decided, then, which driver should be matched with this passenger? In the following designs, we want to maximize the liquidity and the utility respectively.

4.4   How to Make an Optimal Allocation?

Allocating a car to the passenger can be modeled as a bipartite matching naturally. Let G be a dynamic graph with one bipartition designated as the drivers, and the other bipartition designated as the passengers. A new request arrives along with incident edges. An incident edge indicates that a passenger can be matched with a driver. In most of the existing literatures, graphs are studied as static objects, while graphs are subject to discrete changes in a dynamic setting. However, in this paper, this problem is far more complicated than before. With the changing of a driver's schedule, the incident edges with remaining passengers change as well.

Liquidity Maximization: We design the algorithm LIQMAX_GRE to maximize the liquidity.

Theorem 1. *The LIQMAX_GRE algorithm has a competitive ratio of $\frac{1}{\lambda+1}$, where λ is the maximum number of passengers a car can take.*

Algorithm 2. Scheduler

Input : locations $(x_0, x_1, \ldots, x_{u+2v+2})$, request profile of a collaborator tuple
$\{r_1, r_2, \ldots, r_{u+v+1}\}$, remaining budget $(b_1, b_2, \ldots, b_{u+v+1})$

Output: Schedule result

1   $(dis_1, dis_2, \ldots, dis_{u+v+1}) = (\text{Dis}(s_1, d_1), \text{Dis}(s_2, d_2), \ldots, \text{Dis}(s_{u+v+1}, d_{u+v+1}))$;

2   **for** $i = 1$ *to* $u + v + 1$ **do**

3     $active(i) = true$;

4   $Loc = 0 = \sigma(0)$;

5   **for** $i = 1$ *to* $u + 2v + 2$ **do**

6   choose the nearest unvisited feasible neighbour $x_{\sigma(i)}$ of position x_{Loc};

7     $t_{\sigma(i)} = t_{Loc} + T(x_{\sigma(i)}, x_{Loc})$;

8     $Loc = \sigma(i)$;

9     $Dis = Dis(x_{\sigma(i-1)}, x_{\sigma(i)})$;

10   **for** *each active* r_j **do**

11       $c_j = c_j + b_j * \dfrac{dis_j}{\sum\limits_{all\ active\ i} dis_j} \times Dis$;

12   **if** $c_j > b_j$ **then**

13   **return** $false$;

14   **if** x_{Loc} *is the destination of some* r_k **then**

15       $active(k) = false$;

16   **for** $i = 1$ *to* $u + v + 1$ **do**

17   **if** $t_i > t_i^2$ **then**

18   **return** $false$;

19   **for** $i = v$ *to* $2v$ **do**

20   **if** $t_{u+2+i} > t_{u+2+i}^1$ **then**

21   **return** $false$;

22   **return** $\sigma, < x_{\sigma(0)}, x_{\sigma(1)}, \ldots, x_{\sigma(u+2v+2)} >$

Algorithm 3. LIQMAX_GRE

Input : All participants' profiles
Output: Matching Result

1   When the next passenger $P \in \mathbb{P}$ arrives:

2   Match P to any available driver(if any);

3   update the schedules of drivers;

Proof. The competitive ratio is used to compare the performance of online algorithm with the performance of the optimal algorithm. The definition of the competitive ratio is:

$$\min_{\overrightarrow{r}, \sigma} \frac{W(\overrightarrow{r}, \sigma)}{W^*(\overrightarrow{r}, \sigma)}$$

where $\vec{r}$ and σ are the profiles of all requests and their arriving order respectively. W is the result of our algorithm and W^* is the optimal result.

Assume in the maximum matching, the matched passenger set is

$$\mathbb{S}_P = \{P_1, P_2, ..., P_n\}$$

and the matched driver set is

$$\mathbb{S}_D = \{D_1, D_2, ..., D_m\}.$$

So the optimal algorithm returns n. LIQMAX_GRE outputs x. Denote the matched passenger set as $\mathbb{S}_P^1$ and its complement is denoted as $\mathbb{S}_P^2$. The corresponding driver is denoted as $\mathbb{S}_D^1$ and $\mathbb{S}_D^2$, where the drivers in $\mathbb{S}_D^1$ is matched with some passengers in $\mathbb{S}_P^1$. Then we have:

$$\mathbb{S}_P^1 \cup \mathbb{S}_P^2 = \mathbb{S}_P, \ \mathbb{S}_D^1 \cup \mathbb{S}_D^2 = \mathbb{S}_D.$$

Let $|\mathbb{S}_P^1| = x$, then

$$\frac{x}{\lambda} \leq |\mathbb{S}_D^1| \leq x.$$

The inequality holds because a car can take at least one passenger and up to λ passengers. Since $|\mathbb{S}_P^1| + |\mathbb{S}_P^2| = n$, then $|\mathbb{S}_P^2| = n - x$. These passengers fail to be matched because the drivers matched with them in the maximum matching are matched with passengers in $\mathbb{S}_P^1$ in our algorithm. So we have $\mathbb{S}_D^1 \geq \frac{n-x}{\lambda}$, then $x \geq \frac{n-x}{\lambda}$. So we have $x \geq \frac{n}{\lambda+1}$.

We have proved that the competitive ratio is $\frac{1}{\lambda+1}$.

5   Conclusion

In this paper, we have studied the dynamic ridesharing problem, where a passenger requests for a car at any time and any position. We have designed a charging method, a schedule algorithm and two allocation algorithms. In the charging method, active passengers share the driving cost proportional to their distances. It is easy to be understood and accepted by the passengers. In the schedule algorithm, we always choose the nearest position from the current position. Our objectives are maximizing the liquidity. We have proposed an algorithm LIQMAX_GRE, which can achieve a competitive ratio of $\frac{1}{\lambda}$.

References

1. Agatz, N., Erera, A.L., Wang, X.: Dynamic ride-sharing: a simulation study in metro atlanta. Transp. Res. Part B-Methodol. **45**(9), 1450–1464 (2011)
2. Banerjee, S., Johari, R., Riquelme, C.: Pricing in ride-sharing platforms: a queueing-theoretic approach. In: EC, pp. 639–639. ACM (2015)
3. Blum, A., Sandholm, T., Zinkevich, M.: Online algorithms for market clearing. J. ACM **53**(5), 845–879 (2006)

4. Deakin, E., Frick, K.T., Shively, K.: Markets for dynamic ridesharing. Transp. Res. Rec. **2187**, 131–137 (2011)
5. Fang, Z., Huang, L., Wierman, A.: Prices and subsidies in the sharing economy. In: Proceedings of the 26th International Conference on World Wide Web, pp. 53–62. International World Wide Web Conferences Steering Committee (2017)
6. Feuerstein, E., Stougie, L.: On-line single-server dial-a-ride problems. Theor. Comput. Sci. **268**(1), 91–105 (2001)
7. Huang, Y., Bastani, F., Jin, R., Wang, X.S.: Large scale real-time ridesharing with service guarantee on road networks. Proc. VLDB Endow. **7**(14), 2017–2028 (2014)
8. Jacob, J., Roet-Green, R.: Ride solo or pool: the impact of sharing on optimal pricing of ride-sharing services (2017)
9. Kalyanasundaram, B., Pruhs, K.: Online weighted matching. J. Algorithms **14**(3), 478–488 (1993)
10. Kalyanasundaram, B., Pruhs, K.: An optimal deterministic algorithm for online b-matching. Theor. Comput. Sci. **233**, 319–325 (2000)
11. Karp, R.M., Vazirani, U.V., Vazirani, V.V.: An optimal algorithm for on-line bipartite matching. In: STOC, pp. 352–358. ACM (1990)
12. Kleiner, A., Nebel, B., Ziparo, V.A.: A mechanism for dynamic ride sharing based on parallel auctions. In: IJCAI. AAAI (2011)
13. Kubo, M., Kasugai, H.: Heuristic algorithms for the single vehicle dial-a-ride problem. J. Oper. Res. Soc. Jpn. **33**(4), 354–365 (1990)
14. Mehta, A., Saberi, A., Vazirani, U., Vazirani, V.: Adwords and generalized online matching. J. ACM **54**(5) (2007)
15. Neoh, J.G., Chipulu, M., Marshall, A.: What encourages people to carpool? An evaluation of factors with meta-analysis. Transportation **44**(2), 423–447 (2017)
16. Psaraftis, H.N.: Analysis of an $o(N^2)$ heuristic for the single vehicle many-to-many Euclidean dial-a-ride problem. Transp. Res. Part B-Methodol. **17**(2), 133–145 (1983)
17. Santos, D.O., Xavier, E.C.: Dynamic taxi and ridesharing: a framework and heuristics for the optimization problem (2013)
18. Tian, C., Huang, Y., Liu, Z., Bastani, F., Jin, R.: Noah: a dynamic ridesharing system. In: SIGMOD. ACM (2013)
19. Zhang, D., Li, Y., Zhang, F., Lu, M., Liu, Y., He, T.: coRide: carpool service with a win-win fare model for large-scale taxicab networks. In: SenSys. ACM (2013)
20. Zhao, D., Zhang, D., Gerding, E.H., Sakurai, Y., Yokoo, M.: Incentives in ridesharing with deficit control. In: AAMAS. International Foundation for Autonomous Agents and Multiagent Systems (2014)

New LP Relaxations for Minimum Cycle/Path/Tree Cover Problems

Wei Yu[1](✉), Zhaohui Liu[1], and Xiaoguang Bao[2]

[1] Department of Mathematics, East China University of Science and Technology,
Shanghai 200237, China
{yuwei,zhliu}@ecust.edu.cn
[2] College of Information Technology, Shanghai Ocean University, Shanghai 201306,
China
xgbao@shou.edu.cn

Abstract. Given an undirected complete weighted graph $G = (V, E)$ with nonnegative weight function obeying the triangle inequality, a set $\{C_1, C_2, \ldots, C_k\}$ of cycles is called a *cycle cover* if $V \subseteq \bigcup_{i=1}^{k} V(C_i)$ and its cost is given by the maximum weight of the cycles. The Minimum Cycle Cover Problem (MCCP) aims to find a cycle cover of cost at most λ with the minimum number of cycles. We propose new LP relaxations for MCCP as well as its variants, called the Minimum Path Cover Problem (MPCP) and the Minimum Tree Cover Problem, where the cycles are replaced by paths or trees. Moreover, we give new LP relaxations for a special case of the rooted version of MCCP/MPCP and show that these LP relaxations have significantly better integrality gaps than the previous relaxations.

Keywords: Vehicle routing · Cycle cover · Path cover
Approximation algorithm · Integrality gap

1 Introduction

Given an undirected complete graph $G = (V, E)$ with metric weight function $w : V \times V \to \mathbb{N}$ that is nonnegative, symmetric and obeys the triangle inequality, a set $\{C_1, C_2, \ldots, C_k\}$ of cycles is called a *cycle cover* if $V \subseteq \bigcup_{i=1}^{k} V(C_i)$ and the cost of a cycle cover is given by the maximum weight of the cycles. The goal of the Minimum Cycle Cover Problem (MCCP) is to find a cycle cover of cost at most λ with the minimum number of cycles. By replacing the cycles with paths and trees we obtain the Minimum Path Cover Problem (MPCP) and the Minimum Tree Cover Problem (MTCP), respectively.

When the vertices represent customers to be served by a fleet of vehicles and the cycles or paths correspond to the travel routes of the vehicles, MCCP and MPCP are exactly the most fundamental model of vehicle routing problems (VRPs) [8,15]. These problems and their variants (min-max

© Springer Nature Switzerland AG 2018
S. Tang et al. (Eds.): AAIM 2018, LNCS 11343, pp. 221–232, 2018.
https://doi.org/10.1007/978-3-030-04618-7_18

cycle/path/tree cover problem, rooted minimum cycle/path/tree cover problem etc.) have attracted considerable research attention due to their widespread applications in both operations research and computer science communities. Typical applications include mail and newspaper delivery [5], nurse station location [4], disaster relief efforts routing [3], distance-constrained vehicle routing [14], data gathering and wireless recharging in wireless sensor networks [17], multi-vehicle scheduling problem [2,10], political districting [9], and so on.

Due to the NP-Completeness of the Hamiltonian Cycle/Path Problem [7], the problems MCCP/MPCP cannot be approximated within a ratio less than 2 unless P=NP. The results in [14] imply that MTCP has an inapproximability lower bound of 3/2. Therefore, it is unlikely that these problems admit PTASes.

Arkin et al. [1] first presented a 3-approximation algorithm for both MTCP and MPCP, which implies a 6-approximation algorithm for MCCP by a simple edge-doubling strategy. Khani and Salavatipour [11] derived an improved 5/2-approximation algorithm for MTCP, which implies a 5-approximation algorithm for MCCP. Yu and Liu [19] showed that a ρ-approximation algorithm for the well-known Traveling Salesman Problem can be transformed into a 4ρ-approximation algorithm for MCCP. They also proposed a matching-based 14/3-approximation algorithm for MCCP, which was improved to an algorithm with approximation ratio 32/7 by Yu et al. [20].

In the rooted version of MCCP (MTCP), called Rooted MCCP (Rooted MTCP), there is a depot $r \in V$ and each cycle (tree) in the cycle cover (tree cover) has to contain r. The Rooted MPCP can be defined similarly except that this time each path is required not only to contain r but also to start from r. The above-mentioned inapproximability results for MCCP/MPCP/MTCP also apply to their rooted versions. Nagarajan and Ravi [13] showed that Rooted MCCP and Rooted MPCP are within a factor of two in terms of approximability. Nagarajan and Ravi [12,13] developed an $O(\min\{\log n, \log \lambda\})$-approximation algorithm for Rooted MPCP, which was improved to $O(\min\{\log n, \frac{\log \lambda}{\log \log \lambda}\})$ by Friggstad and Swamy [6]. Whether Rooted MCCP/MPCP admits a constant-factor approximation algorithm is a major open problem in this area (see Nagarajan and Ravi [14]).

However, constant-factor approximation algorithms are proposed for a special case, called the Rooted MCCP/MPCP on a tree, where G is the metric closure of a weighted tree graph T. For Rooted MCCP on a tree, Nagarajan and Ravi [14] developed a 2-approximation algorithm showed an inapproximability lower bound of 3/2. For Rooted MPCP on a tree, Nagarajan and Ravi [12] devised a 4-approximation algorithm.

We note that most of the existing algorithms are purely combinatorial algorithms. In contrast, there are few results on algorithms based on the integer programming formulations and their linear programming relaxations. LP-based approaches (e.g. LP-rounding, primal-dual method) are powerful techniques on the design of approximation algorithms for many combinatorial optimization problems [16]. The success of an LP-based algorithm relies crucially on the integrality gap of the LP relaxations on the problem in study. As a result, the first step to derive an efficient LP-based algorithm is to come up with an linear programming relaxation with provable good integrality gaps.

For Rooted MPCP, the above-mentioned results in [12,13] and [6] also imply that the integrality gap of a natural set-covering LP relaxation is at most $O(\min\{\log n, \log \lambda\})$ and $O(\min\{\log n, \frac{\log \lambda}{\log \log \lambda}\})$, respectively. For Rooted MPCP on a tree, Nagarajan and Ravi [12] proved that the integrality gap of the same LP relaxation has an upper bound of 64. Nagarajan and Ravi [13] considered similar set-covering LP relaxations for MPCP and Rooted MCCP on a tree and showed that the integrality gaps are at most 17 and 20, respectively. Therefore, there is a large gap between the best available approximation ratios and the best upper bounds on the integrality gap for (Rooted) MCCP/MPCP.

In this paper, we significantly narrow these gaps by giving new LP relaxations whose integrality gaps almost match the best available approximation algorithms for MCCP/MPCP/MTCP and Rooted MCCP/MPCP on a tree. To be specific, we obtain an LP relaxation for MCCP with integrality gap at most 6 and an LP relaxation for MPCP/MTCP whose integrality gap is bounded by 4. For Rooted MCCP on a tree we derive an LP relaxation with integrality gap at most 5/2 and for Rooted MPCP on a tree we propose an LP relaxation with integrality gap no more than 5. We achieve these results by exploiting the problem structure and adding powerful valid inequalities in the new relaxations.

The rest of the paper is organized as follows. We formally state the problem and give some preliminary results in Sect. 2. In Sect. 3 we treat Rooted MCCP/MPCP on a tree, which is followed by the discussion on MCCP/MPCP/MTCP in Sect. 4.

2 Preliminaries

Given an undirected weighted graph $G = (V, E)$ with vertex set V and edge set E, $w(e)$ denotes the weight or length of edge e. If $e = (u, v)$, we also use $w(u, v)$ to denote the weight of e. For $B > 0$, $G[B]$ denotes the subgraph of G obtained by removing all the edges in E with weight greater than B. For a subgraph G' of G, the graph obtained by adding some copies of the edges in G' is called a multi-subgraph of G. For a (multi-)subgraph H (e.g. tree, cycle, path) of G, let $V(H), E(H)$ be the vertex set and edge set of H, respectively. The weight of H is defined as $w(H) = \sum_{e \in E(H)} w(e)$. If H is a multi-subgraph, $E(H)$ is a multi-set of edges and the edges appearing multiple times contribute multiply to $\sum_{e \in E(H)} w(e)$. If H is connected, let $MST(H)$ be the minimum spanning tree on $V(H)$ and its weight $w(MST(H))$ is simplified to $w_T(H)$.

A cycle C is also called a tour on $V(C)$. A cycle (tree) that contains some special vertex $r \in V$, called the depot, is referred to as an r-cycle (r-tree). An r-path is a path starting from r. A set $\{C_1, \ldots, C_k\}$ of cycles is called an cycle cover if $V \subseteq \bigcup_{i=1}^{k} V(C_i)$. And the cost of this cycle cover is defined as $\max_{1 \leq i \leq k} w(C_i)$, i.e., the maximum weight of the cycles. If each C_i is an r-cycle, $\{C_1, \ldots, C_k\}$ is called an r-cycle cover. By replacing cycles with paths (trees) we can define *path cover* (*tree cover*) or r -*path cover* (r-*tree cover*) and their cost similarly.

We formally state the problems to be studied as follows.

In the Minimum Cycle Cover Problem (MCCP), we are given $\lambda > 0$, an undirected complete graph $G = (V, E)$ and a metric weight function $w : E \rightarrow \mathbb{N}$ that is nonnegative, symmetric and obeys the triangle inequality, the aim is to find a cycle cover of cost at most λ with the minimum number of cycles.

For the Rooted Minimum Cycle Cover Problem (RMCCP), a depot $r \in V$ is specified in addition to the input of MCCP, and the goal is to find an r-cycle cover of cost at most λ with the minimum number of cycles.

By replacing the cycles in MCCP with paths, we obtain the Minimum Path Cover Problem (MPCP). Similarly, the Rooted Minimum Path Cover Problem (RMPCP) is derived by substituting r-paths for r-cycles in RMCCP.

The RMCCP on a tree (RMPCP on a tree) is a special case of RMCCP (RMPCP) where G is the metric closure of a weighted tree $T = (V, E)$. For an edge e (a vertex u) and a vertex v in T, v is called below e (u) if the unique path from r to v passes e (u). A set $V' \subseteq V$ of vertices is called below e (u) if each vertex in V' is below e (u).

Given an instance of MCCP (MPCP) or its rooted version, we call each cycle (path) in the optimal solution an optimum cycle (path). By the triangle inequality, we can assume w.l.o.g that any two optimum cycles (paths) are vertex-disjoint. We use n to denote the number of vertices of G. If (IP) is an integer programming formulation for the MCCP (MPCP) or its rooted version, we denote by OPT_{IP} the optimal value of (IP). OPT_{LP} is defined similarly for an LP relaxation (LP) for the problem.

The following cycle-splitting result on breaking a long cycle into a series of short paths is very useful. The basic idea is to add the edges greedily to a path along the cycle and throw out the last edge once this path has a length more than the target value.

Lemma 1. *[1,5,18] Given a tour C on V' and $B > 0$, we can split the tour into $\lceil \frac{w(C)}{B} \rceil$ paths of length at most B such that each vertex is located at exactly one path in $O(|V'|)$ time.*

3   Tree Metric

In this section we deal with RMCCP/RMPCP on a tree. We present new LP relaxations for both problems with integrality gaps at most 5/2 and 5, respectively. In contrast, the upper bounds on the integrality gaps of the LP relaxations in [12,13] are 20 for RMCCP on a tree and 64 for RMPCP on a tree. Moreover, we also give an example to show that the integrality gaps of both LP relaxations we proposed are at least 2.

3.1   Rooted Minimum Cycle Cover

Given an instance of RMCCP consisting of $G = (V, E)$ with depot r and $\lambda > 0$, let $\mathcal{C}$ be the set of all r-cycles of length at most λ. Nararajan and Ravi [13] investigated the following set-covering integer programming (IPC) for RMCCP, in which a binary variable x_C is associated with each r-cycle $C \in \mathcal{C}$.

$$\min \sum_{C \in \mathcal{C}} x_C$$

$$s.t. \quad \sum_{C \in \mathcal{C}: v \in V(C)} x_C \geq 1, \ \forall v \in V \setminus \{r\} \qquad (IPC)$$

$$x_C \in \{0, 1\}, \ \forall C \in \mathcal{C},$$

where the first constraint is to ensure that each non-depot vertex is covered by at least one r-cycle in $\mathcal{C}$. The corresponding LP relaxation (LPC) is obtained by neglecting the integral constraints on the variables.

$$\min \sum_{C \in \mathcal{C}} x_C$$

$$s.t. \quad \sum_{C \in \mathcal{C}: v \in V(C)} x_C \geq 1, \ \forall v \in V \setminus \{r\} \qquad (LPC)$$

$$x_C \geq 0, \ \forall C \in \mathcal{C},$$

Note that the constraints $x_C \leq 1$ for all $C \in \mathcal{C}$ is unnecessary due to the first constraint and the minimization objective.

Nagarajan and Ravi [13] proved that the integrality gap of (LPC) is at most $O(\min\{\log n, \log \lambda\})$ for the general RMCCP and is bounded by 20 for RMCCP on a tree $T = (V, E)$. As noted by Nagarajan and Ravi [13,14], we can assume without loss of generality that T is a binary tree rooted at r (otherwise one can add some dummy vertices and zero-weight edges).

A crucial concept in their proof on the tree metric is so-called *heavy cluster*, which is a set of vertices $F \subseteq V$ such that the induced subgraph of F is connected and all the vertices in F cannot be covered by a single r-cycle in $\mathcal{C}$. They obtained the following results.

Lemma 2. *[13, 14] (i) There is a polynomial algorithm that finds k disjoint heavy clusters $F_1, \ldots, F_k \subseteq V$ and uses at most $2k + 1$ r-cycles in $\mathcal{C}$ to cover all the vertices in T; (ii) If there exist k disjoint heavy clusters $F_1, \ldots, F_k \subseteq V$ in the tree T, the minimum number of r-cycles in $\mathcal{C}$ required to cover $\bigcup_{i=1}^{k} F_i$ is at least $k + 1$.*

This lemma implies straightforwardly the 2-approximation algorithm for RMCCP on a tree in [13,14]. In what follows we derive a new integer programming formulation for RMCCP on a tree by heavily using the tree structure and show that the integrality gap of the corresponding LP relaxation has an upper bound of 5/2.

First, it can be seen that for RMCCP on a tree, an r-cycle is a multi-subgraph of T that consists of two copies of the edges of some r-tree of T. As a consequence, covering all the vertices in T is equivalent to covering at least twice all the edges. So we can replace the first constraint in (IPC) by

$$\sum_{C \in \mathcal{C}:e \in E(C)} x_C \geq 2, \quad \forall e \in E.$$

(For $e = (r, v)$, $x_C = 2$ for the r-cycle $C = r \, e \, v \, e \, r$.)

Let $F_1, \ldots, F_k \subseteq V$ be the heavy clusters in Lemma 2(i) and n_e be the number of heavy clusters in $F_1, \ldots, F_k$ that are below e. By Lemma 2(ii) we know that there are at least $n_e + 1$ optimum r-cycles each of which contains two copies of the edge e due to the tree structure. Then the above inequality can be strengthened to

$$\sum_{C \in \mathcal{C}:e \in E(C)} x_C \geq 2(n_e + 1), \quad \forall e \in E.$$

Therefore, our new integer programming (IPC-T) and its LP relaxation (LPC-T) for RMCCP on a tree are described below.

$$\min \sum_{C \in \mathcal{C}} x_C$$

$$s.t. \quad \sum_{C \in \mathcal{C}:e \in E(C)} x_C \geq 2(n_e + 1), \ \forall e \in E \qquad (IPC - T)$$

$$x_C \in \{0, 1\}, \ \forall C \in \mathcal{C}$$

$$\min \sum_{C \in \mathcal{C}} x_C$$

$$s.t. \quad \sum_{C \in \mathcal{C}:e \in E(C)} x_C \geq 2(n_e + 1), \ \forall e \in E \qquad (LPC - T)$$

$$0 \leq x_C \leq 1, \ \forall C \in \mathcal{C}$$

This new LP relaxation has much better integrality gap than (LPC) while the proof is also simpler than that in [13].

Theorem 1. *The integrality gap of (LPC-T) is at most 5/2.*

Proof. Suppose $(x_C^*)_{C \in \mathcal{C}}$ is an optimal solution to (LPC-T) and $OPT_{LPC-T} = \sum_{C \in \mathcal{C}} x_C^*$ is the optimal value.

By the first constraint of (LPC-T) corresponding to any edge e we have

$$OPT_{LPC-T} = \sum_{C \in \mathcal{C}} x_C^* \geq \frac{1}{2} \sum_{C \in \mathcal{C}:e \in E(C)} x_C^* \geq n_e + 1 \geq 1, \qquad (1)$$

where the first inequality follows from the fact that each $C \in \mathcal{C}$ contains two copies of edges used by it and the second inequality is due to the constraints of (LPC-T). We distinguish two cases.

Case 1. $n_e = 0$ for any $e \in E$. Since T is a binary tree, the depot r has two possible children u_1, u_2. Set $e_i = (r, u_i)(i = 1, 2)$. Let T_{e_i} be the r-tree of T consisting of e_i and the subtree rooted at u_i. Since $n_{e_i} = 0$, we know that $w(T_{e_i}) \leq \frac{\lambda}{2}$ and T can be covered by at most two r-cycles C_1, C_2, where $C_i (i = 1, 2)$ is obtained by doubling the edges in T_{e_i}. Therefore we have a feasible integral solution to (IPC-T) of objective value at most $2 \leq 2OPT_{LPC-T}$, where the inequality follows from (1).

Case 2. There exists some edge $\tilde{e}$ with $n_{\tilde{e}} \geq 1$, which implies $k \geq 1$. Then the first constraint of (LPC-T) corresponding to $\tilde{e}$ leads to

$$OPT_{LPC-T} \geq \frac{1}{2} \sum_{C \in \mathcal{C}: \tilde{e} \in E(C)} x_C^* \geq n_{\tilde{e}} + 1 \geq 2.$$

Multiply by $w(e)$ in the first constraint of (LPC-T) and take the summation over all $e \in E$, we obtain

$$\sum_{e \in E} 2(n_e + 1)w(e) \leq \sum_{e \in E} \left(\sum_{C \in \mathcal{C}: e \in E(C)} x_C^* \right) w(e)$$

$$= \sum_{C \in \mathcal{C}} x_C^* w(C)$$

$$\leq \lambda \sum_{C \in \mathcal{C}} x_C^* = \lambda OPT_{LPC-T}, \qquad (2)$$

where the first equality holds by exchanging the order of the two summations and the last inequality follows from the definition of $\mathcal{C}$.

On the other hand, for $i = 1, \ldots, k$, by definition the induced subgraph of F_i, denoted by $T[F_i]$, is actually a subtree of T. Let $v_i \in F_i$ be the highest vertex in F_i. We obtain an r-cycle C_i by doubling the edges in the r-tree consisting of $T[F_i]$ and the unique path from r to v_i. Let $E' \subseteq E$ be the set of edges used by $C_1, \ldots, C_k$. For each $e \in E'$, it is used at most $2(n_e + 1)$ times by the cycles $C_1, \ldots, C_k$, where $2n_e$ is due to the heavy clusters below it and if e happens to be in some $T[F_i]$ it appears two more times in C_i. Since F_i is a heavy cluster, we have $w(C_i) \geq \lambda$. Then

$$k\lambda \leq \sum_{i=1}^{k} w(C_i) \leq \sum_{e \in E'} 2(n_e + 1)w(e) \leq \sum_{e \in E} 2(n_e + 1)w(e) \leq \lambda OPT_{LPC-T},$$

where the last inequality follows from (2). This implies $k \leq OPT_{LPC-T}$. Combining this inequality with $OPT_{LPC-T} \geq 2$ and Lemma 2(i), we have a feasible integral solution to (IPC-T) with objective value at most

$$2k + 1 \leq \frac{2k + 1}{\max\{2, k\}} OPT_{LPC-T} \leq \left(2 + \frac{1}{\max\{2, k\}} \right) OPT_{LPC-T} \leq \frac{5}{2} OPT_{LPC-T}.$$

3.2   Rooted Minimum Path Cover

Given an instance of RMPCP, let $\mathcal{P}$ be the set of all r paths of length at most λ. Nagarajan and Ravi [12] considered the following path-version of (IPC) for RMPCP, in which a binary variable x_P is associated with each r-path $P \in \mathcal{P}$.

$$\min \sum_{P \in \mathcal{P}} x_P$$

$$s.t. \quad \sum_{P \in \mathcal{P}: v \in V(P)} x_P \geq 1, \ \forall v \in V \setminus \{r\} \qquad (IPP)$$

$$x_P \in \{0, 1\}, \ \forall P \in \mathcal{P}.$$

The corresponding LP relaxation (LPP) is obtained by dropping the integral constraints on the variables.

$$\min \sum_{P \in \mathcal{P}} x_P$$

$$s.t. \quad \sum_{P \in \mathcal{P}: v \in V(P)} x_P \geq 1, \ \forall v \in V \setminus \{r\} \qquad (LPP)$$

$$x_P \geq 0, \ \forall P \in \mathcal{P}$$

Let (IPC') and (LPC') be the integer programming and its LP relaxation obtained by replacing the parameter λ with 2λ in (IPC) and (LPC), respectively. It follows that $OPT_{IPP} \leq 2OPT_{IPC'}$ since each r-cycle of length at most 2λ can be broken into two r-paths of length at most λ by removing one edge in the middle. On the other hand, it holds that $OPT_{LPC'} \leq OPT_{LPP}$ because each r-paths of length at most λ can be turned into an r-cycle of length at most 2λ by doubling all the edges. Therefore, an upper bound α of the integrality gap for (LPC) implies an upper bound of 2α of the integrality gap for (LPP). By the results in [6,13], the integrality gap of (LPP) is at most $O(\min\{\log n, \frac{\log \lambda}{\log \log \lambda}\})$ for the general RMPCP and can be bounded by a constant (more exactly, 40) for RMPCP on a tree $T = (V, E)$. As before, we assume that T is a binary tree rooted at r.

Next we derive an LP relaxation for RMPCP on a tree with an integrality gap of at most 5. Similar to the tour-version problem, Nagarajan and Ravi [12] define a *heavy cluster* to be a set of vertices $F \subseteq V$ such that the induced subgraph of F is connected and all the vertices in F cannot be covered by a single r-path in $\mathcal{P}$. They proposed a 4-approximation algorithm based on the following results.

Lemma 3. *[12] (i)There is a polynomial algorithm that finds k disjoint heavy clusters $F_1, \ldots, F_k \subseteq V$ and uses at most $2k + 1$ r-paths in $\mathcal{P}$ to cover all the vertices in T; (ii)If there are k disjoint heavy clusters $F_1, \ldots, F_k \subseteq V$ in the tree T, the minimum number of r-paths in $\mathcal{P}$ required to cover $\bigcup_{i=1}^{k} F_i$ is at least $\lfloor \frac{k+1}{2} \rfloor + 1$.*

As in the previous section, our new integer programming replaces the first constraint in (IPP) with

$$\sum_{P \in \mathcal{P}: e \in E(P)} x_P \geq 1, \quad \forall e \in E.$$

Note that in this new constraint the right-hand side is changed to 1, since for RMPCP on a tree an r-path P is obtained by doubling all the edges of the r-tree induced by $V(P)$ except those on the unique path from r to the farthest vertex in $V(P)$.

Let $F_1, \ldots, F_k \subseteq V$ be the heavy clusters in Lemma 3 and n_e be the number of heavy clusters in $F_1, \ldots, F_k$ that are below e. By Lemma 3(ii) we known that there are at least $\lfloor \frac{n_e+1}{2} \rfloor + 1$ optimum r-paths that pass the edge e due to the tree structure. Then the above inequality can be strengthened to

$$\sum_{P \in \mathcal{P}: e \in E(P)} x_P \geq \left\lfloor \frac{n_e + 1}{2} \right\rfloor + 1, \quad \forall e \in E.$$

So our new integer programming (IPP-T) and its LP relaxation (LPP-T) are given as follows.

$$\min \sum_{P \in \mathcal{P}} x_P$$

$$s.t. \quad \sum_{P \in \mathcal{P}: e \in E(P)} x_P \geq \left\lfloor \frac{n_e + 1}{2} \right\rfloor + 1, \quad \forall e \in E \qquad (IPP-T)$$

$$x_P \in \{0, 1\}, \quad \forall P \in \mathcal{P},$$

$$\min \sum_{P \in \mathcal{P}} x_P$$

$$s.t. \quad \sum_{P \in \mathcal{P}: e \in E(P)} x_P \geq \left\lfloor \frac{n_e + 1}{2} \right\rfloor + 1, \quad \forall e \in E \qquad (LPP-T)$$

$$0 \leq x_P \leq 1, \quad \forall P \in \mathcal{P}$$

Now we can show an upper bound on the integrality gap of (LPP-T).

Theorem 2. *The integrality gap of (LPP-T) is at most 5.*

Proof. Suppose $(x_P^*)_{P \in \mathcal{P}}$ is an optimal solution to (LPP-T) and $OPT_{LPP-T} = \sum_{P \in \mathcal{P}} x_P^*$ is the optimal value.

By the first constraint of (LPP-T) corresponding to any edge e we deduce that

$$OPT_{LPP-T} = \sum_{P \in \mathcal{P}} x_P^* \geq \frac{1}{2} \sum_{P \in \mathcal{P}: e \in E(P)} x_P^* \geq \frac{1}{2} \left(\left\lfloor \frac{n_e + 1}{2} \right\rfloor + 1 \right) \geq \frac{1}{2}, \qquad (3)$$

where the first inequality follows from the fact that each $P \in \mathcal{P}$ contains at most two copies of edges used by it. We consider two cases.

Case 1. $n_e = 0$ for any $e \in E$. Since T is a binary tree, we assume that u_1, u_2 be the possible children of the depot r and set $e_i = (r, u_i)(i = 1, 2)$. Let T_{e_i} be the r-tree of T consisting of e_i and the subtree rooted at u_i. Since $n_{e_i} = 0$, we know that all the vertices in T_{e_i} can be covered by an r-path in $\mathcal{P}$ and T can be covered by at most two r-paths. In other words, we derive a feasible integral solution to (IPP-T) of objective value at most $2 \leq 4OPT_{LPP-T}$, where the inequality follows from (3).

Case 2. There exists some edge $\tilde{e}$ with $n_{\tilde{e}} \geq 1$, which implies $k \geq 1$. Then the first constraint of (LPP-T) corresponding to $\tilde{e}$ implies

$$OPT_{LPP-T} \geq \frac{1}{2}\left(\left\lceil \frac{n_e + 1}{2} \right\rceil + 1\right) \geq 1.$$

Multiply by $w(e)$ in the first constraint of (LPP-T) and take the summation over all $e \in E$, we obtain

$$\sum_{e\in E}\left(\frac{n_e}{2} + 1\right)w(e) \leq \sum_{e\in E}\left(\left\lceil \frac{n_e + 1}{2} \right\rceil + 1\right)w(e)$$

$$\leq \sum_{e\in E}\left(\sum_{P\in\mathcal{P}:e\in E(P)} x_P^*\right)w(e)$$

$$= \sum_{P\in\mathcal{P}} x_P^* w(P)$$

$$\leq \lambda \sum_{P\in\mathcal{P}} x_P^* = \lambda OPT_{LPP-T}, \qquad (4)$$

where the first equality holds by exchanging the order of the two summations and the last inequality follows from the definition of $\mathcal{P}$.

On the other hand, for $i = 1, \ldots, k$ let v_i be the highest vertex in F_i and T_i be the r-tree consisting of $T[F_i]$ and the path from r to v_i. By doubling all the edges in T_i except those on the path from r to the farthest vertex in F_i we can generate an r-path P_i. Let $E' \subseteq E$ be the set of edges used by $P_1, \ldots, P_k$. For each $e \in E'$, it is used at most $n_e + 2$ times by the paths $P_1, \ldots, P_k$, where n_e is due to the heavy clusters below it and if e happens to be in some $T[F_i]$ it may appear two more times in P_i. Since F_i is a heavy cluster, we have $w(P_i) \geq \lambda$. Then

$$k\lambda \leq \sum_{i=1}^{k} w(P_i) \leq \sum_{e\in E'}(n_e + 2)w(e) \leq \sum_{e\in E}(n_e + 2)w(e) \leq 2\lambda OPT_{LPP-T},$$

where the last inequality follows from (4). So we have $k \leq 2OPT_{LPP-T}$. Combining this inequality with $OPT_{LPP-T} \geq 1$ and Lemma 3(i), we have a feasible integral solution to (IPP-T) with objective value at most

$$2k + 1 \leq \frac{2k+1}{\max\left\{1, \frac{k}{2}\right\}} OPT_{LPP-T} \leq \left(4 + \frac{2}{\max\{2, k\}}\right)OPT_{LPP-T} \leq 5OPT_{LPP-T}.$$

3.3   A Lower Bound on the Integrality Gap

Consider a star consisting of the depot r and $n + 1$ leaves. All the $n + 1$ edges have unit weight. Set $\lambda = 2n$. It can be seen that $OPT_{IPC-T} = OPT_{IPP-T} = 2$. For $i = 1, \ldots, n + 1$, let C_i be the r-cycle visiting all the leaves except the ith leaf. We have a solution to (LPC-T) by setting $x_{C_1} = \cdots = x_{C_{n+1}} = \frac{1}{n}$ and $x_C = 0$ for all $C \in \mathcal{C} \setminus \{C_1, \ldots, C_{n+1}\}$. This solution is feasible for (LPC-T) since there are exactly n r-cycles from $C_1, \ldots, C_{n+1}$ covering the ith leaf for each $i = 1, \ldots, n+1$. The objective value of this solution is $\frac{n+1}{n} \geq OPT_{LPC-T}$, which implies that the integrality gap of (LPC-T) is at least 2 for sufficiently large n.

Similarly, if we define P_i as the r-path visiting all the leaves except the ith leaf one can show that the integrality gap of (LPP-T) cannot be smaller than 2.

4   General Metric

In this section we give new LP relaxations for the general MCCP/MPCP/MTCP. Our LP relaxation for MCCP has an integrality gap of at most 6 and the integrality gap of the LP relaxations for MPCP/MTCP is bounded by 4. Previously, Nagarajan and Ravi [12,13] showed an upper bound of 17 on the integrality gap of the unrooted version of (LPP) where $\mathcal{P}$ is redefined as the set of all paths of length no more than λ.

The details on the LP relaxations for MCCP/MPCP/MTCP will be presented in the full version of our paper.

Acknowledgements. This research is supported by the National Natural Science Foundation of China under grants numbers 11671135, 11701363 and the Fundamental Research Fund for the Central Universities under grant number 22220184028.

References

1. Arkin, E.M., Hassin, R., Levin, A.: Approximations for minimum and min-max vehicle routing problems. J. Algorithms **59**, 1–18 (2006)
2. Bhattacharya, B., Hu, Y.: Approximation algorithms for the multi-vehicle scheduling problem. In: Cheong, O., Chwa, K.-Y., Park, K. (eds.) ISAAC 2010. LNCS, vol. 6507, pp. 192–205. Springer, Heidelberg (2010). https://doi.org/10.1007/978-3-642-17514-5_17
3. Campbell, A.M., Vandenbussche, D., Hermann, W.: Routing for relief efforts. Transp. Sci. **42**, 127–145 (2008)
4. Even, G., Garg, N., Koemann, J., Ravi, R., Sinha, A.: Min-max tree covers of graphs. Oper. Res. Lett. **32**, 309–315 (2004)
5. Frederickson, G.N., Hecht, M.S., Kim, C.E.: Approximation algorithms for some routing problems. SIAM J. Comput. **7**(2), 178–193 (1978)
6. Z. Friggstad, C. Swamy, Approximation algorithms for regret-bounded vehicle routing and applications to distance-constrained vehicle routing. In: The Proceedings of the 46th Annual ACM Symposium on Theory of Computing, pp. 744–753 (2014)
7. Garey, M.R., Johnson, D.S.: Computers and Intractability: A Guide to the Theory of NP-Completeness. Freeman, San Francisco (1979)
8. Golden, B.L., Raghavan, S., Wasil, F. A. (eds.): The Vehicle Routing Problem: Latest Advances and New Challenges. Springer, Heidelberg (2008). https://doi.org/10.1007/978-0-387-77778-8

9. Karakawa, S., Morsy, E., Nagamochi, H.: Minmax tree cover in the Euclidean space. J. Graph Algorithms Appl. **15**, 345–371 (2011)
10. Karuno, Y., Nagamochi, H.: 2-Approximation algorithms for the multi-vehicle scheduling problem on a path with release and handling times. Discret. Appl. Math. **129**, 433–447 (2003)
11. Khani, M.R., Salavatipour, M.R.: Approximation algorithms for min-max tree cover and bounded tree cover problems. Algorithmica **69**, 443–460 (2014)
12. Nagarajan, V., Ravi, R.: Minimum vehicle routing with a common deadline. In: Díaz, J., Jansen, K., Rolim, J.D.P., Zwick, U. (eds.) APPROX/RANDOM -2006. LNCS, vol. 4110, pp. 212–223. Springer, Heidelberg (2006). https://doi.org/10. 1007/11830924_21
13. Nagarajan, V., Ravi, R.: Approximation algorithms for distance constrained vehicle routing problems. Carnegie Mellon University, Pittsburgh, Tepper School of Business (2008)
14. Nagarajan, V., Ravi, R.: Approximation algorithms for distance constrained vehicle routing problems. Networks **59**(2), 209–214 (2012)
15. Toth, P., Vigo, D. (eds.): The Vehicle Routing Problem. SIAM, Philadelphia (2002)
16. Vazirani, V.V.: Approximation Algorithms. Springer, Heidelberg (2001). https:// doi.org/10.1007/978-3-662-04565-7
17. Xu, W., Liang, W., Lin, X.: Approximation algorithms for min-max cycle cover problems. IEEE Trans. Comput. **64**(3), 600–613 (2015)
18. Xu, Z., Xu, L., Zhu, W.: Approximation results for a min-max location-routing problem. Discret. Appl. Math. **160**, 306–320 (2012)
19. Yu, W., Liu, Z.: Improved approximation algorithms for some min-max cycle cover problems. Theor. Comput. Sci. **654**, 45–58 (2016)
20. Yu, W., Liu, Z., Bao, X.: New approximation algorithms for the minimum cycle cover problem. In: Chen, J., Lu, P. (eds.) FAW 2018. LNCS, vol. 10823, pp. 81–95. Springer, Cham (2018). https://doi.org/10.1007/978-3-319-78455-7_7

Computation of Kullback-Leibler Divergence Between Labeled Stochastic Systems with Non-identical State Spaces

Krishnendu Ghosh[✉]

College of Charleston, Charleston, SC 29401, USA
ghoshk@cofc.edu

Abstract. Model checking of biological systems is computational intensive because of state explosion. Model reduction is one of the directions that has been addressed for state explosion. Formal modeling of
biological pathways leads to additional challenges given that biological
pathways are multiscale and stochastic. Model abstractions incorporating multiscale biological processes are represented as labeled stochastic
systems. Kullback-Leibler divergence is computed to measure the close-
ness of stochastic systems. A fixed point polynomial time algorithm is
presented to compute Kullback-Leibler divergence with an approximation when comparing labeled stochastic systems with non-identical state
spaces.

Keywords: Preorder relation · Algorithm
Kullback-Leibler divergence

1 Introduction

Formal modeling in systems biology has been an active research area in recent
years [4,6,22]. Formal methods such as model checking have been used as a
querying mechanism by posing biological queries in temporal logics to the model,
a finite state machine (FSM) representing biological processes. One of the challenges in model checking [8] is the state explosion problem. Recent advances
in research have addressed the state explosion problem [12,20]. Biological processes are multiscale and stochastic. Multiscale processes execute at different
orders of time scale. For example, communication between molecular processes
with cellular processes. Modeling biological processes with a system of differential equations is unstable given there is imprecise and incomplete information
with regards to the concentration of the chemicals (biochemicals) in a system.
The interactions of biological processes operating at different time scales create
a large state space if the lowest time scale is the reference of the system. We
motivate the construction of a multiscale system that incorporates stochasticity
with the following example.

© Springer Nature Switzerland AG 2018
S. Tang et al. (Eds.): AAIM 2018, LNCS 11343, pp. 233–243, 2018.
https://doi.org/10.1007/978-3-030-04618-7_19

Motivational example [14]: Consider four biochemical pathways, $\mathcal{A}, \mathcal{B}, \mathcal{C}$ and $\mathcal{D}$ with chemicals V, W, X, Y and Z:

$$\mathcal{A} : X \xrightarrow{\epsilon} Y.$$
$$\mathcal{B} : Y \xrightarrow{M} X + Z.$$
$$\mathcal{C} : Y \xrightarrow{N} X + V.$$
$$\mathcal{D} : Y \xrightarrow{P} X + W.$$

The notation, $C_s \xrightarrow{\alpha} C_p$ denotes a set of substrates, C_s in the presence of a catalyst (chemical), α produces a set of products, C_p. Also, $\alpha \in \{M, N, P, \epsilon\}$ where M, N, P denotes catalysts and ϵ represents the absence of a catalyst.

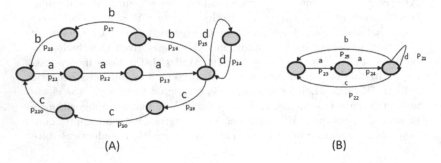

(A)   (B)

Fig. 1. Finite state machine representing identical partial ordering of pathways $\mathcal{A}, \mathcal{B}, \mathcal{C}$ and $\mathcal{D}$ represented by edge labels a,b,c and d, respectively. (A) System of pathways with 9 states (B) System of pathways with 3 states.

Figure 1 shows two FSMs with different state space sizes for pathways, The FSMs represent pathways $\mathcal{A}, \mathcal{B}, \mathcal{C}$ and $\mathcal{D}$. The concentrations of the chemicals are in moles. The initial state, S contains one mole of the chemicals X, M, N and P. In Fig. 1(A), the transition with label a represents pathway $\mathcal{A}$ being executed and consuming 0.25 mol of chemical, X. Similarly, transitions with labels b, c, and d represent execution of pathways, $\mathcal{B}, \mathcal{C}$ and $\mathcal{D}$, respectively. Each transition represents consumption of 0.25 mol of the substrate during a reaction. Pathways, $\mathcal{B}, \mathcal{C}$ and $\mathcal{D}$ execute nondeterministically after completion of pathway $\mathcal{A}$ executing three times successively. Also, the exact concentration of the chemicals is often not known initially. Assume, the concentration of X is 0.75 mol and 0.5 mol in Fig. 1(A) and (B). The concentration of $X \in \mathcal{A}$ in Fig. 1(A) is higher than in Fig. 1(B) and the rate of reaction is faster in $\mathcal{A}$ than $\mathcal{B}, \mathcal{C}$ and $\mathcal{D}$. Modeling chemical reactions is challenging because data of the rate of reactions and initial concentrations of chemicals are often imprecise. Assigning probabilities on the edges of the FSMs is a way to quantify imprecise data in the model. The states store concentrations of the chemicals produced or consumed. Each transition represents a time step for execution of a pathway. The order of execution of pathways without the repeats of execution in a path in Fig. 1(A) and (B) is identical. For example, the three successive transitions labeled, a in Fig. 1(A) can be collapsed to one transition similar to collapsing two successive labels of a in Fig. 1(B). Hence, the state between successive transitions labeled with a in Fig. 1(A) is not present in Fig. 1(B). The FSMs of Fig. 1(A) and (B) represent identical partial ordering of the pathways but the number of states is different. The exact concentrations of the chemicals are not known, so the edges have probabilities in the form p_{1x} and p_{2y} where $x \in \{1, 2, 3, 4, 5, 6, 7, 8, 9, 10\}$ and $y \in \{1, 2, 3, 4, 5\}$. It is clear that the FSMs represent the same order of execution in a path but the details are different because the size of state spaces of the FSMs are different. In this work, we seek to answer the query, "Are the probabilistic FSMs representing the multiscale pathways (dis-)similar by a numerical quantity, x?". The question is addressed by computing the Kullback-Leibler Divergence (KLD) on the two FSMs. KLD is computed on same state space. The state space are different because of the repetitions of edge labels, *repeats* in the FSM. A notion of *read equivalence* is introduced for computation of KLD on the same state space. A preorder relation is constructed to identify the partial ordering of the pathways represented by the edge labels of the FSMs in Fig. 1(A)–(B). We consider only the edge labels of the FSM to evaluate the identifiability of the partial ordering of pathways. Stochastic structures are created for the comparison and computation of KLD. The objectives of this work are: (i) Identification of partial ordering of pathways (without repeats) on the FSMs representing a system of pathways and, (ii) compute the KLD on the probabilities stored in the edge labels of the FSMs. The goal of this work is to identify the model that has the least number of states among several models that represent the biochemical pathways. The model with the least number of states will not be detailed but computationally, it will be less intensive. To the best of our knowledge, this is the first work that has addressed computation of KLD using a preorder relation on probabilistic structures. KLD as a metric can be used on structures that have identical state space. This work addresses constructing an identical state space on two probabilistic structures with different state spaces by defining a preorder relation.

2   Background and Related Work

In this section, we review the literature on algorithms computing bisimulations, asynchronous modeling, temporal logics on probabilistic systems, distances or approximations on stochastic systems and related formal modeling in systems biology. These different theories form the foundations of our work. Computing bisimulations on finite state machines has been an active research area. The Partition algorithm [19] constructed equivalence classes and addressed computation of the states that are bisimilar. Identically labeled states occurring successively in a path of finite state machines has been referred as *stuttering* and computation of stuttering bisimulations have been reported [16]. A $O(m \log n)$ algorithm, where n is the number of states and m is the transitions, have been constructed to compute stuttering equivalence and branching bismulations [17].

An polynomial time algorithm for computating equivalence of labeled Markov chain had been constructed [11] but trace refinement on Markov decision process has been proved to be undecidable [13]. A general class of metrics between Markov chains based on behaviour is introduced [9]. A survey of approximating metrics on probabilistic bisimulations is published [1]. Model reduction of continuous time stochastic systems using approximations based on Wasserstein metrics [25] has been performed. Bisimilarity of probabilistic systems was computed by **P**-hard reduction of the monotone circuit problem [7]. Metric-based [10] state space reduction for Markov chains, when solved as a bilinear program and the threshold problem, have been proven to be in PSPACE and NP-hard [2].

A synctactic Markovian bisimulation based probabilistic bisimulation over the structure of chemical reaction network in polynomial time has been constructed [5]. A model representing multiscale processes and a fixed point algorithm for computation of preorder relationship on FSM have been reported [14]. Construction of model abstractions for study of the processes at difference levels of granularity have been stated [24]. A survey on model reduction for large scale biological models has been reported [23].

3   Preliminaries

Definition 1. *(Labeled transition system (LTS))* Given a set of propositions, AP being the set of labels for states and $\mathcal{E}$, a set of labels for edges, a *labeled state transition system* is defined as $\mathcal{M} = \langle S_0, S, E, L_e, L \rangle$ where,

1. S is the set of states.
2. $S_0 \subseteq S$ is the initial set of states.
3. $E \subseteq S \times S$ is the transition relation.
4. $L : S \to 2^{AP}$ where L is the labeling function that labels each state with a subset from the set, AP.
5. $L_e : E \to \mathcal{E}$ is an edge-labeling function.

The state based definition of the stochastic structures such as discrete time Markov chain [3] is:

Definition 2. *(Discrete Time Markov Chain (DTMC))* a *discrete-time Markov chain is a tuple:* $\mathcal{M}_m \langle S, S_0, \iota_{init}, P, L \rangle$ *where:*

- S *is a finite set of states.*
- S_0 *is the set of initial states.*
- $P : S \times S \to [0,1]$, *where* P *represents the probability matrix and* $\sum_{s,s' \in S} \mathcal{P}(s, s') = 1$.
- $\iota_{init} : S \to [0,1]$ *where* $\sum_{s \in S} \iota_{init}(s) = 1$ *is the initial distribution.*
- $L : S \to 2^{AP}$, *where* L *is a labeling function and* AP *the set of atomic propositions.*

Definition 3. *(Labeled Probabilistic System (LPS)) a LPS is a tuple,* $\mathcal{W} =$ $\langle S, S_0, \iota_{init}, P, L_e, L, E \rangle$ *where:*

- $\langle S, S_0, \iota_{init}, P, L \rangle$ *is DTMC.*
- $L_e : S \times S \rightarrow E$ *where, E is the set of edge labels.*

For simplicity, each edge label on the probabilitic structure will be a pair, $\langle p, a \rangle$ where p is the probability of the action, $a \in \mathcal{A}$. The reading is an action, a from a state, $s \in S$ is assigned a probability, p.

Kullback-Leibler divergence [18] or relative entropy is a non-symmetric measure between two probability distributions.

Definition 4. *(Kullback-Leibler Divergence (KLD)) Kullback-Leibler divergence [18] or relative entropy is a non-symmetric measure between two probability distributions. Formally, [21]: Given P and Q be two probability distributions over the random variable X, the KLD is denoted by $H(P\|Q)$ of P with respect to Q is, $H(P\|Q) =$*

$$\sum_{x \in X} P(x) log \frac{P(x)}{Q(x)}$$

$H(P\|Q)$ is not a metric because $H(P\|Q) \neq H(Q\|P)$.

4 Formalization of Stochastic Multiscale Processes

The description of a mechanistic formal model for a sequence of chemical reactions is represented by a LTS. The following is the model stated [15]: Given (1) a set $\mathcal{C}$ of chemicals, (2) for each chemical $C \in \mathcal{C}$ a finite set of numbers, $0, 1, \ldots, k$ where $k \in \mathbb{N}$ represent the number of moles for each chemical C and (3) a set of edge labels, $\mathcal{E} \cup \{\epsilon\}$. An edge label is of the form, $\langle \hat{A}_1, \hat{A}_2, \ldots, \hat{A}_{ks}, \hat{B}_1, \hat{B}_2, \ldots, \hat{B}_{kp}, Rate, \boldsymbol{Cat}, \boldsymbol{Inh}, RType \rangle$ where $\hat{A}_1, \hat{A}_2, \ldots, \hat{A}_{ks}$ and $\hat{B}_1, \hat{B}_2, \ldots, \hat{B}_{kp}$ represent concentration of substrates and products for a reaction, respectively.

Rate is represents the rate of reaction. $\boldsymbol{Cat}$ and $\boldsymbol{Inh}$ are the set of chemicals representing catalyst and inhibitor, respectively. *RType* is the type of reaction such as endothermic and exothermic. $ks, kp \in \mathbb{N}$ and $e \in \mathcal{E}$ represents a reaction. ϵ represents no reaction taking place. It is used to maintain the totality of the LTS. We assume that a reaction with forward and backward rate of reactions are modeled as two different reactions.

The LTS, $\mathcal{M} = \langle S, S_0, E, L_e, L \rangle$ representation of a system of chemical reactions is given by: The labels of the state: AP is the set of all the atomic formulas $c_0 = 0$, $c_1 = 1$, or $c_i = k$ for all $C \in \mathcal{C}$, $i \in \mathbb{N}$. The atomic formula represent the molar concentration of a chemical. The states contain concentrations of all the chemicals in the system and are represented by the atomic formulas of each chemical that are true in the state.

S_0 is the set of initial states of the LTS. An initial state contains concentration of all the chemicals before any reaction. Hence, $\mid S_0 \mid = 1$.

A labeled transition is represented, $s \xrightarrow{e} s'$ where $e \in \mathcal{E}$ and $s, s' \in S$. A labeled transition is represented, $s \xrightarrow{e} s'$. A transition represents a reaction is taking place and the consumption(production) of substrates(products) is based on the x, number of moles to be consumed in a unit of time. A transition also represents a time step in a reaction. The number of moles consumed or produced is computed using conservation of mass action. As stated earlier, edge label ϵ represents no reaction. A reaction takes places only if there is minimal amount of substrates available. The label on state, s' contains the concentration of the substrates and products after the reaction. There is no self loop on the states that have atleast one reaction that is not ϵ.

The conversion of the LTS representing the system of chemical reactions to a LPS is the following: The state labels are identical, but the edge labels on the LPS have additional information in the form of probabilities. The edge label,e of LTS is mapped to an edge label in the form of a pair, $\langle e, p \rangle$. In the LPS, the set of edge labels which are pairs is denoted by E. The pair corresponds the reaction represented by $e \in E$ and executes with a probability of p. If s is the state from where the reactions begin, the sum of probabilities in all the outgoing edge labels,$s \xrightarrow{e} s_k$, where $k \in \mathbb{N}$, in LPS is 1. The reading is there can be multiple reactions,$e_1, e_2, \ldots e_k$ from s to states, s_k that can take place from the chemicals from state, s and each reaction can occur with probability of $p_1, p_2, \ldots, p_k$, respectively. The LPS may not be total because the probability of ϵ transition of LTS is zero.

A path in a LPS is a finite or infinite sequence $\sigma = s_0 \xrightarrow{\alpha_0} s_1 \xrightarrow{\alpha_1} \ldots$. The sequence of reaction is the sequence of edge labels. For an edge, w in a LPS $\mathcal{W}$, let $\Pi(w)$ be the set of paths starting with w, and let $\Pi(\mathcal{W})$ be the set of all paths in LPS $\mathcal{W}$. We use variable π_w for an element of $\Pi(w)$. A prefix of length m of a path π_{w_1}, beginning from edge, w_1 is a finite sequence, $\pi_{w_1}^m = w_{0,1}, w_{1,1}, \ldots, w_{m-1,1}$ where $m \in \mathbb{N}$.

5   Computation of KLD on LPSs

We use the following notation for the computation of KLD on LPSs. A LPS is stated as $\mathcal{W}_i = \langle S, S_0, \iota_{init}, P, L_e, L, \mathcal{E} \rangle$ where $i = 1, 2$. The edge labels of $\mathcal{W}$ is w,a pair $\langle e, p \rangle$ where p is the probability, e is the label representing reaction and E is the set of reaction labels in $\mathcal{W}$ unless stated otherwise.

A path in LPS is a finite or infinite sequence $\sigma = s_0 \xrightarrow{\alpha_0} s_1 \xrightarrow{\alpha_1} \ldots$ where $s_0, s_1, \ldots$ are the states. The sequence of reactions is given by $e_0, e_1, \ldots$ from the sequence of edge labels,$w_0, w_1, \ldots$. In the description, we will focus on the reaction labels and the corresponding probability will be the corresponding index. For example, the probabilites for the sequence of reactions, $e_0, e_1, \ldots$ is $p_0, p_1, \ldots$. Notation, an edge, $w \in \mathcal{W}$ will be the reaction label in w which is e. Therefore, for an edge $w \in \mathcal{W}$, let $\Pi(e)$ be the set of paths starting with e, and let $\Pi(\mathcal{W})$ be the set of all paths in LPS $\mathcal{W}$. S_i, E_i and let $\Pi(\mathcal{W}_i)$ denote the set of states, edges and paths in $\mathcal{W}_i$ for $i = 1, 2$. We use variable π_e for an element of $\Pi(e)$. A prefix of length m of a path π_{e_1}, beginning from edge, e_1 is a finite sequence, $\pi_{e_1}^m = e_{0,1}, e_{1,1}, \ldots, e_{m-1,1}$ where $m \in \mathbb{N}$.

The *outline* for the computation of KLS on two LPSs is:

1. Given two LTSs, $\mathcal{M}_1$ and $\mathcal{M}_2$, construct LPSs, $\mathcal{W}_1$ and $\mathcal{W}_2$, respectively.
2. Compute preorder on the edge labels (without successive repetitions) of the LPSs
3. While computing the preoder, compute the KLD on the same sequence of reactions (paths) of the LPSs.

Definition 5. *(Read)* For an infinite path, $\pi = e_0, e_1, e_2, e_3, \ldots$ in a LPS $\mathcal{W}$, $\alpha_0, \alpha_1, \alpha_2, \ldots$ denotes the sequence of reaction labels in π. The read of a path is the subsequence of *reaction* labels $\tilde{\pi} = \alpha_0, \alpha_{i_1}, \alpha_{i_2}$ where $0 \leq i_1 \leq i_2 \leq \ldots, \alpha_{i_j}$ is in $\tilde{\pi}$ iff $\alpha_{i_j} \neq \alpha_{i_{j-1}}$ and $\alpha_0 \neq \alpha_{i_1}$.

A finite path segment $\sigma = e_0 \rightarrowtail e_1 \rightarrowtail e_2 \rightarrowtail e_3 \ldots \rightarrow e_m \rightarrowtail \ldots$, is identically labeled *(il)* if the reactions are identical. We explicitly allow $m = 0$; in that case we write $e_0 \rightsquigarrow e_0$. Notation $e_0 \overset{+}{\rightsquigarrow} e'$ means that for some $m \geq 0, e_0 \rightsquigarrow e_m \rightarrowtail e'$, and $L_e(e_0) \neq L_e(e')$.

Definition 6. *The* compact probability, $P_c(e, e')$ *between two edges is computed by the following equations dependent on the label of the successive edges.*

$$
P_c(e, e') = \begin{cases} P(e, e') \text{ if, } e \neq e' \\ P(e) \times P(e_1) \times \cdots P(e_k) \text{if } e \rightarrowtail e_1 \rightarrowtail \ldots, e_k \rightarrowtail e' \end{cases}
$$

The compact probability for an *il* path fragment is computed by the products of the probabilities.

Definition 7. *(Read equivalence on paths)* Paths $\pi_1 \in \Pi(\mathcal{W}_1), \pi_2 \in \Pi(\mathcal{W}_2)$ are said to be read equivalent iff their reads are identical. This is denoted by $\pi_1 \equiv_r \pi_2$.

Definition 8. *(Read equivalence on reactions)* Given two LPSs, $\mathcal{W}_1$ and $\mathcal{W}_2$, the relation read on edges $(\equiv_r)$ is defined on reaction labels, $e_1 \in E_1$ and $e_2 \in E_2$. $e_1 \equiv_r e_2$ if and only if the following conditions hold:

1. $L_e(e_1) = L_e(e_2)$.
2. For all paths, $\pi_{e_1} \in \Pi(e_1) \exists$ a path $\pi_{e_2} \in \Pi(e_2)$ such that $\pi_{e_1} \equiv_r \pi_{e_2}$.
3. For all paths, $\pi_{e_2} \in \Pi(e_2) \exists$ a path $\pi_{e_1} \in \Pi(e_1)$ such that $\pi_{e_1} \equiv_r \pi_{e_2}$.

The compact probabilities computed from the probabilities of *il* path on $\mathcal{W}_1$ and $\mathcal{W}_2$ by Definitions 6, 5 and 8.

Definition 9. A relation, R_e defined on the edges of $\mathcal{W}_1$ and $\mathcal{W}_2$ is given by $(e_1, e_2) \in R_e, e_1 \in E_1$ and $e_2 \in E_2$ where, $L_e(e_1) = L_e(e_2)$.

Definition 10. *(Predecessor)* The subset of ordered pairs, Predecessor $Pred^r(Y)$ is defined from the set of ordered pairs, $(e_1, e_2) \in R_e$ represented by the Y is: $Pred^r(Y) = \{(e_1, e_2) \in Y \mid \forall e'_1, e_1 \rightarrowtail e'_1$ implies $\exists$ an *il*-path fragment $e_2 \rightarrowtail \ldots \rightarrowtail e_{m,2} \rightarrowtail c'_2, \forall i \leq m, (e_1, e_{i,2}) \in Y \wedge (e'_1, e'_2) \in Y$, and $\forall e'_2, e_2 \rightarrowtail e'_2$ implies $\exists$ an *il* path-fragment $e_1 \rightarrowtail \ldots \rightarrowtail e_{m,1} \rightarrowtail e'_1, \forall i \leq m(e_{i,1}, e_2) \in Y \wedge (e'_1, e'_2) \in Y\}$.

The operators, Fst and Snd are defined on R_e to extract the abscissa and ordinates. Pre is a function defined on the reaction label,e and returns the non-identical predecessors of a reaction label, e' in the sequence of reactions labels. An example: $Fst(e_1, e_2) = e_1$ and $Snd(e_1, e_2) = e_2$. The algorithm computes the KLD on the edge labels of W_1 and W_2. The edge labels contain probabilities. The algorithm is based on fixed point computation on two probabilistic labeled structures that have identical read. A requirement for computation of KLD is that the state space is the same for the two probabilistic structures. The input of the algorithm is R_e as stated earlier. $H(W_1\|W_2)$ is the KLD value of W_1 with respect to W_2. The algorithm shows the computation of $H(W_1\|W_2)$ but the same algorithm can be used to compute $H(W_2\|W_1)$.

Algorithm 1. Fixed Point Computation of KL Divergence on LPSs

---

Input: Set of Ordered Pairs,R_e
Output: Set of ordered pairs in the greatest fixed point,Y_∞.
1: $Y := R_e$;
2: $Y' := 0$;
3: $H(W_1\|W_2) = 0$;
4: while $(Y \neq Y')$
5:    {
6:        $Y' := Y$;
7:        $Y := Y \cap Pred^r(Y)$;
8:        $H(W_1\|W_2) =$
    $H(W_1 \| W_2) + P_c(Fst(Y), Pre(Fst(Y))) log(\frac{P_c(Fst(Y), Pre(Fst(Y)))}{P_c(Snd(Y), Pre(Snd(Y)))})$
9:    }
10: $Y_\infty = Y'$

Algorithm 1 terminates: The loop that begins in line (4) takes a finite number of steps,$i \in \mathbb{N}$ for the algorithm to terminate because there is a finite number of ordered pairs of edges in R_e.

Claim: The algorithm computes the fixed point, i.e. $Y = Pred^r(Y)$. Let Y_∞ be the set of ordered pairs at the end of the loop and $Y_\infty = Y' = Y$. By definition of the set, $Y' = \{(e_1, e_2) \mid e_1 \in E_1, e_2 \in E_2, L_e(e_1) = L_e(e_2)\}$. For every $(e_1, e_2) \in Y'$ implies $(e_1, e_2) \in Y$ because at the end of the loop, $Y_\infty = Y' = Y$. According to the statement in line (6) in the algorithm, every $(e_1, e_2) \in Y$ implies $(e_1, e_2) \in Pred^r(Y)$. Each $(e_1, e_2) \in Pred^r(Y)$ implies through Definition 10 that $(e_1, e_2) \in Y$. Therefore, $Y = Pred^r(Y)$.

The time complexity of the algorithm is $O(m^2)$ where $m = | R_e |$. In the worst case, the set of ordered pairs in $Pred^r(Y)$ is constructed by removing a pair (e_1, e_2) at a time. The *while* loop iterates m times over m computations in $Pred^r(Y)$. The above algorithm computing the greatest fixed point is based on the following recursive relation, Y_i defined on the ordered pairs (e_1, e_2):

$Y_{i+1} = Y_i \cap Pred^r(Y_i)$, where $Y_0 = \{(e_1, e_2) \mid L_e(e_1) = L_e(e_2)\}$. The greatest fixed point is the first $i \in \mathbb{N}$ such that $Y_\infty = Y_{i+1} = Y_i$. The algorithm is constructed inductively for read equivalence on the paths of the LPSs. The errors in the computation of (dis-)similarity by the fixed point algorithm are defined as: Auto-Path (A-Path) error and Co-Path(C-Path) error. Errors occur in the computation of the probabilities when two paths are compared in the algorithm.

A-Path error: occurs when KLD on two read equivalent paths are computed in the algorithm with one of the paths, being an *il* path and the other not an *il* path. The A-Path error is added to the KLD value and is computed by $plog\frac{p}{P_c(p_1,...,p_k)}$ where p is the probability of the non-*il* path and $P_c(p_1,...,p_k)$ represents the compact probability of the *il* path. A-path error quantifies the error for substituting the *il* path in the computation of read equivalence.

C-Path error: occurs when two *il* paths are read equivalent. The error is computed by $P_c(p_1,...,p_k)log\frac{P_c(p_1,...,p_k)}{P_c(p'_1,...,p'_m)}$ where $P_c(p_1,...,p_k)$ and $P_c(p'_1,...,p'_m)$ are the compact probabilities for *il* paths, π and π'.

The total error for a read of length one, ξ = A-Path error + C-Path error.

Theorem 1. *For every read in W_1 and W_2, the computation of $H(W_1\|W_2)$ is within the total error of ξ.*

Proof. By computation of $H(W_1\|W_2)$ in the Algorithm 1 for a read and definition of total error, the maximum error is ξ.

Theorem 2. *Given a total error, ξ there exists a read between W_1 and W_2 such that $H(W_1\|W_2) = P_c(Fst(Y), Pre(Fst(Y))log(\frac{P_c(Fst(Y),Pre(Fst(Y)))}{P_c(Snd(Y),Pre(Snd(Y)))}).$*

Proof. By construction and definition of ξ, there exists a read between W_1 and W_2 which has a total error of ξ. The KLD of the read is
$$H(W_1\|W_2) = P_c(Fst(Y), Pre(Fst(Y))log(\frac{P_c(Fst(Y),Pre(Fst(Y)))}{P_c(Snd(Y),Pre(Snd(Y)))}).$$

By similar reasoning, Theorems 1 and 2 is true for $H(W_2\|W_1)$.

6   Conclusion

The work focussed on the mechanistic approach to construct a multiscale and stochastic model of a system of chemical reactions. The formalism addressed uncertainty in data and in the model. The approximation in the form of products of probabilities of identically labeled fragments in the KLD computation was computed to fulfil the computation of KLD of two LPS with an identical state space. Error analysis of the algorithm provided quantification of the approximation of the algorithm in the comparing the two LPSs. The formalism provided a methodology to compare two models with similar ordering of reactions.

References

1. Abate, A.: Approximation metrics based on probabilistic bisimulations for general state-space Markov processes: a survey. Electron. Notes Theor. Comput. Sci. **297**, 3–25 (2013)
2. Bacci, G., Bacci, G., Larsen, K.G., Mardare, R.: On the metric-based approximate minimization of Markov Chains. J. Log. Algebraic Methods Program. **100**, 36–56 (2018)

3. Baier, C., Katoen, J.P., Larsen, K.G.: Principles of Model Checking. MIT press, Cambridge (2008)
4. Barbuti, R., Caravagna, G., Maggiolo-Schettini, A., Milazzo, P., Tini, S.: Foundational aspects of multiscale modeling of biological systems with process algebras. Theor. Comput. Sci. **431**, 96–116 (2012)
5. Cardelli, L., Tribastone, M., Tschaikowski, M., Vandin, A.: Syntactic Markovian bisimulation for chemical reaction networks. In: Aceto, L., Bacci, G., Bacci, G., Ingólfsdóttir, A., Legay, A., Mardare, R. (eds.) Models, Algorithms, Logics and Tools. LNCS, vol. 10460, pp. 466–483. Springer, Cham (2017). https://doi.org/10.1007/978-3-319-63121-9_23
6. Chabrier, N., Fages, F.: Symbolic model checking of biochemical networks. In: Priami, C. (ed.) CMSB 2003. LNCS, vol. 2602, pp. 149–162. Springer, Heidelberg (2003). https://doi.org/10.1007/3-540-36481-1_13
7. Chen, D., van Breugel, F., Worrell, J.: On the complexity of computing probabilistic bisimilarity. In: Birkedal, L. (ed.) FoSSaCS 2012. LNCS, vol. 7213, pp. 437–451. Springer, Heidelberg (2012). https://doi.org/10.1007/978-3-642-28729-9_29
8. Clarke, E.M., Grumberg, O., Peled, D.: Model Checking. MIT press, Cambridge (1999)
9. Daca, P., Henzinger, T.A., Kretinsky, J., Petrov, T.: Linear distances between Markov Chains. In: Desharnais, J., Jagadeesan, R. (eds.) 27th International Conference on Concurrency Theory (CONCUR 2016). Leibniz International Proceedings in Informatics (LIPIcs), vol. 59, pp. 20:1–20:15. Schloss Dagstuhl-Leibniz-Zentrum fuer Informatik, Dagstuhl, Germany (2016). http://drops.dagstuhl.de/opus/volltexte/2016/6182
10. Desharnais, J., Jagadeesan, R., Gupta, V., Panangaden, P.: Approximating labeled Markov processes. In: Proceedings of 15th Annual IEEE Symposium on Logic in Computer Science 2000, pp. 95–106. IEEE (2000)
11. Doyen, L., Henzinger, T.A., Raskin, J.F.: Equivalence of labeled Markov Chains. Int. J. Found. Comput. Sci. **19**(03), 549–563 (2008)
12. Feret, J., Henzinger, T., Koeppl, H., Petrov, T.: Lumpability abstractions of rule-based systems. Theor. Comput. Sci. **431**, 137–164 (2012)
13. Fijalkow, N., Kiefer, S., Shirmohammadi, M.: Trace refinement in labelled Markov decision processes. In: Jacobs, B., Löding, C. (eds.) FoSSaCS 2016. LNCS, vol. 9634, pp. 303–318. Springer, Heidelberg (2016). https://doi.org/10.1007/978-3-662-49630-5_18
14. Ghosh, K.: Computing equivalences on model abstractions representing multiscale processes. Nano Commun. Netw. **6**(3), 118–123 (2015)
15. Ghosh, K., Schlipf, J.: Formal modeling of a system of chemical reactions under uncertainty. J. Bioinf. Comput. Biol. **12**(05), 1440002 (2014)
16. Groote, J.F., Vaandrager, F.: An efficient algorithm for branching bisimulation and stuttering equivalence. In: Paterson, M.S. (ed.) ICALP 1990. LNCS, vol. 443, pp. 626–638. Springer, Heidelberg (1990). https://doi.org/10.1007/BFb0032063
17. Groote, J.F., Jansen, D.N., Keiren, J.J., Wijs, A.J.: An o (m log n) algorithm for computing stuttering equivalence and branching bisimulation. ACM Trans. Comput. Logic (TOCL) **18**(2), 13 (2017)
18. Kullback, S., Leibler, R.A.: On information and sufficiency. Ann. Math. Stat. **22**(1), 79–86 (1951)
19. Paige, R., Tarjan, R.E.: Three partition refinement algorithms. SIAM J. Comput. **16**(6), 973–989 (1987)

20. Paulevé, L.: Reduction of qualitative models of biological networks for transient dynamics analysis. IEEE/ACM Trans. Comput. Biol. Bioinf. **15**, 1167–1179 (2017, in press)
21. Pham, T.D., Zuegg, J.: A probabilistic measure for alignment-free sequence comparison. Bioinformatics **20**(18), 3455–3461 (2004)
22. Shin, S.W., Thachuk, C., Winfree, E.: Verifying chemical reaction network implementations: a pathway decomposition approach. Theor. Comput. Sci. (2017)
23. Snowden, T.J., van der Graaf, P.H., Tindall, M.J.: Methods of model reduction for large-scale biological systems: a survey of current methods and trends. Bull. Math. Biol. **79**, 1–38 (2017)
24. Sunnåker, M., Schmidt, H., Jirstrand, M., Cedersund, G.: Zooming of states and parameters using a lumping approach including back-translation. BMC Syst. Biol. **4**(1), 28 (2010)
25. Thorsley, D., Klavins, E.: Model reduction of stochastic processes using wasserstein pseudometrics. In: 2008 American Control Conference, pp. 1374–1381. IEEE (2008)

Order Preserving Barrier Coverage
with Weighted Sensors on a Line

Robert Benkoczi[1]([✉]), Daya Gaur[1], and Xiao Zhang[2]

[1] Department of Mathematics and Computer Science, University of Lethbridge,
Lethbridge, Alberta, Canada
`robert.benkoczi@uleth.ca`
[2] Singapore University of Technology and Design, Engineering Systems and Design,
Singapore, Singapore

Abstract. We consider a barrier coverage problem with heterogeneous
mobile sensors where the sensors are located on a line and the goal is
to move the sensors so that a given line segment, called the barrier, is
covered by the sensors. We focus on an important generalization of the
classical model where the cost of moving a sensor equals the weighted
travel distance and the sensor weights are arbitrary. For the objective of
minimizing the maximum cost of moving a sensor, it was recently shown
that the problem is NP-hard when the sensing ranges are arbitrary. In
contrast, Chen et al. give an $O(n^2 \log n)$ algorithm for the problem with
uniform weights and arbitrary sensing ranges. Xiao shows that restrict-
ing the problem to uniform sensing ranges but allowing arbitrary sensor
weights can also be solved exactly in time $O(n^2 \log n \log \log n)$, raising
the question whether other restrictions can be solved in time polynomial
in the size of the instance. In this paper, we show that a natural restriction in which sensors must preserve their relative ordering (the sensors
move on rails, for example) but the sensors have arbitrary sensing ranges
and weights can be solved in time $(n^2 \log^3 n)$. Due to the combinatorially rich set of configurations of the optimal solution for our problem, our
algorithm uses the general parametric search method of Megiddo which
parameterizes a feasibility test algorithm. Interestingly, it is not easy
to design an efficient feasibility test algorithm for the order preserving
problem. We overcome the difficulties using the concept of critical budget
values and employing standard computational geometry techniques.

1 Introduction and Problem Definition

Barrier coverage is an effective approach to ensure that access to a certain area
is monitored by a set of sensors. Rather than deploying a large number of simple
sensors in the area of interest, barrier coverage deploys a small number of mobile
sensors that monitor the perimeter around the area.

Kumar et al. [7,8] were among the first to investigate barrier coverage problems. Their work has motivated many researchers who have looked at a broad
range of problems in different topologies and with different objective functions.

© Springer Nature Switzerland AG 2018
S. Tang et al. (Eds.): AAIM 2018, LNCS 11343, pp. 244–255, 2018.
https://doi.org/10.1007/978-3-030-04618-7_20

Perhaps one of the simplest geometric settings is to cover a barrier represented by a line segment, using mobile sensors that can travel only along the line containing the segment. But even in this setting, the problem is not trivial. Czyzowicz et al. [4] were the first to propose this problem with the aim of minimizing the maximum travel distance for the sensors. They describe a natural algorithm that outputs an optimal solution in time $O(n^2)$ for the problem where sensors have identical covering ranges. Chen et al. later improved this complexity to $O(n \log n)$ [3]. They were the first to show that the version with sensors that have arbitrary covering ranges can also be solved in polynomial time, in $O(n^2 \log n)$ time. Their contribution is significant since for a long time, no polynomial time algorithm for this problem was known. Their solution uses the powerful general technique of parametric optimization that we have also employed here.

A different variation of barrier coverage was proposed in a paper by Bar-Noy et al. [1], where the goal is not only to optimize the movement of the sensors, but also their coverage range which is variable. We note that in the Bar-Noy et al. paper, the cost of moving sensors is uniform. Other geometric settings have also been proposed. For example, Bhattacharya et al. consider the problem of covering the perimeter of a planar region and sensors move from the interior of the region to the boundary [2].

In this paper, we extend the results known so far for a natural generalization of the problem that was recently proposed in the PhD thesis of Zhang [10], where the moving cost for a sensor equals the weighted travel distance for the sensor. Unlike the problem in which the weight for all sensors is one, Xiao Zhang shows in his thesis that the weighted barrier coverage problem is NP-hard. He then shows that weighted barrier coverage can be solved in time polynomial in the problem size if the sensors have the same covering range, and he asks whether a different constraint on weighted barrier coverage in which sensors must maintain their relative ordering on the line but the sensing ranges are arbitrary, can also be solved exactly in time polynomial in the size of the input.

In this paper, we answer this question in the affirmative. We first propose a greedy feasibility test for the problem with arbitrary sensor weights and covering ranges, in which we are given a value λ and we need to determine if there exists a feasible solution that covers the barrier so that no sensor needs to pay a moving cost larger than λ. The constraint on the relative ordering of sensors helps to prove that the feasibility test is correct, but otherwise it introduces a number of technical difficulties which we were able to overcome using simple and elegant computational geometry concepts.

We use the feasibility test algorithm to give a polynomial time algorithm for the optimization version of the order preserving weighted barrier coverage problem using the parametric optimization method of Megiddo. The feasibility test receives for input a value λ and returns "yes" if it is possible to cover the barrier in such a way that no sensor has a moving cost larger than λ, otherwise it returns "no". We execute the feasibility test in parametric form (the input λ is an unknown parameter). We simulate the execution of the feasibility test on the optimal solution λ^* of the problem at every step of the algorithm that uses the unknown parameter λ^* in a comparison with a deterministic value by calling the feasibility test with a an appropriate constant value for λ.

However, the application of the parametric optimization method on the weighted barrier problem is not straightforward. In particular, the feasibility test does not always correctly emulate the decisions that the feasibility test would take if λ^* would have been known. This problem appears in the case of uniform sensor weights as well [3]. The authors overcome this problem by extending their feasibility test to determine whether the optimal cost is strictly smaller than a given value. We argue that, for the weighted version, the feasibility test can also be extended to distinguish strict inequalities, however the solution is different than for the problem with uniform weights. The parametric search procedure is a general method that has been applied for many problems. For our problem as well as for many of the variations we mentioned earlier, the currently known characterizations of the optimal solution are too rich combinatorially and parametric search seems the only approach that given efficient algorithms. It would be interesting to study whether other properties of the optimal solution are discovered so that we can extend our algorithmic toolbox for minmax barrier coverage.

To summarize, we feel that our contributions in this paper are significant in several ways.

- We contribute to an important yet little studied problem. Barrier coverage problems with weighted sensors and arbitrary covering ranges are very natural extensions of the barrier coverage problems studied so far, but very few algorithms exists in the literature concerning such problems.
- We solve the feasibility test efficiently and we give a first parametric optimization algorithm that runs in time polynomial in the number of sensors.
- We propose an effective data structure that allows us to handle constraints on the relative ordering of sensors when both the weight and the sensing ranges are arbitrary; this data structure may be useful in other similar contexts.

Problem definition:

Input:
- Barrier represented by the line segment $[0, L]$ with $L > 0$.
- A set of n sensors located on the line. Each sensor i has a known initial location x_i on the line, a covering range r_i, and a weight w_i. The sensors are indexed in the order they appear on the line, in other words, $x_i \leq x_{i+1}$ for all $i \in \{1, \ldots, n-1\}$.

Output:
- Final positions y_i for each sensor i so that every point on the barrier is covered by some sensor and the relative order of the sensors on the line is preserved. More precisely, $\forall x \in [0, L] \exists i \in \{1, \ldots, n\} | x - y_i| \leq r_i$ and $y_i \leq y_{i+1}$ for all $i \in \{1, \ldots, n-1\}$.

Measure: Minimize the maximum weighted travel distance, or

$$\min(\max_{1 \leq i \leq n} w_i |x_i - y_i|).$$

2 The Decision Version of the Order Preserving Barrier Coverage Problem

Solving the decision version of the order preserving barrier coverage problem with weighted sensors plays a central role in our approach to solve the problem. In the decision problem, we test the feasibility of the optimization version of the problem. We are given a weighted barrier problem instance consisting of a set of sensors and a barrier as defined in Sect. 1 and a positive value λ. We need to determine whether it is possible to move the sensors to cover the barrier so that the weighted travel distance for any sensor is not larger than λ.

We propose a natural greedy algorithm for this feasibility test that runs in time $O(n \log n)$ after a preprocessing step taking $O(n \log n)$ time, where n represents the number of sensors. After preprocessing is completed, the algorithm runs n iterations attempting to move the sensors one by one starting with sensor 1, in order to cover points on the barrier from left to right.

Consider iteration i of the algorithm, where $1 \leq i \leq n$. The algorithm maintains a sub-interval containing the origin which consists of points on the barrier that are covered by sensors strictly from set $\{1, \ldots, i\}$. We note that there may exist sensors with an index larger than i that cover points in this sub-interval, but we ignore such sensors at this stage. We denote by $q_i(\lambda)$ the rightmost point of the sub-interval of covered points obtained at the end of iteration i. The sub-interval is maximal, therefore $q_i(\lambda)$ is maximum. We define $q_0(\lambda) = 0$, thus when the algorithm starts, the entire barrier is considered uncovered.

Figure 1 illustrates a configuration of sensors and of covered barrier points at the start of iteration i. The set of covered barrier points is interval $[0, q_{i-1}(\lambda)]$ which was obtained in previous iterations. The figure shows the initial position x_i of sensor i and the final positions and covering ranges of sensors denoted $i_1, i_2, \ldots, i_l$ that were used by the algorithm to extend the covered region in previous iterations. At iteration i, the algorithm determines if sensor i can extend the coverage of the barrier to interval $[0, q_i(\lambda)]$ where $q_i(\lambda) \geq q_{i-1}(\lambda)$ and $q_i(\lambda)$ is maximum.

If the barrier coverage can be extended, sensor i may be moved to the left or to the right, depending on the value r_i of the covering range of sensor i. We note that the movement of sensor i is constrained by the relative order of the sensors and this order must be maintained. Figure 1 depicts three regions on the barrier.

- Region A may contain sensors with an index larger that i_l for which the cost λ was insufficient to allow the covered area of the barrier to be extended. Possibly, some of these sensors have been moved to the final position y_{i_l} of sensor i_l, the last sensor used by the algorithm to extend the barrier coverage, only to allow sensor i_l to reach point y_{i_l}. Other sensors in region A may have not been moved at all from their initial positions. These sensors cannot constrain the movement of sensor i even if i moves to the left because sensor i does not need to travel to the left of point $q_{i-1}(\lambda)$.

- Region B may contain sensors with an index larger that i_l but smaller than i. Again, such sensors were not moved during previous iterations of the algorithm. However, these sensors may constrain the movement of sensor i if i needs to move to the left.
- Finally, region C contains sensors with index larger than i which may constrain the movement of sensor i when i needs to move to the right.

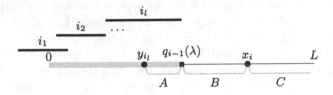

Fig. 1. Sensors and covered barrier points at the start of iteration i of the feasibility test algorithm

We can interpret the cost value λ in the feasibility test as a budget that allows sensor i to be moved. Let $Y_i(\lambda)$ be the set of points that sensor i can reach when given budget λ with the constraint that the relative order of the sensors is maintained while sensor i moves. Thus, part of budget λ is used to pay for moving other sensors in regions B or C if necessary. Let $Q_i(\lambda)$ be the set of locations for sensor i that can extend the interval $[0, q_{i-1}(\lambda)]$ of covered barrier points at the start of iteration i. Formally, $Q_i(\lambda) = \{y_i : y_i(\lambda) - r_i \leq q_{i-1}(\lambda) \leq y_i(\lambda) + r_i\}$. Thus sensor i can extend the coverage of the barrier in iteration i if $Y_i(\lambda) \cap Q_i(\lambda) \neq \emptyset$. The following relation defines the rightmost endpoint of the portion of the barrier covered by the sensors from set $\{1, \ldots, i\}$.

$$q_i(\lambda) = \begin{cases} \max\limits_{y_i \in Y_i(\lambda) \cap Q_i(\lambda)} (y_i + r_i), & \text{if } Y_i(\lambda) \cap Q_i(\lambda) \neq \emptyset \\ q_{i-1}(\lambda), & \text{otherwise.} \end{cases} \tag{1}$$

To complete the description of the feasibility test algorithm, we need to explain how to compute the interval $Q_i(\lambda)$ of points reachable by sensor i. We first present a procedure which executes in $O(n)$ time and, in Sect. 2.2, we show that the complexity of this procedure can be reduced to $O(\log^2 n)$ using preprocessing.

2.1 Computing the Order Preserving Sensor Movement

Without loss of generality, consider that sensor i moves to the right, and suppose there exits another sensor j to the right of i. In order to preserve the relative ordering of the sensors, if sensor i moves to a location $x > x_j$, sensor j needs to maintain the same location with sensor i, therefore the true cost of moving sensor i to a location $x > x_j$ is $\max\{w_i(x - x_i), w_j(x - x_j)\}$. Notice that, if $w_j \leq w_i$, the cost of moving sensor i dominates that of moving sensor j.

When $w_j > w_i$, the cost of moving sensor i can be dominated by the cost of moving sensor j if budget λ is sufficiently large. We can compute the critical value for the budget, denoted $\lambda_{i,j}$, so that if $\lambda \geq \lambda_{i,j}$, then the cost of moving sensor j dominates the cost of moving sensor i. We have,

$$\lambda_{i,j} = |x_j - x_i|\frac{w_j w_i}{w_j - w_i}. \tag{2}$$

We can generalize this example to consider all sensors to the right of sensor i. Then the true cost of moving sensor i to the right on the point with coordinate x, which we denote $Z_i(x)$, is given by the following relation.

$$Z_i(x) = \max\{w_j(x - x_j) : i \leq j \leq n \text{ and } x_j \leq x\}. \tag{3}$$

Given a value λ for the moving cost of sensor i, we can compute the farthest point $y_i^+(\lambda)$ to the right of i that sensor i can move to, possibly displacing other sensors to preserve order so that the moving cost of i is no larger than λ.

$$y_i^+(\lambda) = \min_{i \leq j \leq n} \left(\frac{\lambda}{w_j} + x_j\right). \tag{4}$$

Similarly, we can compute the farthest point to the left that sensor i can move given budget λ. We can therefore state the following theorem.

Theorem 1. *The set $Q_i(\lambda)$ of points that can be reached by sensor i while preserving the relative order of sensors so that none of the sensor moved incurs a moving cost larger than λ can be obtained, in a straightforward way, in time $O(n)$.*

We describe the algorithm for feasibility test in Algorithm 1. It is not difficult to see that the complexity of this algorithm is $O(n^2)$ if we compute $Q_i(\lambda)$ according to Theorem 1.

Before we discuss a more efficient implementation of the feasibility test algorithm, we argue its correctness. We state the following theorem.

Theorem 2. *Algorithm 1 returns YES if and only if there exists a feasible cover of the barrier with cost no larger than λ.*

Proof. ($\Rightarrow$) Suppose the algorithm returns YES. Since the algorithm computes a cover in this case, it follows that such a cover must exist.

($\Leftarrow$) The proof of this statement relies on the following lemma which can be shown by induction over i. The proof of the lemma is omitted for brevity. The lemma formalizes the fact that the greedy step in the feasibility test is actually optimal.

Lemma 1. *Let $Y' = (y_j')_{1 \leq j \leq n}$ be an arbitrary vector of final sensor positions so that the moving cost of any sensor is no larger than λ. For any $i \in \{1, \ldots, n\}$, let $[0, q_i']$ be the interval containing 0 covered by the set $\{1, \ldots, i\}$ of sensors positioned according to Y'. Let $[0, q_i(\lambda)]$ be the interval containing 0 covered by the same set of sensors according to Algorithm 1. Then $[0, q_i'] \subseteq [0, q_i(\lambda)]$.*

Algorithm 1. Feasibility test

Data: n sensors with initial positions x_i, weights w_i, and sensor ranges r_i,
   $1 \leq i \leq n$,
Barrier $[0, L]$,
Cost λ.
Result: YES if the barrier can be completely covered by moving each sensor i
   to point y_i so that $y_i \leq y_{i+1}$ $\forall 1 \leq i < n$ and the moving cost for each
   sensor is no larger than λ
   NO otherwise

1 $q_0(\lambda) \leftarrow 0$
2 $S \leftarrow \emptyset$ // Set S contains sensors that have not been moved and may
    constrain the movement of the current sensor to the left
3 **for** $i \leftarrow 1$ **to** n **do**
4 | Compute the set $Q_i(\lambda)$ of points reachable by i
5 | **if** *interval i can extend the coverage according to Eq. (1)* **then**
6 | | Compute $q_i(\lambda)$ and y_i
7 | **else**
8 | | $S \leftarrow S \cup \{i\}$
9 | | $y_i = x_i$, $q_i(\lambda) = q_{i-1}(\lambda)$.
10 | **end**
11 **end**
12 **if** $q_n(\lambda) \geq L$ **then**
13 | **return** *YES; output y_i for $1 \leq i \leq n$*
14 **else**
15 | **return** *NO*
16 **end**

2.2 Computing the Order Preserving Sensor Movement Efficiently

Consider Fig. 2 representing the cost of moving sensor i to the right at the
location represented by the x axis. The order preserving cost of moving sensor
i to the right is determined by the upper envelope of the linear costs for every
sensor $j \geq i$ (see Eq. (3)).

Using the concept of critical budget values in Eq. (2) from the previous
section, and a technique similar to that used by Tamir et al. [6], we can compute this upper envelope in $O(n)$ time. We notice that, if the upper envelope is
determined by sensors $i, j_1, j_2, \ldots$, then the critical budget values are ordered as
follows: $\lambda_{i,j_1} < \lambda_{j_1,j_2} < \ldots$. If we store this ordered sequence of critical budget
values, we can obtain the farthest point $y_i^+(\lambda)$ reachable by sensor i in $O(\log n)$
time rather than $O(n)$ time as in Sect. 2.1. Unfortunately, the construction of
the upper envelope needs to be performed for each choice of i and the complexity
of the feasibility test algorithm remains $O(n^2)$.

However, we can precompute a segment tree data structure [5] to store this
upper envelope at each node of the structure. The segment tree consists of a
balanced binary tree with n leaves, with one leaf for each sensor (see Fig. 3).
Each internal node of the structure contains the upper envelope of the moving

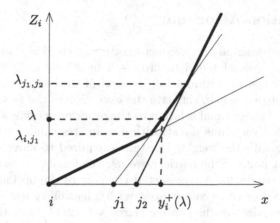

Fig. 2. The cost $Z_i(x)$ of moving sensor i to the right on location x while preserving the relative order of all sensors encountered by i

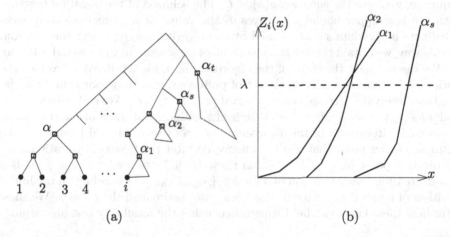

Fig. 3. (a) A segment tree storing the upper envelope of sensor moving costs; (b) Upper envelopes stored at nodes $\alpha_1, \alpha_2, \dots$

costs for the sensors that are descendants of the node. For example, in Fig. 3, node α contains the upper envelope of moving costs for sensors $\{1, \dots, 4\}$.

It can be shown that constructing this data structure takes $O(n \log n)$ time and space. Consider now computing the farthest point reachable by sensor i using this structure. Let $\alpha_1, \alpha_2, \dots, \alpha_t$ be the $O(\log n)$ nodes in the segment tree covering the set $\{i, \dots, n\}$ of leaves. The upper envelope function stored at every α_s node where $s \in \{1, \dots, t\}$, starts always on the x axis, but may interact with the envelopes from adjacent nodes. We compute the intersection of line λ with each envelope using binary search as described earlier, then we select the leftmost intersection point as answer. Since there are $O(\log n)$ such envelopes, the entire procedure takes $O(\log^2 n)$ time. We state the following theorem.

Theorem 3. *Algorithm 1 uses $O(n \log^2 n)$ time and $O(n \log n)$ space.*

3   Optimization Algorithm

In this section, we describe our parameterization of the feasibility test in Algorithm 1, using the general template proposed by Megiddo [9]. The idea is to run the feasibility test algorithm using the cost of the cover, λ, as an unknown parameter. The objective is to simulate the execution of the feasibility test as if $\lambda = \lambda^*$, the cost of the optimal solution to the order preserving weighted barrier coverage problem. Every time the algorithm compares value λ with some quantity (for example with the weighted distance required to move some sensor to cover the leftmost point of the barrier), we use the feasibility test to resolve the comparison as if parameter λ represented the cost of the optimal solution, λ^*. We will show how to resolve comparisons with a feasibility test shortly.

During the iterations of the parameterized version of Algorithm 1, we maintain a range of candidate values for λ^*, $\lambda^* \in (\lambda_{\min}, \lambda_{\max}]$. This interval may be refined by subsequent feasibility tests. Once we process the last interval in the sequence, we solve the equation $q(\lambda) = L$. The solution of the equation together with the best upper bound $\lambda_{\max}$ gives us the value for λ^*, which we then back-substitute in the data structure used by the algorithm to represent the solution. In this way, we obtain the actual movement of all sensors in the optimal solution.

We discuss next the steps of the algorithm in more detail. We introduce the notion of *interval* i to refer to the set of points covered by sensor i; in its initial position, interval i corresponds to interval $[x_i - r_i, x_i + r_i]$. We call points $x_i - r_i$ and $x_i + r_i$ the left endpoint and the right endpoint of interval i respectively. We consider iteration i of the algorithm when we test if interval i can cover the leftmost barrier point that was left uncovered after the first $i - 1$ iterations. We denote this point by $q_{i-1}(\lambda)$. We can show by induction over i that $q_i(\lambda)$ is a non-decreasing linear function of λ for $\lambda \in (\lambda_{\min}, \lambda_{\max}]$. Depending on the initial position of interval i relative to this point, we distinguish three cases. We show later how these cases can be distinguished using the feasibility test algorithm.

3.1   Case A: Interval i Is to the Left of $q_{i-1}(\lambda)$

In this case, interval i in its original position is to the left of the leftmost uncovered point on the barrier, $x_i + r_i < q_{i-1}(\lambda)$ (see Fig. 4). We will use the feasibility test algorithm with actual values for λ to determine whether interval i covers some points on the barrier in the optimal solution or not.

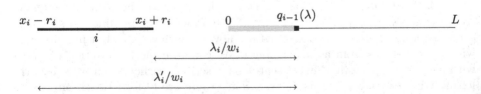

Fig. 4. Covering interval i is to the left of barrier point $q_{i-1}(\lambda)$

Recall that in Sect. 2.2, we describe a segment tree and an ordered list of critical budget values which allows us to answer the feasibility test in $O(n \log^2 n)$ time. Since we may use the feasibility test at least once per iteration, we do not need to use the segment tree in the parameterized version of the feasibility test. We can simply build the complete upper envelope for the sensors in set $\{i, \ldots, n\}$ in time $O(n)$ as argued at the start of Sect. 2.2. We can now identify the linear piece on the upper envelope that intersects λ^* as follows.

We select a critical budget value λ_{j_s, j_s+1} following the binary search routine, and we test whether $\lambda^* \leq \lambda_{j_s, j_s+1}$. The test can be resolved by comparing λ_{j_s, j_s+1} with the intervals $\lambda_{\min}$ and $\lambda_{\max}$. In case $\lambda_{\min} \leq \lambda_{j_s, j_s+1} \leq \lambda_{\max}$, we run Algorithm 1 with $\lambda = \lambda_{j_s, j_s+1}$. If the test is positive, then we know that $\lambda^* \leq \lambda_{j_s, j_s+1}$ and we select a smaller critical value in the binary search procedure, otherwise we select a larger critical value. Every time we run the feasibility test, we update the ranges $\lambda_{\min}$ or $\lambda_{\max}$.

At the end of this procedure which uses $O(\log n)$ feasibility tests with actual values, we have identified the linear function that determines how far sensor i can move given budget λ^*. We denote this function $y_i(\lambda)$. We now need to decide whether interval i can reach point $q_{i-1}(\lambda)$. This can be determined easily from the two functions linear in λ, y_i and q_{i-1}.

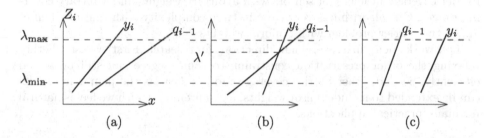

Fig. 5. Deciding if sensor i covers point $q_{i-1}(\lambda)$ in the optimal solution

Figure 5 illustrates the three possible situations. In case (a), sensor i does not cover point $q_{i-1}(\lambda)$ in the optimal solution, therefore sensor i is not moved. We update $q_i(\lambda) = q_{i-1}(\lambda)$. In case (c), sensor i covers point $q_{i-1}(\lambda)$. We update $q_i(\lambda) = y_i(\lambda)$ and we record the new position of sensor i as $y_i(\lambda) - r_i$. In Case (b), we run another feasibility test with value λ' and based on the answer, we reduce the case to either alternative (a) or (c).

To save space, the description for cases B and C is omitted, but will appear in the full version of the paper.

There is one important difference from the procedure executed in Case A. Consider the case when the linear functions y_i and q_{i-1} intersect precisely at line $\lambda_{\max}$. In this situation, we do not know if $\lambda^* = \lambda_{\max}$, in which case sensor i is moved to cover the barrier, or if $\lambda^* < \lambda_{\max}$, in which case sensor i is not moved.

We can use a procedure similar to that of Chen et al. [3] to determine the strict inequality above. When we determine, for example, the furthest reachable point of sensor i at iteration i of the feasibility test, we can test for equality or strict inequality by comparing the farthest reachable point by sensor i with the rightmost uncovered barrier point.

Based on the comments we made so far about our optimization algorithm, we state the following theorem.

Theorem 4. *The time complexity of the algorithm to solve the optimization version of the order preserving weighted barrier coverage problem is $O(n^2 \log^3 n)$ and the space complexity is $O(n \log n)$.*

4   Conclusion

We study an interesting generalization of the barrier coverage problem on a line where sensors are not identical. Besides having arbitrary covering ranges, sensors have different weights that define their moving cost. However, the mobility of the sensors is restricted: their original relative order must be maintained. In practice, such a constraint is motivated by applications where the mobile agents travel on rails. We give the first polynomial time algorithm for the order preserving barrier coverage problem for sensors with arbitrary weights and arbitrary covering ranges. Our algorithm is very close in time complexity with that of Chen et al. for the barrier problem with uniform weights [3].

Our work opens up a several new directions for research. First, we ask whether relaxing the order preservation constraint to some degree can still be solved efficiently. Second, other barrier coverage models such as the planar model [2] can be extended to include sensor weights, a requirement which we feel is natural for many practical applications.

Acknowledgement. The first author acknowledges the support received for this research from an NSERC Discovery Grant. We thank the anonymous reviewers for their comments which have strengthened this paper.

References

1. Bar-Noy, A., Rawitz, D., Terlecky, P.: Maximizing barrier coverage lifetime with mobile sensors. SIAM J. Discret. Math. **31**(1), 573–596 (2017)
2. Bhattacharya, B., Burmester, M., Hu, Y., Kranakis, E., Shi, Q., Wiese, A.: Optimal movement of mobile sensors for barrier coverage of a planar region. Theor. Comput. Sci. **410**(52), 5515–5528 (2009). Combinatorial Optimization and Applications
3. Chen, D.Z., Gu, Y., Li, J., Wang, H.: Algorithms on minimizing the maximum sensor movement for barrier coverage of a linear domain. Discret. Comput. Geom. **50**(2), 374–408 (2013)
4. Czyzowicz, J., et al.: On minimizing the maximum sensor movement for barrier coverage of a line segment. In: Ruiz, P.M., Garcia-Luna-Aceves, J.J. (eds.) ADHOC-NOW 2009. LNCS, vol. 5793, pp. 194–212. Springer, Heidelberg (2009). https://doi.org/10.1007/978-3-642-04383-3_15

5. de Berg, M., Cheong, O., van Kreveld, M., Overmars, M.: Computational Geometry: Algorithms and Applications. Springer, Heidelberg (2008). https://doi.org/10.1007/978-3-540-77974-2
6. Hassin, R., Tamir, A.: Improved complexity bounds for location problems on the real line. Oper. Res. Lett. **10**(7), 395–402 (1991)
7. Kumar, S., Lai, T.H., Arora, A.: Barrier coverage with wireless sensors. In: Proceedings of the 11th Annual International Conference on Mobile Computing and Networking, pp. 284–298. ACM (2005)
8. Kumar, S., Lai, T.H., Arora, A.: Barrier coverage with wireless sensors. Wirel. Netw. **13**(6), 817–834 (2007)
9. Megiddo, N.: Applying parallel computation algorithms in the design of serial algorithms. J. ACM (JACM) **30**(4), 852–865 (1983)
10. Zhang, X.: Algorithms for Barrier Coverage with Wireless Sensors. Ph.D. thesis, City University of Hong Kong (2016)

Achieving Location Truthfulness in Rebalancing Supply-Demand Distribution for Bike Sharing

Hongtao Lv[1], Fan Wu[1(✉)], Tie Luo[2], Xiaofeng Gao[1], and Guihai Chen[1]

[1] Department of Computer Science and Engineering, Shanghai Jiao Tong University, Shanghai, China
lvhongtao@sjtu.edu.cn, {fwu,gao-xf,gchen}@cs.sjtu.edu.cn
[2] Institute for Infocomm Research, A*STAR, Singapore, Singapore
luot@i2r.a-star.edu.sg

Abstract. Recently, station-free Bike sharing as an environment-friendly transportation alternative has received wide adoption in many cities due to its flexibility of allowing bike parking at anywhere. How to incentivize users to park bikes at desired locations that match bike demands - a problem which we refer to as a rebalancing problem - has emerged as a new and interesting challenge. In this paper, we propose a solution under a crowdsourcing framework where users report their original destinations and the bike sharing platform assigns proper relocation tasks to them. We first prove two impossibility results: (1) finding an optimal solution to the bike rebalancing problem is NP-hard, and (2) there is no approximate mechanism with bounded approximation ratio that is both truthful and budget-feasible. Therefore, we design a two-stage heuristic mechanism which selects an independent set of locations in the first stage and allocates tasks to users in the second stage. We show analytically that the mechanism satisfies location truthfulness, budget feasibility and individual rationality. In addition, extensive experiments are conducted to demonstrate the effectiveness of our mechanism. To the best of our knowledge, we are the first to address 2-D location truthfulness in the perspective of mechanism design.

Keywords: Location truthfulness · Bike sharing · Mechanism design

1   Introduction

Bike sharing as a convenient, health-promoting, and eco-friendly form of transportation, has been widely adopted in more than 1000 cities across the world [1].

This work was supported in part by the National Key R&D Program of China 2018YFB1004703, in part by China NSF grant 61672348, 61672353, and 61472252. The opinions, findings, conclusions, and recommendations expressed in this paper are those of the authors and do not necessarily reflect the views of the funding agencies or the government.

S. Tang et al. (Eds.): AAIM 2018, LNCS 11343, pp. 256–267, 2018.
https://doi.org/10.1007/978-3-030-04618-7_21

It substantially contributes to the reduction of traffic congestion and air pollution. In recent years, a new type of bike sharing, called station-free bike sharing, has been deployed in many cities[1] and attracted increasing attention.

Compared with traditional bike sharing, users of a station-free bike sharing system can pick up and drop off bikes at any valid locations rather than at designated stations. This new system brings new challenges. The foremost challenge is a more serious imbalance of bike distribution as compared to the traditional bike sharing, due to the much less restriction on parking locations and the asymmetry of bike demand. For instance, suppose a hospital is short of bikes while a nearby shopping mall has many redundant bikes. Without a proper rebalancing mechanism, subsequent shoppers would still go to the shopping mall to park for convenience, leading to a more and more serious imbalance.

To tackle this problem, a plausible solution is to design an incentive mechanism to motivate users to park their bikes in desirable locations. However, there are two challenges. First, there is a limited budget for the bike sharing platform to use as the incentive, and hence it should be used to the maximal efficiency. Second, there is a continuum of possible parking locations and a large number of bikes, making computation tractability a practical issue.

This paper addresses the bike rebalancing problem and our main contributions are as follows:

- We characterize the imbalance between bike demand and supply using the Kullback-Leibler (KL) divergence, and formulate an optimization problem under a crowdsourcing framework.
- Pertaining to this model, we prove two impossibility results: (1) the optimization problem is NP-hard, and the traditional VCG mechanism cannot be applied; (2) there is no truthful and budget-feasible mechanism for this problem that can achieve a bounded approximation ratio.
- Thus, we propose a two-stage heuristic mechanism as an alternative solution, which achieves both location truthfulness, budget feasibility, and individual rationality. To the best of our knowledge, we are the first to study the 2-D location truthfulness in the perspective of mechanism design.
- We conduct experiments using real-world data, and demonstrate the effectiveness of our mechanism as a viable solution.

2 Related Work

Optimizing bike sharing systems has attracted much research effort [2–4]. For station-based bike sharing, Singla et al. [5] proposed a crowdsourcing mechanism that incentivizes users in the bike repositioning process, where users report their destination stations and the system provides an offer that consists of recommended stations and corresponding incentives. Ghosh et al. [1] generated repositioning tasks with trailers using an optimization method. For station-free bike sharing, a deep reinforcement learning algorithm is proposed in [6]. In that work,

---

[1] https://mobike.com/cn/about/.

the platform learns to determine the payment based on their behaviors. It takes spatial and temporal features into consideration, but the proposed mechanism does not guarantee truthfulness. In contrast, our work achieves truthfulness and budget feasibility simultaneously.

In the field of crowdsourcing [7] and crowdsensing [8], a large body of works study the allocation and payment of spatial tasks [9,10], and especially some papers take the quality into consideration [8,11] which are similar to our work in a sense. In these works, users report their cost for tasks directly, but in reality, users may not know their exact cost. In our work, users only need to report their respective destinations, which would be a more practical approach.

3   The Model

In the bike rebalancing problem as illustrated in Fig. 1, there is a set of n users $N = \{1, 2, 3, \cdots, n\}$, and a set of m discrete locations $M = \{1, 2, 3, \cdots, m\}$. We assumed that the demand distribution $D(l)$ at all the locations $l \in M$ and the current bike distribution A_0 are known to the system (e.g., through the mobile apps and GPS), where A_0 means the set of the existing parked bikes and their respective locations. In this model, each user i who uses a bike needs to indicate or report her intended destination d_i on the map. The destinations of users are continuous in the 2-D area, but locations of tasks M are discrete points, each of which indexes a grid (see Fig. 1). We focus on bikes that are being in use and have not been parked (the parked ones are accounted for by A_0).

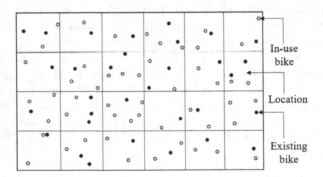

Fig. 1. Bike rebalancing problem: existing (parked) bikes, in-use bikes (to be parked), and locations (grids) for parking tasks.

As explained earlier, serious imbalance of bike distribution can happen if all the users park their bikes exactly at their destinations. Hence, the bike sharing platform would like to allocate a system-desired location l_i (rather than d_i) to user i for her to park her bike in order to match demand of bikes. In return, the platform offers an incentive p_i to user i if she takes that task. We employ an crowdsourcing framework as follows. Each location $l \in M$ corresponds to a task and each user $i \in N$ reports her destination as her bid. The distance between any two points x and y is denoted by H_{xy} and can be retrieved by the platform. The cost of user i for parking her bike at location l_i rather than her intended destination d_i is denoted by C_i or $C_{d_i l_i} = c * H_{d_i l_i}$, where the constant c is the unit travel cost. Hence, the utility of user i who takes a task of location l_i is $u_i = p_i - C_i$. In addition, we assume that a user does not accept a task whose location is outside range h of d_i, where h is a constant.

The platform has a budget B, within which it aims to design a mechanism to allocate desirable locations to users for balancing the demand and supply of bikes. The mechanism should satisfy the following properties:

- *Location truthfulness*: the utility of each user bidding truthfully should be no less than the utility of misreporting, i.e., $u_i(d_i, d_{-i}) \geq u_i(d_i', d_{-i}), \forall d_i' \neq d_i$.
- *Budget feasibility*: the payment to all users should not exceed the budget limitation, $\sum_i p_i \leq B$.
- *Computational efficiency*: the algorithm should terminate in polynomial time.
- *Individual rationality*: the utility of any user should be nonnegative, i.e., $p_i \geq C_i$.

3.1 Problem Formulation

We characterize the imbalance of bike distribution using KL divergence, which measures the expected logarithmic difference between two probability distribution X and Y, as defined by

$$KL(X\|Y) = \sum_i X(i) log \frac{X(i)}{Y(i)}.$$

The smaller the KL divergence is, the smaller the gap between X and Y is, and $KL(X\|Y) = 0$ means that X and Y are identical probability distributions. In our case, we substitute $Q(l) = \frac{D(l)}{\sum_{l' \in M} D(l')}$ for $X(i)$ (demand), and $\frac{|A(l)|}{|A|}$ for $Y(i)$ (supply), where A is the set of all the parked bikes including existing bikes and bikes with allocated tasks, and $A(l)$ is defined the same way but for location l only. We assume $|A_0(l)| > 0$ for all locations[2] to avoid singularity.

Thus, the KL-divergence is

$$KL(A) = \sum_l Q(l) log \frac{Q(l)|A|}{|A(l)|} \tag{1}$$

where we omit Q on the left hand side for notational convenience. Now, let A_i denote the set of all the parked bikes before user i parks her bike. If user i takes the task of parking a bike at location l_i, then we have

$$KL(A_i \cup (i, l_i)) = \sum_{l \neq l_i} Q(l) log \frac{Q(l)(|A_i| + 1)}{|A_i(l)|} + Q(l_i) log \frac{Q(l_i)(|A_i| + 1)}{|A_i(l_i)| + 1} \tag{2}$$

---

[2] This is generally ensured as long as a grid is not too small.

In this work, our goal is to minimize the imbalance of bike distribution, namely the KL divergence, so we define the contribution of user i as the difference between $KL(A_i)$ and $KL(A_i \cup (i, l_i))$. Based on Eq. (1) and (2), we have

$$\xi_i = KL(A_i) - KL(A_i \cup (i, l_i))$$
$$= log \frac{|A_i|}{|A_i| + 1} + Q(l_i)log \frac{|A_i(l_i)| + 1}{|A_i(l_i)|}.$$

Denote

$$\xi_i^1 = log \frac{|A_i|}{|A_i| + 1}, \quad \xi_i^2 = Q(l_i)log \frac{|A_i(l_i)| + 1}{|A_i(l_i)|}.$$

We can observe that the sum of the first item only depends on the total number of users $|N|$. Since our objective is to minimize the KL divergence, which is the total contribution of all the users, we can omit the first term ξ_i^1 because the sum of ξ_i^1 is a constant. Thus, we let $\xi_i = \xi_i^2$ in the following. Moreover, note that the sequence of task allocation influences users' contribution, because $A_i(l_i)$ and A_i are evolving when we sequentially calculate each user's contribution.

Based on the above, the bike rebalancing problem can be formulated as:

$$\begin{aligned} \textbf{max} \quad & \xi = \sum_{i \in U} \xi_i & (3) \\ \textbf{s.t.} \quad & \sum_{i \in U} p_i \leq B \\ & p_i \geq C_{d_i l_i} \quad \forall i \in U \end{aligned}$$

where U is the subset of users that are chosen to park in particular locations (namely, to perform parking tasks), p_i is the payment given to user i, which should be no less than her cost of performing the task. For users who are not selected (i.e., $N \setminus U$), they can just park at their intended destinations and the system does not allocate tasks to them.

3.2   NP-hardness

We prove that the problem (3) is NP-hard.

Theorem 1. *The bike rebalancing problem is NP-hard.*

Proof. We prove the decision version of the bike rebalancing problem is NP-hard. In the decision version, the question is whether there exists a subset of items U that satisfies both $\sum_{i \in U} \xi_i \geq K$ and $\sum_{i \in U} p_i \leq B$ for a given constant K.

We use reduction to NP-hardness from the *0-1 knapsack problem* which is a classic NP-complete problem, and is defined as follows.

Definition 1 (An Instance of 0-1 Knapsack Problem). *Given a set of n items, each with a positive weight w_i and a positive value v_i. Given a maximum weight capacity W and a constant K, the question is whether there exists a subset of items U that satisfies $\sum_{i \in U} v_i \geq K$ and $\sum_{i \in U} w_i \leq W$.*

We simplify the decision version of our problem to an instance where the acceptable range h is small enough such that there is only one choice of $\hat{l}_i$ for each user i and all the $\hat{l}_i$'s are non-overlapping. Thus, the quantities v_i, w_i, and W in the 0-1 knapsack problem correspond to ξ_i, p_i, and B in our case, respectively. Hence, the solution to the instance of the 0-1 knapsack problem is exactly the solution to the instance of our problem. In addition, the above reduction ends in polynomial time, which completes the proof.   □

4   Impossibility of Approximate Mechanisms

Theorem 1 shows that VCG mechanism is unusable due to the exponential time complexity of finding an optimal solution. One possible direction is to make use of the available results in [14,15] where the authors proposed budget-feasible approximate mechanisms for *submodular* functions which are defined as follows:

Definition 2 [17]. *A function* $V : 2^{[n]} \to \mathcal{R}_+$ *is submodular if* $V(S \cup \{i\}) - V(S) \geq V(T \cup \{i\}) - V(T), \forall S \subseteq T.$

In short, it means that the marginal contribution of a user decreases when the chosen user set becomes larger. However, our problem does not satisfy submodularity: when the set of chosen users expands from S to T, the (additional) user i's marginal contribution may *increase* because the user i may have multiple choices of tasks and the task allocated to her (and hence her contribution) may change when S changes to T. Therefore, the mechanisms introduced in [14,15] cannot be directly used. In fact, we prove that there does not exist an approximate mechanism with bounded approximation ratio for our problem.

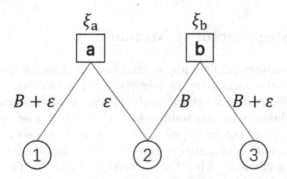

Fig. 2. An example showing impossibility where circles denote users and boxes denote tasks.

Theorem 2. *There is no approximate mechanism with a bounded approximation ratio that is truthful, budget-feasible and individually rational simultaneously for the bike rebalancing problem.*

Proof. Let us consider an example shown in Fig. 2, where the bid of location has been easily converted to the bid of cost by calculating the distance. The bidding profile is $x = \{(B+\epsilon, \infty), (\epsilon, B), (\infty, B+\epsilon)\}$, where ϵ can be any positive number less than B. ξ_a and ξ_b are the contribution of fulfilling task a and b respectively, and $\frac{\xi_b}{\xi_a}$ can be arbitrarily large. In the optimal solution, location b should be allocated to user 2, leading to a total contribution of ξ_b. We now show that any truthful, budget feasible and individually rational mechanism can achieve at most a total contribution of ξ_a.

Assume for the purpose of contradiction that there exists a mechanism f that satisfies these properties and guarantees a bounded approximation ratio. Let's consider the case of bidding profile $y = \{(B+\epsilon, \infty), (\epsilon, B+\epsilon), (\infty, B+\epsilon)\}$, where user 2 declares $B + \epsilon$ instead for location b. In this case, the optimal solution will allocate location a to user 2, so does the mechanism f. The reason is that (1) if f allocates location b to user 2, the cost of user 2 is above B, so it is neither budget feasible nor individually rational, and (2) if none of the locations a and b is allocated to user 2, then the total contribution is 0, and thus f can not guarantee a bounded approximation ratio. Given this allocation, to achieve truthfulness, the payment to user 2 for parking at location a has to be B because, otherwise, user 2 can misreport B for location a. Now, we can compare the bidding profiles x and y. In the case of y, the utility of user 2 is $B - \epsilon$. In the case of x, if mechanism f allocates location b to user 2, the utility of user 2 is at most 0, so she has incentive to misreport $B + \epsilon$ for location b to change the bidding profile into y to get better utility. Therefore, to ensure truthfulness, mechanism f has two choices in the case of x: allocating location a to user 2 or allocating nothing. In either case, the approximation ratio of total contribution is $\frac{OPT}{\sum_{l \in M} \xi_i} \geq \frac{\xi_b}{\xi_a}$ which can be arbitrarily large. Therefore, the mechanism cannot guarantee bounded approximation ratio, which constitutes the contradiction. $\square$

5   A Two-Stage Incentive Mechanism

Due to the impossibility result of approximate mechanisms, we propose a heuristic mechanism in this section for the bike rebalancing problem.

The main idea is to convert the problem into a submodular problem and then employ techniques for submodular functions. We choose some representative locations that are not overlapping, and restrict each user to choose one of these locations or none (not participating). This way, the function of total contribution becomes a submodular function. Note that this method is not impractical because in the real world there are typically some sparse locations that are short of bikes, such as subway stations or residential areas.

However, there are still two challenges in designing a heuristic mechanism: (1) the selection of locations is a maximum weighted independent set problem, which is an NP-complete problem [16], and (2) allocating tasks to users to achieve truthfulness and budget feasibility simultaneously is a difficult problem.

We propose a two-stage incentive mechanism. In the first stage, we construct a conflict network among locations by adding an edge of two locations if the

---

Algorithm 1. The Two-stage Mechanism

Input: N, M, A_0, $Q = \{Q(1), Q(2), \cdots, Q(m)\}$, set of bids $d = \{d_1, d_2, \cdots d_n\}$, and the conflict network $G = (V, E, W)$.

Output: set of winning allocation $(i, l_i) \in U$, and payment p_i, for winning user i.

1: $U_l \leftarrow \emptyset$, $L \leftarrow \emptyset$, $TC \leftarrow 0$, $N_l \leftarrow 0$;
2: **for** $l \in V$, $i \in N$ **do**
3:     **if** $H_{d_i l} \leq h$ **then**
4:         $N_l \leftarrow N_l \cup i$;
5:     **end if**
6: **end for**
7: **for** $l \in V$ **do**
8:     $\overline{\xi_l(A_0)} = Q(l) \log \frac{|A_0(l)|+1}{|A_0(l)|}$;
9: **end for**
10: Sort locations based on the contribution $\overline{\xi_l(A_0)}$ into a list M' in descending order;
11: **while** $M' \neq \emptyset$ **do**
12:     Let l' be the head of the list, and $G_{l'}$ be the neighbor set of l';
13:     $L \leftarrow L \cup \{l'\}$, $M' \leftarrow M' \backslash \{G_{l'} \cup l'\}$;
14: **end while**
15: **for** $l \in L$ **do**
16:     $B_l = \frac{|N_l| \cdot B}{\sum_{l' \in L} |N_{l'}|}$;
17:     Sort users in set N_l into a list N_l' based on $C_{d_i l}$ in nondecreasing order, and let j be the head of N_l';
18:     **while** $C_{d_j l} \leq \frac{B_l}{|U_l|+1}$ **do**
19:         $U_l \leftarrow U_l \cup (j, l)$, $N_l' \leftarrow N_l' \backslash i$;
20:         Let j be the new head of N_l';
21:     **end while**
22:     **for** $(i, l) \in U_l$ **do**
23:         $p_i = \min\{C_{d_j l}, \frac{B_l}{|U_l|}\}$;
24:     **end for**
25: **end for**

---

distance between them is no more than $2h$, and we assign the weight of each location to be the contribution of the first user who parks at the location. Then, we use a greedy method to find the maximum weighted independent set of locations. In the second stage, the budget is divided for selected locations, and users are chosen for each location using the critical price mechanism. The complete procedure is presented in Algorithm 1.

In Algorithm 1, line 2–6 is to determine the candidates that are adjacent to each location. Line 7–14 determines a maximum weighted independent set of locations and proportionally divides the budget to each location based on the number of users. Line 15–25 is to find the optimal set of winners in a greedy manner for each location, where the critical price $\min\{C_{d_j l}, \frac{B_l}{|U_l|}\}$ is used as the payment for the first unselected user j.

In the following, we prove four important properties of our proposed mechanism: truthfulness, individual rationality, budget feasibility and computation efficiency. For proving truthfulness, we give a definition of symmetric modular function and a lemma presented below.

Definition 3 [17]. *A function $V : 2^{[n]} \to \mathcal{R}_+$ is symmetric submodular if there exist $r_1 \geq \cdots \geq r_n \geq 0$, such that $V(S) = \sum_{i=1}^{|s|} r_i$.*

Intuitively, a function is *symmetric* if the value of the function is only determined by the cardinality of the set, and it is *submodular* if the marginal value is monotonously non-increasing.

Lemma 1 [14]. *For a symmetric submodular function with a given budget, the above mechanism of determining winners (line 17–24) is truthful.*

Theorem 3. *The two-stage incentive mechanism is location truthful.*

Proof. In the first stage, it's obvious that users cannot manipulate the selected locations because the sorting of locations only relies on the condition of locations rather than the bids of users. So, let user i be a candidate of location l, if she misreports her destination $d_i' \neq d_i$, it must fall into one of following cases:

Case 1: $H_{d_i'l} > h$. In this case, user i either becomes a candidate of another location $l' \neq l$, or fails to be a candidate. In the former scenario, based on the non-overlapping characteristic between different selected locations, we have $H_{d_il'} > h$, so it's beyond the acceptable range of user i. In the latter scenario, we easily have that user i's utility $u_i(d_i', d_{-i}) = 0 \leq u_i(d_i, d_{-i})$.

Case 2: $H_{d_i'l} \leq h$ and $d_i' \neq d_i$. In this case, we use Lemma 1. Due to the monotonicity of function $\log \frac{x+1}{x}$, if $S_l \subseteq T_l$, we have

$$\xi_l(S_l \cup \{i\}) - \xi_l(S_l) = Q(l) \log \frac{|A_0(l)| + |S_l| + 1}{|A_0(l)| + |S_l|}$$

$$\geq Q(l) \log \frac{|A_0(l)| + |S_l| + |T_l \backslash S_l| + 1}{|A_0(l)| + |S_l| + |T_l \backslash S_l|}$$

$$= Q(l) \log \frac{|A_0(l)| + |T_l| + 1}{|A_0(l)|} - Q(l) \log \frac{|A_0(l)| + |T_l|}{|A_0(l)|}$$

$$= \xi_l(T_l \cup \{i\}) - \xi_l(T_l).$$

Moreover, the function of total contribution $\xi_l = Q(l) \log \frac{|A_0(l)| + |U_l|}{|A_0(l)|}$ depends on cardinality only. Therefore, the contribution of a single location is a symmetric submodular function, and by Lemma 1, the above mechanism is truthful.   □

Theorem 4. *The two-stage incentive mechanism satisfies individual rationality.*

Proof. For an unselected user i, her payment and cost are both zero, so the utility $u_i = p_i - C_i = 0$. For a winning user i, by the line 18 in the algorithm, we have $C_i \leq \frac{B_l}{|U_l|}$, and by the nondecreasing order of N_l', we can get $C_i \leq C_j$, where j is the first unselected user. Therefore, we have that $u_i = \min\{C_j, \frac{B_l}{|U_l|}\} - C_i \geq 0$.□

Theorem 5. *The algorithm of the two-stage incentive mechanism has a polynomial-time computation complexity.*

Proof. The complexity of allocating users to adjacent locations (line 3–7) is $O(|V| \cdot |N|)$. The operation of sorting locations (line 10) is $O(|V| \cdot \log |V|)$. The computation complexity of determining winners for single location (line 17–24) is $O(|N_l| \cdot \log |N_l|)$, so for all selected locations, it's at most $O(|N| \cdot \log |N|)$. Since we have $|N| > \log |V|$ and $|V| > \log |N|$ in reality, the overall complexity of the two-stage mechanism is $O(|V| \cdot |N|)$.   □

Theorem 6. *The two-stage incentive mechanism is budget feasible.*

Proof. In the mechanism, the given budget is divided for each selected location, so we only need to prove the mechanism for each single location is budget feasible. For location l and the set of selected users A_l, the price is $\min\{C_{d_j l}, \frac{B_l}{|U_l|}\}$ where j is the first unselected user, so we have

$$\sum_{i \in N} p_i = \sum_{l \in L} \min\{C_{d_j l}, \frac{B_l}{|U_l|}\} \cdot |U_l|$$
$$\leq \sum_{l \in L} \frac{B_l}{|U_l|} \cdot |U_l|$$
$$= B$$

which proves the budget feasibility.   □

6   Performance Evaluation

We evaluate the effectiveness of our proposed mechanism using a real-world dataset from Mobike[3], which is a popular bike sharing company in China. We build a simulator that generates parking users and demand users based on the dataset of Beijing city from 10th to 14th May 2017.

The parameter values are set as follows. The cost of unit distance for each user is $c = 1RMB/km$, and the maximum acceptable range $h = 2\,km$. Unless otherwise specified, the number of existing bikes is 4000, the number of parking users is 1700, and the number of demand is 5000. We perform each experiment for 30 times and present the average value.

Three mechanisms are compared: our proposed two-stage heuristic mechanism (TSH), a randomized mechanism (RAN) and a randomized mechanism with selected locations (RAN-SL). In RAN, one user is chosen randomly in each round, and the platform picks all of nearby locations with higher demand than her affiliated location, then randomly chooses one to allocate to the user and pays her the maximum possible cost $p_i = c * h$ for performing that task. RAN-SL is similar to TSH in that it selects an independent set of locations the same way as in our mechanism. However, the platform randomly chooses a location and a candidate user for that location in each round, and the payment for each task is also the maximum cost $p_i = c * h$.

---

[3] https://mobike.com/global/.

We use *successful service ratio* (SSR) as the evaluation metric, which is defined as the proportion of demand that is satisfied, formally,

$$SSR = \frac{\sum_{l \in M} \min\{D(l), |A_0(l)| + |U_l|\}}{\sum_{l \in M} D(l)}.$$

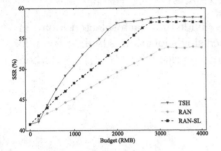

Fig. 3. Comparison on SSR with varying budget.

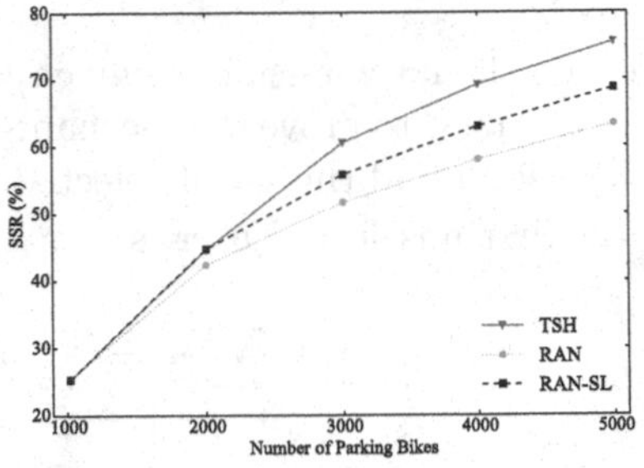

Fig. 4. The effect of the number of parking users.

The comparison of the successful service ratio (SSR) with varying budgets is illustrated in Fig. 3. We observe that SSR of all the three methods increases with the increase of budget until a threshold value. This is because all the candidate users have been selected and there is a remaining budget. Our method TSH outperforms the other methods in general. In addition, we see that the threshold of our method is about 2000 whereas the threshold of RAN and RAN-SL is about 3000, which indicates the *budget-saving* advantage of our mechanism (Fig. 4).

7   Conclusion

In this paper, we have studied the bike rebalancing problem in station-free bike sharing. We have proved two impossibility results for optimal and approximate mechanisms, respectively. Therefore, we have proposed a two-stage heuristic mechanism as an alternative solution, and showed that it is effective and outperforms other choices through our extensive experiments based on a real-world dataset. In future work, we plan to explicitly incorporate the temporal factor into an online model, and conduct pilot experiments in a real city.

References

1. Ghosh, S., Varakantham, P.: Incentivizing the use of bike trailers for dynamic repositioning in bike sharing systems. In: Proceedings of the Twenty-Seventh International Conference on Automated Planning and Scheduling, ICAPS 2017, Pittsburgh, Pennsylvania, USA, 18–23 June 2017, pp. 373–381 (2017)
2. Raviv, T., Tzur, M., Forma, I.A.: Static repositioning in a bike-sharing system: models and solution approaches. Euro J. Transp. Logist. **2**(3), 187–229 (2013)
3. Dell'Amico, M., Hadjicostantinou, E., Iori, M., Novellani, S.: The bike sharing rebalancing problem: mathematical formulations and benchmark instances. Omega **45**(2), 7–19 (2014)
4. Laporte, G., Meunier, F., Calvo, R.W.: Shared mobility systems. 4OR **13**(4), 341–360 (2015)
5. Singla, A., Santoni, M., Bartók, G., Mukerji, P., Meenen, M., Krause, A.: Incentivizing users for balancing bike sharing systems. In: Proceedings of the Twenty-Ninth AAAI Conference on Artificial Intelligence, 25–30 January 2015, Austin, Texas, USA, pp. 723–729 (2015)
6. Pan, L., Cai, Q., Fang, Z., Tang, P., Huang, L.: Rebalancing dockless bike sharing systems. arXiv preprint arXiv:1802.04592 (2018)
7. Luo, T., Kanhere, S.S., Das, S.K., Tan, H.-P.: Incentive mechanism design for heterogeneous crowdsourcing using all-pay contests. IEEE Trans. Mob. Comput. **15**(9), 2234–2246 (2016)
8. Peng, D., Wu, F., Chen, G.: Pay as how well you do: a quality based incentive mechanism for crowdsensing. In: Proceedings of the 16th ACM International Symposium on Mobile Ad Hoc Networking and Computing, pp. 177–186. ACM (2015)
9. Feng, Z., Zhu, Y., Zhang, Q., Ni, L.M., Vasilakos, A.V.: TRAC: truthful auction for location-aware collaborative sensing in mobile crowdsourcing. In: INFOCOM, 2014 Proceedings IEEE, pp. 1231–1239. IEEE (2014)
10. Wang, Q., Zhang, Y., Xiao, L., Wang, Z., Qin, Z., Ren, K.: Real-time and spatio-temporal crowd-sourced social network data publishing with differential privacy. IEEE Trans. Dependable Secur. Comput. **15**(4), 591–606 (2018)
11. Yang, S., Fan, W., Tang, S., Gao, X., Yang, B., Chen, G.: On designing data quality-aware truth estimation and surplus sharing method for mobile crowdsensing. IEEE J. Sel. Areas Commun. **35**(4), 832–847 (2017)
12. Procaccia, A.D., Tennenholtz, M.: Approximate mechanism design without money. In: Proceedings of the 10th ACM Conference on Electronic Commerce, pp. 177–186. ACM (2009)
13. Serafino, P., Ventre, C.: Truthful mechanisms without money for non-utilitarian heterogeneous facility location. In: AAAI, pp. 1029–1035 (2015)
14. Singer, Y.: Budget feasible mechanisms. In: 2010 51st Annual IEEE Symposium on Foundations of Computer Science (FOCS), pp. 765–774. IEEE (2010)
15. Chen, N., Gravin, N., Lu, P.: On the approximability of budget feasible mechanisms. In: Proceedings of the Twenty-second Annual ACM-SIAM Symposium on Discrete Algorithms, pp. 685–699. Society for Industrial and Applied Mathematics (2011)
16. Sakai, S., Togasaki, M., Yamazaki, K.: A note on greedy algorithms for the maximum weighted independent set problem. Discret. Appl. Math. **126**(2–3), 313–322 (2003)
17. Vickrey, W.: Counterspeculation, auctions, and competitive sealed tenders. J. Financ. **16**(1), 8–37 (1961)

Approximation Algorithms and a Hardness Result for the Three-Machine Proportionate Mixed Shop

Longcheng Liu[1,4], Guanqun Ni[2,4], Yong Chen[3,4], Randy Goebel[4], Yue Luo[1], An Zhang[3,4], and Guohui Lin[4(✉)]

[1] School of Mathematical Sciences, Xiamen University, Xiamen, China
longchengliu@xmu.edu.cn
[2] College of Management, Fujian Agriculture and Forestry University, Fuzhou, China
guanqunni@163.com
[3] Department of Mathematics, Hangzhou Dianzi University, Hangzhou, China
{chenyong,anzhang}@hdu.edu.cn
[4] Department of Computing Science, University of Alberta, Edmonton, AB, Canada
{rgoebel,guohui}@ualberta.ca

Abstract. A mixed shop is to process a mixture of a set of flow-shop jobs and a set of open-shop jobs. Mixed shops are in general much harder than flow-shops and open-shops, and have been studied since the 1980's. We consider the three machine proportionate mixed shop problem denoted as $M3 \mid prpt \mid C_{\max}$, in which each job has equal processing times on all three machines. Koulamas and Kyparisis (Eur J Oper Res 243:70–74, 2015) showed that the problem is solvable in polynomial time in some very special cases; for the non-solvable case, they proposed a 5/3-approximation algorithm. In this paper, we present an improved 4/3-approximation algorithm and show that this ratio of 4/3 is asymptotically tight; when the largest job is a flow-shop job, we present a fully polynomial-time approximation scheme (FPTAS). On the negative side, while the $F3 \mid prpt \mid C_{\max}$ problem is polynomial-time solvable, we show an interesting hardness result that adding one open-shop job to the job set makes the problem NP-hard if this open-shop job is larger than any flow-shop job.

Keywords: Scheduling · Mixed shop · Proportionate
Approximation algorithm
Fully polynomial-time approximation scheme

1 Introduction

We study in this paper the following three-machine proportionate mixed shop, denoted as $M3 \mid prpt \mid C_{\max}$ in the three-field notation [4]. Given three machines M_1, M_2, M_3 and a set $\mathcal{J} = \mathcal{F} \cup \mathcal{O}$ of jobs, where $\mathcal{F} = \{J_1, J_2, \ldots, J_\ell\}$ and $\mathcal{O} = \{J_{\ell+1}, J_{\ell+2}, \ldots, J_n\}$, each job $J_i \in \mathcal{F}$ needs to be processed non-preemptively

© Springer Nature Switzerland AG 2018
S. Tang et al. (Eds.): AAIM 2018, LNCS 11343, pp. 268–280, 2018.
https://doi.org/10.1007/978-3-030-04618-7_22

through M_1, M_2, M_3 sequentially with a processing time p_i on each machine and each job $J_i \in \mathcal{O}$ needs to be processed non-preemptively on M_1, M_2, M_3 in any machine order with a processing time q_i on each machine. The scheduling constraint is usual in that at every time point a job can be processed by at most one machine and a machine can process at most one job. The objective is to minimize the maximum job completion time, i.e., the makespan.

The jobs of $\mathcal{F}$ are referred to as *flow-shop jobs* and the jobs of $\mathcal{O}$ are called *open-shop jobs*. The mixed shop is to process such a mixture of a set of flow-shop jobs and a set of open-shop jobs. We assume without loss of generality that $p_1 \geq p_2 \geq \ldots \geq p_\ell$ and $q_{\ell+1} \geq q_{\ell+2} \geq \ldots \geq q_n$.

Mixed shops have many real-life applications and have been studied since the 1980's. The scheduling of medical tests in an outpatient health care facility and the scheduling of classes/exams in an academic institution are two typical examples, where the patients (students, respectively) must complete a number of medical tests (academic activities, respectively); some of these activities must be done in the same sequential order while the others can be finished in any order; and the time-spans for all these activities should not overlap with each other. The *proportionate* shops were also introduced in the 1980's [11] and they are one of the most specialized shops with respect to the job processing times which have received many studies [12].

Masuda et al. [10] and Strusevich [16] considered the two-machine mixed shop problem to minimize the makespan, i.e., $M2 \parallel C_{\max}$; they both showed that the problem is polynomial time solvable. Shakhlevich and Sotskov [14] studied mixed shops for processing two jobs with an arbitrary regular objective function. Brucker [1] surveyed the known results on the mixed shop problems either with two machines or for processing two jobs. Shakhlevich et al. [13] studied the mixed shop problems with more than two machines for processing more than two jobs, with or without preemption. Shakhlevich et al. [15] reviewed the complexity results on the mixed shop problems with three or more machines for processing a constant number of jobs.

When $\mathcal{O} = \emptyset$, the $M3 \mid prpt \mid C_{\max}$ problem reduces to the $F3 \mid prpt \mid C_{\max}$ problem, which is solvable in polynomial time [2]. When $\mathcal{F} = \emptyset$, the problem reduces to the $O3 \mid prpt \mid C_{\max}$ problem, which is ordinary (or called weakly) NP-hard [8]. It follows that the $M3 \mid prpt \mid C_{\max}$ problem is at least ordinary NP-hard. Recently, Koulamas and Kyparisis [7] showed that for some very special cases, the $M3 \mid prpt \mid C_{\max}$ problem is solvable in polynomial time; for the non-solvable case, they showed an absolute performance bound of $2 \max\{p_1, q_{\ell+1}\}$ and presented a 5/3-approximation algorithm.

In this paper, we design an improved 4/3-approximation algorithm for (the non-solvable case of) the $M3 \mid prpt \mid C_{\max}$ problem, and show that the performance ratio of 4/3 is asymptotically tight. When the largest job is a flow-shop job, that is $p_1 \geq q_{\ell+1}$, we present a *fully polynomial-time approximation scheme* (FPTAS). On the negative side, while the $F3 \mid prpt \mid C_{\max}$ problem is polynomial-time solvable, we show an interesting hardness result that adding one single open-shop job to the job set makes the problem NP-hard if this open-shop job is larger than any flow-shop job. We construct the reduction from the well-known PARTITION problem [3].

The rest of the paper is organized as follows. In Sect. 2, we introduce some notations and present a lower bound on the optimal makespan $C_{\max}^*$. We present in Sect. 3 the FPTAS for the $M3 \mid prpt \mid C_{\max}$ problem when $p_1 \geq q_{\ell+1}$. The 4/3-approximation algorithm for the case where $p_1 < q_{\ell+1}$ is presented in Sect. 4, and the performance ratio of 4/3 is shown to be asymptotically tight. We show in Sect. 5 that, when there is only one open-shop job J_n and $p_1 < q_n$, the $M3 \mid prpt \mid C_{\max}$ problem is NP-hard, through a reduction from the PARTITION problem. We conclude the paper with some remarks in Sect. 6.

2   Preliminaries

For any subset of jobs $\mathcal{X} \subseteq \mathcal{F}$, the *total processing time* of the jobs of $\mathcal{X}$ on one machine is denoted as

$$P(\mathcal{X}) = \sum_{J_i \in \mathcal{X}} p_i.$$

For any subset of jobs $\mathcal{Y} \subseteq \mathcal{O}$, the *total processing time* of the jobs of $\mathcal{Y}$ on one machine is denoted as

$$Q(\mathcal{Y}) = \sum_{J_i \in \mathcal{Y}} q_i.$$

The set minus operation $\mathcal{J} \setminus \{J\}$ for a single job $J \in \mathcal{J}$ is abbreviated as $\mathcal{J} \setminus J$ throughout the paper.

Given that the *load* (*i.e.*, the total job processing time) of each machine is $P(\mathcal{F}) + Q(\mathcal{O})$, the job $J_{\ell+1}$ has to be processed by all three machines, and one needs to process all the flow-shop jobs of $\mathcal{F}$, the following lower bound on the optimum $C_{\max}^*$ is established [2,7]:

$$C_{\max}^* \geq \max\{P(\mathcal{F}) + Q(\mathcal{O}), \; 3q_{\ell+1}, \; 2p_1 + P(\mathcal{F})\}. \tag{1}$$

3   An FPTAS for the Case Where $p_1 \geq q_{\ell+1}$

In this section, we design an approximation algorithm $A(\epsilon)$ for the $M3 \mid prpt \mid C_{\max}$ problem when $p_1 \geq q_{\ell+1}$, for any given $\epsilon > 0$. The algorithm $A(\epsilon)$ produces a schedule π with its makespan $C_{\max}^\pi < (1 + \epsilon)C_{\max}^*$, and its running time is polynomial in both n and $1/\epsilon$.

Consider a bipartition $\{\mathcal{A}, \mathcal{B}\}$ of the job set $\mathcal{O} = \{J_{\ell+1}, J_{\ell+2}, \ldots, J_n\}$, *i.e.*, $\mathcal{A} \cup \mathcal{B} = \mathcal{O}$ and $\mathcal{A} \cap \mathcal{B} = \emptyset$. Throughout the paper, a part of the bipartition is allowed to be empty. The following *procedure* PROC$(\mathcal{A}, \mathcal{B}, \mathcal{F})$ produces a schedule π:

1. the jobs of $\mathcal{F}$ are processed in the *longest processing time* (LPT) order on all three machines, and every job is processed first on M_1, then on M_2, lastly on M_3;
2. the jobs of $\mathcal{A}$ are processed in the LPT order on all three machines, and every one is processed first on M_2, then on M_3, lastly on M_1;

3. the jobs of $\mathcal{B}$ are processed in the LPT order on all three machines, and every one is processed first on M_3, then on M_1, lastly on M_2; and
4. the machine M_1 processes (the jobs of) $\mathcal{F}$ first, then $\mathcal{B}$, lastly $\mathcal{A}$, denoted as $\langle \mathcal{F}, \mathcal{B}, \mathcal{A} \rangle$;
5. the machine M_2 processes $\mathcal{A}$ first, then $\mathcal{F}$, lastly $\mathcal{B}$, denoted as $\langle \mathcal{A}, \mathcal{F}, \mathcal{B} \rangle$;
6. the machine M_3 processes $\mathcal{B}$ first, then $\mathcal{A}$, lastly $\mathcal{F}$, denoted as $\langle \mathcal{B}, \mathcal{A}, \mathcal{F} \rangle$.

$\text{PROC}(\mathcal{A}, \mathcal{B}, \mathcal{F})$ runs in $O(n \log n)$ time to produce the schedule π, of which an illustration is shown in Fig. 1.

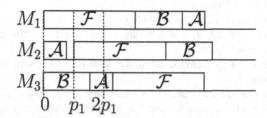

Fig. 1. An illustration of the schedule π produced by $\text{PROC}(\mathcal{A}, \mathcal{B}, \mathcal{F})$, where $\{\mathcal{A}, \mathcal{B}\}$ is a bipartition of the set $\mathcal{O}$ and the jobs of each of $\mathcal{A}, \mathcal{B}, \mathcal{F}$ are processed in the LPT order on all three machines.

The following two lemmas state that if both $Q(\mathcal{A}) \leq p_1$ and $Q(\mathcal{B}) \leq p_1$, or both $Q(\mathcal{A}) \geq p_1$ and $Q(\mathcal{B}) \geq p_1$, then the schedule π produced by $\text{PROC}(\mathcal{A}, \mathcal{B}, \mathcal{F})$ is optimal. Due to the space limit, we refer the readers to our arXiv submission [9] for the detailed proofs.

Lemma 1 [9]. *If both $Q(\mathcal{A}) \leq p_1$ and $Q(\mathcal{B}) \leq p_1$, then the schedule π produced by $\text{PROC}(\mathcal{A}, \mathcal{B}, \mathcal{F})$ is optimal, with its makespan $C_{\max}^\pi = C_{\max}^* = 2p_1 + P(\mathcal{F})$.*

Lemma 2 [9]. *If both $Q(\mathcal{A}) \geq p_1$ and $Q(\mathcal{B}) \geq p_1$, then the schedule π produced by $\text{PROC}(\mathcal{A}, \mathcal{B}, \mathcal{F})$ is optimal, with its makespan $C_{\max}^\pi = C_{\max}^* = P(\mathcal{F}) + Q(\mathcal{O})$.*

Now we are ready to present the approximation algorithm $A(\epsilon)$, for any $\epsilon > 0$.

In the first step, we check whether $Q(\mathcal{O}) \leq p_1$ or not. If $Q(\mathcal{O}) \leq p_1$, then we run $\text{PROC}(\mathcal{O}, \emptyset, \mathcal{F})$ to construct a schedule π and terminate the algorithm. The schedule π is optimal by Lemma 1.

In the second step, the algorithm $A(\epsilon)$ constructs an instance of the KNAP-SACK problem [3], in which there is an item corresponding to the job $J_i \in \mathcal{O}$, also denoted as J_i. The item J_i has a profit q_i and a size q_i. The capacity of the knapsack is p_1. The MIN-KNAPSACK problem is to find a subset of items of minimum profit that *cannot* be packed into the knapsack, and it admits an FPTAS [6]. The algorithm $A(\epsilon)$ runs a $(1 + \epsilon)$-approximation algorithm for the MIN-KNAPSACK problem to obtain a job subset $\mathcal{A}$. It then runs $\text{PROC}(\mathcal{A}, \mathcal{O} \setminus \mathcal{A}, \mathcal{F})$ to construct a schedule, denoted as π^1.

The MAX-KNAPSACK problem is to find a subset of items of maximum profit that can be packed into the knapsack, and it admits an FPTAS, too [5]. In the third step, the algorithm $A(\epsilon)$ runs a $(1 - \epsilon)$-approximation algorithm for the MAX-KNAPSACK problem to obtain a job subset $\mathcal{B}$. Then it runs $\text{PROC}(\mathcal{O} \setminus \mathcal{B}, \mathcal{B}, \mathcal{F})$ to construct a schedule, denoted as π^2.

The algorithm $A(\epsilon)$ outputs the schedule with a smaller makespan between π^1 and π^2. A high-level description of the algorithm $A(\epsilon)$ is provided in Fig. 2.

ALGORITHM $A(\epsilon)$:

1. If $Q(\mathcal{O}) \leq p_1$, then run $\text{PROC}(\mathcal{O}, \emptyset, \mathcal{F})$ to produce a schedule π; output the schedule π.
2. Construct an instance of KNAPSACK, where an item J_i corresponds to the job $J_i \in \mathcal{O}$; J_i has a profit q_i and a size q_i; the capacity of the knapsack is p_1.
   2.1. Run a $(1+\epsilon)$-approximation for MIN-KNAPSACK to obtain a job subset $\mathcal{A}$.
   2.2. Run $\text{PROC}(\mathcal{A}, \mathcal{O} \setminus \mathcal{A}, \mathcal{F})$ to construct a schedule π^1.
3. 3.1. Run a $(1 - \epsilon)$-approximation for MAX-KNAPSACK to obtain a job subset $\mathcal{B}$.
   3.2. Run $\text{PROC}(\mathcal{O} \setminus \mathcal{B}, \mathcal{B}, \mathcal{F})$ to construct a schedule π^2.
4. Output the schedule with a smaller makespan between π^1 and π^2.

Fig. 2. A high-level description of the algorithm $A(\epsilon)$.

In the following performance analysis, we assume without of loss of generality that $Q(\mathcal{O}) > p_1$. We have the following (in-)equalities inside the algorithm $A(\epsilon)$:

$$\text{OPT}^1 = \min\{Q(\mathcal{X}) \mid \mathcal{X} \subseteq \mathcal{O},\ Q(\mathcal{X}) > p_1\}; \tag{2}$$

$$p_1 < Q(\mathcal{A}) \leq (1 + \epsilon)\text{OPT}^1; \tag{3}$$

$$\text{OPT}^2 = \max\{Q(\mathcal{Y}) \mid \mathcal{Y} \subseteq \mathcal{O},\ Q(\mathcal{Y}) \leq p_1\}; \tag{4}$$

$$p_1 \geq Q(\mathcal{B}) \geq (1 - \epsilon)\text{OPT}^2, \tag{5}$$

where OPT^1 (OPT^2, respectively) is the optimum to the constructed MIN-KNAPSACK (MAX-KNAPSACK, respectively) problem.

Lemma 3. *In the algorithm $A(\epsilon)$, if $Q(\mathcal{O} \setminus \mathcal{A}) \leq p_1 - \epsilon\text{OPT}^1$, then for any bipartition $\{\mathcal{X}, \mathcal{Y}\}$ of the job set $\mathcal{O}$, $Q(\mathcal{X}) > p_1$ implies $Q(\mathcal{Y}) \leq p_1$.*

Proof. Note that the job subset $\mathcal{A}$ is computed in Step 2.1 of the algorithm $A(\epsilon)$, and it satisfies Eq. (3). By the definition of OPT^1 in Eq. (2) and using Eq. (3), we have $Q(\mathcal{X}) \geq \text{OPT}^1 \geq Q(\mathcal{A}) - \epsilon\text{OPT}^1$. Furthermore, from the fact that $Q(\mathcal{O}) = Q(\mathcal{X}) + Q(\mathcal{Y}) = Q(\mathcal{A}) + Q(\mathcal{O} \setminus \mathcal{A})$ and the assumption that $Q(\mathcal{O} \setminus \mathcal{A}) \leq p_1 - \epsilon\text{OPT}^1$, we have

$$Q(\mathcal{Y}) = Q(\mathcal{A}) + Q(\mathcal{O} \setminus \mathcal{A}) - Q(\mathcal{X})$$
$$\leq Q(\mathcal{A}) + Q(\mathcal{O} \setminus \mathcal{A}) - (Q(\mathcal{A}) - \epsilon \text{OPT}^1)$$
$$= Q(\mathcal{O} \setminus \mathcal{A}) + \epsilon \text{OPT}^1$$
$$\leq p_1 - \epsilon \text{OPT}^1 + \epsilon \text{OPT}^1$$
$$= p_1.$$

This finishes the proof of the lemma.   □

Lemma 4. *In the algorithm $A(\epsilon)$, if $Q(\mathcal{O} \setminus \mathcal{A}) \leq p_1 - \epsilon \text{OPT}^1$, then $C_{\max}^* \geq P(\mathcal{F}) + Q(\mathcal{O}) + p_1 - \text{OPT}^2$.*

Proof. Consider an arbitrary optimal schedule π^* that achieves the makespan $C_{\max}^*$. Note that the flow-shop job J_1 is first processed on the machine M_1, then on machine M_2, and last on machine M_3.

In the schedule π^*, let S_i and C_i be the start processing time and the finish processing time of the job J_1 on the machine M_i, respectively, for $i = 1, 2, 3$. On the machine M_2, let $\mathcal{J}^1 = \mathcal{O}^1 \cup \mathcal{F}^1$ denote the subset of jobs processed before J_1, and $\mathcal{J}^2 = \mathcal{O}^2 \cup \mathcal{F}^2$ denote the subset of jobs processed after J_1, where $\{\mathcal{O}^1, \mathcal{O}^2\}$ is a bipartition of the job set $\mathcal{O}$ and $\{\mathcal{F}^1, \mathcal{F}^2\}$ is a bipartition of the job set $\mathcal{F} \setminus J_1$. Also, let δ_1 and δ_2 denote the total amount of machine idle time for M_2 before processing J_1 and after processing J_1, respectively (see Fig. 3 for an illustration).

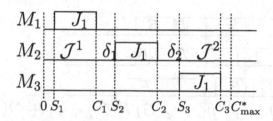

Fig. 3. An illustration of an optimal schedule π^*, in which $\mathcal{J}^1$ and $\mathcal{J}^2$ are the subsets of jobs processed on M_2 before J_1 and after J_1, respectively; δ_1 and δ_2 are the total amount of machine idle time for M_2 before processing J_1 and after processing J_1, respectively.

Note that $\mathcal{F} = J_1 \cup \mathcal{F}^1 \cup \mathcal{F}^2$ is the set of flow-shop jobs. The job J_1 and the jobs of $\mathcal{F}^1$ should be finished before time S_2 on the machine M_1, and the job J_1 and the jobs of $\mathcal{F}^2$ can only be started after time C_2 on the machine M_3. That is,

$$p_1 + P(\mathcal{F}^1) \leq S_2 \tag{6}$$

and

$$p_1 + P(\mathcal{F}^2) \leq C_{\max}^* - C_2. \tag{7}$$

If $Q(\mathcal{O}^1) \leq p_1$, then we have $Q(\mathcal{O}^1) \leq \mathrm{OPT}^2$ by the definition of OPT^2 in Eq. (4). Combining this with Eq. (6), we achieve that $\delta_1 = S_2 - P(\mathcal{F}^1) - Q(\mathcal{O}^1) \geq p_1 - \mathrm{OPT}^2$.

If $Q(\mathcal{O}^1) > p_1$, then we have $Q(\mathcal{O}^2) \leq p_1$ by Lemma 3. Hence, $Q(\mathcal{O}^2) \leq \mathrm{OPT}^2$ by the definition of OPT^2 in Eq. (4). Combining this with Eq. (7), we achieve that $\delta_2 = C^*_{\max} - C_2 - P(\mathcal{F}^2) - Q(\mathcal{O}^2) \geq p_1 - \mathrm{OPT}^2$.

The last two paragraphs prove that $\delta_1 + \delta_2 \geq p_1 - \mathrm{OPT}^2$. Therefore,

$$\begin{aligned} C^*_{\max} &= Q(\mathcal{O}^1) + P(\mathcal{F}^1) + \delta_1 + p_1 + Q(\mathcal{O}^2) + P(\mathcal{F}^2) + \delta_2 \\ &= P(\mathcal{F}) + Q(\mathcal{O}) + \delta_1 + \delta_2 \\ &\geq P(\mathcal{F}) + Q(\mathcal{O}) + p_1 - \mathrm{OPT}^2. \end{aligned}$$

This finishes the proof of the lemma. □

Lemma 5. *In the algorithm* $A(\epsilon)$, *if* $Q(\mathcal{O} \setminus \mathcal{A}) \leq p_1 - \epsilon \mathrm{OPT}^1$, *then* $C^{\pi^2}_{\max} < (1 + \epsilon)C^*_{\max}$.

Proof. Denote $\overline{\mathcal{B}} = \mathcal{O} \setminus \mathcal{B}$. Note that the job set $\mathcal{B}$ computed in Step 3.1 of the algorithm $A(\epsilon)$ satisfies $p_1 \geq Q(\mathcal{B}) \geq (1 - \epsilon)\mathrm{OPT}^2$, and the schedule π^2 is constructed by $\mathrm{PROC}(\overline{\mathcal{B}}, \mathcal{B}, \mathcal{F})$. We distinguish the following two cases according to the value of $Q(\overline{\mathcal{B}})$.

Case 1. $Q(\overline{\mathcal{B}}) \leq p_1$. In this case, the schedule π^2 is optimal by Lemma 1.

Case 2. $Q(\overline{\mathcal{B}}) > p_1$. The schedule π^2 constructed by $\mathrm{PROC}(\overline{\mathcal{B}}, \mathcal{B}, \mathcal{F})$ has the following properties (see Fig. 4 for an illustration):

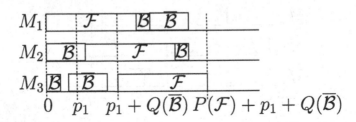

Fig. 4. An illustration of the schedule π^2 constructed by $\mathrm{PROC}(\overline{\mathcal{B}}, \mathcal{B}, \mathcal{F})$ in Case 2, where $Q(\mathcal{B}) \leq p_1$ and $Q(\overline{\mathcal{B}}) > p_1$. The machines M_1 and M_2 do not idle; the machine M_3 may idle between processing the job set $\mathcal{B}$ and the job set $\overline{\mathcal{B}}$ and may idle between processing the job set $\overline{\mathcal{B}}$ and the job set $\mathcal{F}$. M_3 starts processing the job set $\mathcal{F}$ at time $p_1 + Q(\overline{\mathcal{B}})$.

1. The jobs are processed consecutively on the machine M_1 since J_1 is the largest job. The completion time of M_1 is thus $C^{\pi^2}_1 = Q(\mathcal{O}) + P(\mathcal{F})$.
2. The jobs are processed consecutively on the machine M_2 due to $Q(\mathcal{B}) \leq p_1$ and $Q(\overline{\mathcal{B}}) > p_1$. The completion time of M_2 is thus $C^{\pi^2}_2 = Q(\mathcal{O}) + P(\mathcal{F})$.
3. The machine M_3 starts processing the job set $\mathcal{F}$ consecutively at time $p_1 + Q(\overline{\mathcal{B}})$ due to $Q(\mathcal{B}) \leq p_1$. The completion time of M_3 is $C^{\pi^2}_3 = P(\mathcal{F}) + p_1 + Q(\overline{\mathcal{B}})$.

Note that $C_3^{\pi^2} = P(\mathcal{F}) + p_1 + Q(\overline{\mathcal{B}}) \geq P(\mathcal{F}) + Q(\mathcal{B}) + Q(\overline{\mathcal{B}}) = Q(\mathcal{O}) + P(\mathcal{F})$, implying $C_{\max}^{\pi^2} = P(\mathcal{F}) + p_1 + Q(\overline{\mathcal{B}})$. Combining Eq. (5) with Lemma 4, we have

$$
\begin{aligned}
C_{\max}^{\pi^2} &= P(\mathcal{F}) + p_1 + Q(\overline{\mathcal{B}}) \\
&= P(\mathcal{F}) + Q(\mathcal{O}) + p_1 - Q(\mathcal{B}) \\
&\leq P(\mathcal{F}) + Q(\mathcal{O}) + p_1 - (1 - \epsilon)\mathrm{OPT}^2 \\
&\leq C_{\max}^* + \epsilon \mathrm{OPT}^2 \\
&< (1 + \epsilon) C_{\max}^*,
\end{aligned}
$$

where the last inequality is due to $\mathrm{OPT}^2 \leq p_1 < C_{\max}^*$. This finishes the proof of the lemma. □

Lemma 6. *In the algorithm $A(\epsilon)$, if $p_1 - \epsilon \mathrm{OPT}^1 < Q(\mathcal{O} \setminus \mathcal{A}) < p_1$, then $C_{\max}^{\pi^1} < (1 + \epsilon) C_{\max}^*$.*

Proof. Denote $\overline{\mathcal{A}} = \mathcal{O} \setminus \mathcal{A}$. Note that the job set $\mathcal{A}$ computed in Step 2.1 of the algorithm $A(\epsilon)$ satisfies $p_1 < Q(\mathcal{A}) \leq (1 + \epsilon)\mathrm{OPT}^1$, and the schedule π^1 is constructed by $\mathrm{PROC}(\mathcal{A}, \overline{\mathcal{A}}, \mathcal{F})$.

By a similar argument as in Case 2 in the proof of Lemma 5, replacing the two job sets $\mathcal{B}, \overline{\mathcal{B}}$ by the two job sets $\overline{\mathcal{A}}, \mathcal{A}$, we conclude that the makespan of the schedule π^1 is achieved on the machine M_3, $C_{\max}^{\pi^1} = P(\mathcal{F}) + Q(\mathcal{O}) + p_1 - Q(\overline{\mathcal{A}})$. Combining Eq. (1) with the assumption that $p_1 - \epsilon \mathrm{OPT}^1 < Q(\overline{\mathcal{A}})$, we have

$$
C_{\max}^{\pi^1} < P(\mathcal{F}) + Q(\mathcal{O}) + \epsilon \mathrm{OPT}^1 \leq C_{\max}^* + \epsilon \mathrm{OPT}^1 < (1 + \epsilon) C_{\max}^*,
$$

where the last inequality follows from $\mathrm{OPT}^1 \leq Q(\mathcal{O}) \leq C_{\max}^*$. This finishes the proof of the lemma. □

Theorem 1. *The algorithm $A(\epsilon)$ is a $\mathrm{Poly}(n, 1/\epsilon)$-time $(1 + \epsilon)$-approximation for the problem $M3 \mid prpt \mid C_{\max}$ when $p_1 \geq q_{\ell+1}$.*

Proof. First of all, the procedure $\mathrm{PROC}(\mathcal{X}, \mathcal{Y}, \mathcal{F})$ on a bipartition $\{\mathcal{X}, \mathcal{Y}\}$ of the job set $\mathcal{O}$ takes $O(n \log n)$ time. Recall that the job set $\mathcal{A}$ is computed by a $(1 + \epsilon)$-approximation for the MIN-KNAPSACK problem, which takes a polynomial time in both n and $1/\epsilon$; the other job set $\mathcal{B}$ is computed by a $(1 - \epsilon)$-approximation for the MAX-KNAPSACK problem, which also takes a polynomial time in both n and $1/\epsilon$. The total running time of the algorithm $A(\epsilon)$ is thus polynomial in both n and $1/\epsilon$ too.

When $Q(\mathcal{O}) \leq p_1$, or the job set $\mathcal{O} \setminus \mathcal{A}$ computed in Step 2.1 of the algorithm $A_1(\epsilon)$ has total processing time not less than p_1, the schedule constructed in the algorithm $A(\epsilon)$ is optimal by Lemmas 1 and 2. When $Q(\mathcal{O} \setminus \mathcal{A}) < p_1$, the smaller makespan between the two schedules π^1 and π^2 constructed by the algorithm $A(\epsilon)$ is less than $(1 + \epsilon)$ of the optimum by Lemmas 5 and 6. Therefore, the algorithm $A(\epsilon)$ has a worst-case performance ratio of $(1 + \epsilon)$. This finishes the proof of the theorem. □

4   A 4/3-Approximation for the Case Where $p_1 < q_{\ell+1}$

In this section, we present a 4/3-approximation algorithm for the $M3 \mid prpt \mid C_{\max}$ problem when $p_1 < q_{\ell+1}$, and we show that this ratio of 4/3 is asymptotically tight.

Theorem 2. *When $p_1 < q_{\ell+1}$, the $M3 \mid prpt \mid C_{\max}$ problem admits an $O(n \log n)$-time 4/3-approximation algorithm.*

Proof. Consider first the case where there are at least two open-shop jobs. Construct a permutation schedule π in which the job processing order for M_1 is $\langle J_{\ell+3}, \ldots, J_n, \mathcal{F}, J_{\ell+1}, J_{\ell+2} \rangle$, where the jobs of $\mathcal{F}$ are processed in the LPT order; the job processing order for M_2 is $\langle J_{\ell+2}, J_{\ell+3}, \ldots, J_n, \mathcal{F}, J_{\ell+1} \rangle$; the job processing order for M_3 is $\langle J_{\ell+1}, J_{\ell+2}, J_{\ell+3}, \ldots, J_n, \mathcal{F} \rangle$. See Fig. 5 for an illustration, where the start processing time for $J_{\ell+3}$ on M_2 is $q_{\ell+1}$, and the start processing time for $J_{\ell+3}$ on M_3 is $2q_{\ell+1}$. One can check that the schedule π is feasible when $p_1 < q_{\ell+1}$, and it can be constructed in $O(n \log n)$ time.

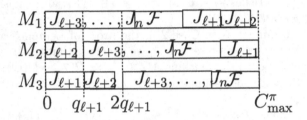

Fig. 5. A feasible schedule π for the $M3 \mid prpt \mid C_{\max}$ problem with $p_1 < q_{\ell+1}$.

The makespan of the schedule π is $C_{\max}^{\pi} = P(\mathcal{F}) + Q(\mathcal{O}) + q_{\ell+1} - q_{\ell+2}$. Combining this with Eq. (1), we have

$$C_{\max}^{\pi} \leq P(\mathcal{F}) + Q(\mathcal{O}) + q_{\ell+1} \leq \frac{4}{3} C_{\max}^{*}.$$

When there is only one open-shop job $J_{\ell+1}$, construct a permutation schedule π in which the job processing order for M_1 is $\langle \mathcal{F}, J_{\ell+1} \rangle$, where the jobs of $\mathcal{F}$ are processed in the LPT order; the job processing order for M_2 is $\langle \mathcal{F}, J_{\ell+1} \rangle$; the job processing order for M_3 is $\langle J_{\ell+1}, \mathcal{F} \rangle$. If $P(\mathcal{F}) \leq q_{\ell+1}$, then π has makespan $3q_{\ell+1}$ and thus is optimal. If $P(\mathcal{F}) > q_{\ell+1}$, then π has makespan $C_{\max}^{\pi} \leq 2q_{\ell+1} + P(\mathcal{F}) \leq \frac{4}{3} C_{\max}^{*}$. This finishes the proof of the theorem.   □

Remark 1. Construct an instance in which $p_i = \frac{1}{\ell-1}$ for all $i = 1, 2, \ldots, \ell$, $q_{\ell+1} = 1$ and $q_i = \frac{1}{n-\ell-2}$ for all $i = \ell+2, \ell+3, \ldots, n$. Then for this instance, the schedule π constructed in the proof of Theorem 2 has makespan $C_{\max}^{\pi} = 4 + \frac{1}{\ell-1}$; an optimal schedule has makespan $C_{\max}^{*} = 3 + \frac{1}{\ell-1} + \frac{1}{n-\ell-2}$ (see for an illustration in Fig. 6). This suggests that the approximation ratio of 4/3 is asymptotically tight for the algorithm in the proof of Theorem 2.

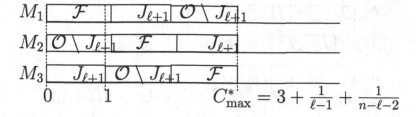

$$C^*_{\max} = 3 + \frac{1}{\ell-1} + \frac{1}{n-\ell-2}$$

Fig. 6. An optimal schedule for the constructed instance of the $M3 \mid prpt \mid C_{\max}$ problem, in which $p_i = \frac{1}{\ell-1}$ for all $i = 1, 2, \ldots, n$, $q_{\ell+1} = 1$ and $q_i = \frac{1}{n-\ell-2}$ for all $i = \ell+2, \ell+3, \ldots, n$.

5   NP-Hardness for the Case Where $\mathcal{O} = \{J_n\}$ and $p_1 < q_n$

In this section, we show that the $M3 \mid prpt \mid C_{\max}$ problem with only one open-shop job is already NP-hard if this open-shop job is larger than any flow-shop job. We prove the NP-hardness through a reduction from the PARTITION problem [3], which is a well-known NP-complete problem.

Theorem 3. *The $M3 \mid prpt \mid C_{\max}$ problem with only one open-shop job is NP-hard if this open-shop job is larger than any flow-shop job.*

Proof. An instance of the PARTITION problem consists of a set $S = \{a_1, a_2, a_3, \ldots, a_m\}$ where each a_i is a positive integer and $a_1 + a_2 + \ldots + a_m = 2B$, and the query is whether or not S can be partitioned into two parts such that each part sums to exactly B.

Let $x > B$, and we assume that $a_1 \geq a_2 \geq \ldots \geq a_m$.

We construct an instance of the $M3 \mid prpt \mid C_{\max}$ problem as follows: there are in total $m + 2$ flow-shop jobs, and their processing times are $p_1 = x, p_2 = x$, and $p_{i+2} = a_i$ for $i = 1, 2, \ldots, m$; there is only one open-shop job with processing time $q_{m+3} = B + 2x$. Note that the total number of jobs is $n = m + 3$, and one sees that the open-shop job is larger than any flow-shop job.

If the set S can be partitioned into two parts S_1 and S_2 such that each part sums to exactly B, then we let $\mathcal{J}^1 = J_1 \cup \{J_i \mid a_i \in B_1\}$ and $\mathcal{J}^2 = J_2 \cup \{J_i \mid a_i \in B_2\}$. We construct a permutation schedule π in which the job processing order for M_1 is $\langle \mathcal{J}^1, \mathcal{J}^2, J_{m+3} \rangle$, where the jobs of $\mathcal{J}^1$ and the jobs of $\mathcal{J}^2$ are processed in the LPT order, respectively; the job processing order for M_2 is $\langle \mathcal{J}^1, J_{m+3}, \mathcal{J}^2 \rangle$; the job processing order for M_3 is $\langle J_{m+3}, \mathcal{J}^1, \mathcal{J}^2 \rangle$. See Fig. 7 for an illustration, in which J_1 starts at time 0 on M_1, starts at time x on M_2, and starts at time $B + 2x$ on M_3; J_2 starts at time $B + x$ on M_1, starts at time $2B + 4x$ on M_2, and starts at time $2B + 5x$ on M_3; J_{m+3} starts at time 0 on M_3, starts at time $B + 2x$ on M_2, and starts at time $2B + 4x$ on M_1. The feasibility is trivial and its makespan is $C^\pi_{\max} = 3B + 6x$, suggesting the optimality.

Conversely, if the optimal makespan for the constructed instance is $3B + 6x = 3q_{m+3}$, then we will show next that S admits a partition into two equal parts.

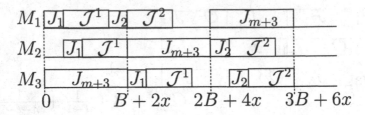

Fig. 7. A feasible schedule π for the constructed instance of the $M3 \mid prpt \mid C_{max}$ problem, when the set S can be partitioned into two equal parts S_1 and S_2. The partition of the flow-shop jobs $\{\mathcal{J}^1, \mathcal{J}^2\}$ is correspondingly constructed. In the schedule, the jobs of $\mathcal{J}^1$ and the jobs of $\mathcal{J}^2$ are processed in the LPT order, respectively.

Firstly, we see that the second machine processing the open-shop job J_{m+3} cannot be M_1, since otherwise M_1 has to process all the jobs of $\mathcal{F}$ before J_{m+3}, leading to a makespan greater than $3B + 6x$; the second machine processing the open-shop job J_{m+3} cannot be M_3 either, since otherwise M_3 has no room to process any job of $\mathcal{F}$ before J_{m+3}, leading to a makespan larger than $3B + 6x$ too. Therefore, the second machine processing the open-shop job J_{m+3} has to be M_2, see Fig. 8 for an illustration.

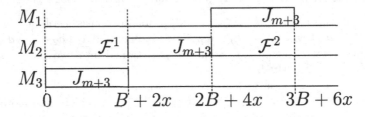

Fig. 8. An illustration of an optimal schedule for the constructed instance of the $M3 \mid prpt \mid C_{max}$ problem with $\mathcal{O} = \{J_{m+3}\}$ and $q_{m+3} = B + 2x$. Its makespan is $3B + 6x = 3q_{m+3}$.

Denote the job subsets processed before and after the job J_{m+3} on M_2 as $\mathcal{F}^1$ and $\mathcal{F}^2$, respectively. Since $x > B$, neither of $\mathcal{F}^1$ and $\mathcal{F}^2$ may contain both J_1 and J_2, which have processing times x. It follows that $\mathcal{F}^1$ and $\mathcal{F}^2$ each contains exactly one of J_1 and J_2, and subsequently $P(\mathcal{F}^1) = P(\mathcal{F}^2) = B + x$. Therefore, the jobs of $\mathcal{J}^1 \setminus \{J_1, J_2\}$ have a total processing time of exactly B, suggesting a subset of S sums to exactly B. This finishes the proof of the theorem. $\square$

6   Concluding Remarks

In this paper, we studied the three-machine proportionate mixed shop problem $M3 \mid prpt \mid C_{max}$. We presented first an FPTAS for the case where $p_1 \geq q_{\ell+1}$; and then proposed a 4/3-approximation algorithm for the other case where $p_1 < q_{\ell+1}$, for which we also showed that the performance ratio of 4/3 is asymptotically tight. The $F3 \mid prpt \mid C_{max}$ problem is polynomial-time solvable; we showed an interesting hardness result that adding only one open-shop job to the job set makes the problem NP-hard if the open-shop job is larger than any flow-shop job.

We believe that when $p_1 < q_{\ell+1}$, the $M3 \mid prpt \mid C_{max}$ problem can be better approximated than 4/3, and an FPTAS is perhaps possible. Nevertheless, a first step towards such an FPTAS is to design an FPTAS for the special case where there is only one open-shop job and the open-shop job is larger than any flow-shop job.

Acknowledgements. LL is supported by the CSC Grant 201706315073 and the Fundamental Research Funds for the Central Universities Grant No. 20720160035. GN is supported by the NSFC Grant 71501045, the NSF of Fujian Province Grant 2016J01332 and the Education Department of Fujian Province. YC and AZ are supported by the NSFC Grants 11771114 and 11571252; YC is also supported by the CSC Grant 201508330054. RG and GL are supported by the NSERC Canada; GL is also supported by the NSFC Grant 61672323.

References

1. Brucker, P.: Scheduling Algorithms. Springer, Berlin (1995). https://doi.org/10.1007/978-3-662-03088-2
2. Chin, F.Y., Tsai, L.L.: On j-maximal and j-minimal flow shop schedules. J. ACM **28**, 462–476 (1981)
3. Garey, M.R., Johnson, D.S.: Computers and Intractability: A Guide to the Theory of NP-Completeness. W. H. Freeman and Company, San Francisco (1979)
4. Graham, R.L., Lawler, E.L., Lenstra, J.K., Kan, R.: Optimization and approximation in deterministic sequencing and scheduling: a survey. Annu. Discret. Math. **5**, 287–326 (1979)
5. Kellerer, H., Pferschy, U.: Improved dynamic programming in connection with an FPTAS for the knapsack problem. J. Comb. Optim. **8**, 5–11 (2004)
6. Kellerer, H., Pferschy, U., Pisinger, D.: Knapsack Problems. Springer, Berlin (2004). https://doi.org/10.1007/978-3-540-24777-7
7. Koulamas, C., Kyparisis, G.J.: The three-machine proportionate open shop and mixed shop minimum makespan problems. Eur. J. Oper. Res. **243**, 70–74 (2015)
8. Liu, C.Y., Bulfin, R.L.: Scheduling ordered open shops. Comput. Oper. Res. **14**, 257–264 (1987)
9. Liu, L., et al.: Approximation algorithms for the three-machine proportionate mixed shop scheduling. CoRR 1809.05745 (2018)
10. Masuda, T., Ishii, H., Nishida, T.: The mixed shop scheduling problem. Discret. Appl. Math. **11**, 175–186 (1985)
11. Ow, P.S.: Focused scheduling in proportionate flowshops. Manag. Sci. **31**, 852–869 (1985)
12. Panwalkar, S., Smith, M.L., Koulamas, C.: Review of the ordered and proportionate flow shop scheduling research. Nav. Res. Logist. **60**, 46–55 (2013)
13. Shakhlevich, N., Sotskov, Y.N., Werner, F.: Shop-scheduling problems with fixed and non-fixed machine orders of the jobs. Ann. Oper. Res. **92**, 281–304 (1999)

14. Shakhlevich, N.V., Sotskov, Y.N.: Scheduling two jobs with fixed and nonfixed routes. Computing **52**, 17–30 (1994)
15. Shakhlevich, N.V., Sotskov, Y.N., Werner, F.: Complexity of mixed shop scheduling problems: a survey. Eur. J. Oper. Res. **120**, 343–351 (2000)
16. Strusevich, V.A.: Two-machine super-shop scheduling problem. J. Oper. Res. Soc. **42**, 479–492 (1991)

A New Algorithm Design Technique for Hard Problems, Building on Methods of Complexity Theory

András Faragó[(✉)]

Department of Computer Science, The University of Texas at Dallas,
Richardson, TX, USA
farago@utdallas.edu

Abstract. Our goal is to develop a general algorithm design technique for a certain type of heuristic algorithms, so that it *provably* applies to a large class of hard problems. A heuristic algorithm provides a correct decision for most inputs, but may fail on some. We focus on the case when failure means that the algorithm does not return any answer, rather than returning a wrong result. This type of failure is represented by a "don't know" answer. Such algorithms are called **errorless heuristics.** Their advantage is that whenever the algorithm returns any answer (other than "don't know"), the answer is *guaranteed* to be correct.

A reasonable quality measure for heuristics is the failure rate over the set of n-bit instances. When no efficient exact algorithm is available for a problem, then, ideally, we would like one with vanishing failure rate. We show, however, that this is hard to achieve: unless a complexity theoretic hypothesis fails (albeit less standard than $\mathbf{P} \neq \mathbf{NP}$), some $\mathbf{NP}$-complete problems cannot have a polynomial-time errorless heuristic algorithm with any vanishing failure rate.

On the other hand, as a key result, we prove that vanishing, even exponentially small, failure rate is achievable, if we use a somewhat different accounting scheme to count the failures. This is based on special sets, that we call α-spheres, which are the images of the n-bit strings under a bijective, polynomial-time computable and polynomial-time invertible encoding function α. The α-spheres form a partition of all binary strings, with similar properties as the sets of n-bit strings.

Our main result is that polynomial-time errorless heuristic algorithms exist, with exponentially low failure rates on the α-spheres, for a large class of decision problems. *This class includes, surprisingly, all known intuitively natural NP-complete problems.* Furthermore, the proof of the main theorem actually supplies a general scheme to *construct* the desired encoding and the errorless heuristic.

Keywords: Algorithm design · Errorless heuristics
Complexity theory

© Springer Nature Switzerland AG 2018
S. Tang et al. (Eds.): AAIM 2018, LNCS 11343, pp. 281–292, 2018.
https://doi.org/10.1007/978-3-030-04618-7_23

1   Introduction and Motivation

When a hard task is intractable to solve exactly, it is reasonable to try a heuristic algorithm, hoping that it will provide a correct decision for most inputs, but possibly not for all.

The quality of such heuristics can be measured by the success rate, that is, the fraction of inputs on which correct decision is guaranteed. On the rest of the instances the algorithm may return an incorrect decision, or no decision at all. If an algorithm never returns a wrong answer, only correct decisions, or possibly no decision at all (failure), then it is called *errorless heuristic*. In this paper we focus on this type of polynomial-time heuristics for decision problems.

The standard way to measure the success rate is to count the number of correct decisions on the set of n-bit strings. If the algorithm fails on r_n inputs, out of all n-bit strings, then the failure rate is $r_n/2^n$, or, equivalently, the success rate is $1 - r_n/2^n$. If no exact polynomial time algorithm is available for a decision problem, then, ideally, we look for an approximation with exponentially small failure rate, that is, $r_n/2^n \leq c^n$, for some constant $c < 1$.

Unfortunately, this ideal case is unlikely to be achievable, even if we are satisfied with *any* vanishing failure rate, which does not have to be exponentially small. We show in Sect. 4 that not all **NP**-complete problems have polynomial-time errorless heuristics with vanishing failure rates, unless a complexity theoretic hypothesis fails, albeit a less standard one than **P**$\neq$**NP**.

On the other hand, we prove that vanishing, even exponentially small, failure rate is achievable, if we use a somewhat different accounting scheme to count the failures. Specifically, we define sets, called α-spheres, which are defined by a bijective encoding $\alpha : \{0,1\}^* \mapsto \{0,1\}^*$ of strings, such that both α and α^{-1} are computable in polynomial time. An α-sphere, denoted by $S^{(\alpha)}$, is the image of the n-bit strings under α, that is, $S_n^{(\alpha)} = \{\alpha(x) \mid |x| = n\}$. This set system has many similarities to the set system of n bit strings: the α-sphere $S_n^{(\alpha)}$ has size 2^n, they form a partition of $\{0,1\}^*$, the mapping between n-bit strings and the elements of $S_n^{(\alpha)}$ is 1–1, it does not change string lengths more than polynomially, and the mapping is computable and invertible in polynomial time. Therefore, $\alpha(x)$ can be viewed as a feasible one-to-one re-encoding of the input x.

Our main result is that polynomial-time errorless heuristic algorithms do exist, with exponentially low failure rates on α-spheres, for a large class of decision problems. This class, surprisingly, includes all known intuitively natural **NP**-complete problems. Furthermore, the proof of the main theorem actually supplies a general method to construct the desired encoding and the errorless heuristic, thus providing a novel algorithm design technique.

To further motivate our approach, let us briefly review the area of heuristic algorithms. These algorithms come in two primary flavors:

1. Algorithms that may err on some inputs. These algorithms are required to run in polynomial time, provide an answer to all inputs, but may return a wrong answer on some inputs. The key issue here is the error frequency: on how many instances can the answer be wrong, out of the total of 2^n n-bit instances? (*Note:* we distinguish this error frequency from the *error rate*, by which we mean the *relative frequency* of errors.) Unfortunately, aiming at low error frequencies runs into conflict with widely accepted hypotheses in complexity theory. For a survey, see Hemaspaandra and Williams [7]. For example, it has been known for a long time that achieving polynomially bounded error frequency is impossible, unless $\mathbf{P} = \mathbf{NP}$. Subexponentially bounded error frequency is still known to imply highly unlikely complexity class collapses.

How about then *exponential* error frequency? Note that it can still yield an exponentially low error *rate*. For instance, a $2^{n/2}$ error frequency yields an error rate of $2^{n/2}/2^n = 2^{-n/2}$. Is that not good enough? The answer is that this task already turns "too easy:" it allows meaningless trivial heuristics. For example, if we pad an n-bit input x to $x0^n$, so that its length becomes $N = 2n$, and apply the trivial heuristic that accepts *all* inputs, then the error rate on the padded language is at most $2^{N/2}/2^N = 2^{-N/2}$. Of course, it does not produce the same error rate when mapped back to the *original* problem. But often just the strong asymmetry of yes- or no-instances in the original language can already lead to similar trivial cases, without the need for padding, which occurs even in natural tasks. For example, regarding the well known HAMILTONIAN CIRCUIT problem in graphs, one can prove[1] that all but an exponentially small fraction of n-vertex graphs have a Hamiltonian circuit. Thus, the "accept everything" trivial heuristic works with exponentially low error rate for this natural problem. Such a trivial heuristic is not meaningful, as it ignores the very structure we are looking for. This motivates a stronger class of heuristics, which excludes any error, but allows "don't know" answers, as explained below.

2. Errorless heuristics. These polynomial time algorithms never output a wrong decision, but may *fail* on some inputs (returning "don't know"). The error rate is zero, since no error is allowed, but there may be a nonzero *failure rate*. Observe that in the errorless case one cannot simply use a trivial heuristic, capitalizing on the strong asymmetry of yes- or no-instances. It would unavoidably lead to errors, which are not allowed here at all. That is, the algorithm has to correctly know when to say "don't know," which may be nontrivial to achieve. These schemes have intimate connections to average-case complexity, for a survey see Bogdanov and Trevisan [3].

Note, however, that the failure rate can depend on which sets of strings are used for reference. The traditional metric is to count the failures relative to all 2^n bit strings of length n. Let us call the latter sets the *spheres* of radius n, denoted by S_n. Nothing forces us, however, to use the S_n as reference sets. If α is a *bijection* on all strings, then we may just as well count the failures on the same sized sets $\alpha(S_n)$. If both α and α^{-1} are computable in polynomial time, then we call it a *p-isomorphic encoding*. Observe that such a transformation cannot hide much complexity. It preserves the sphere sizes, the sets $\alpha(S_n)$ form a partition of $\{0,1\}^*$, the mapping between n-bit strings and the elements of $S_n^{(\alpha)}$ is 1-1, it does not change string lengths more than polynomially, and the mapping is computable and invertible in polynomial time. Therefore, $\alpha(x)$ can

---

[1] Non-trivially, using methods from random graph theory, see, e.g., Bollobas [4].

be viewed as a feasible one-to-one re-encoding of the input x. But it may still alter the failure rate, because $|\alpha(S_n)| = |S_n|$ does not imply that the two sets have the same number of "don't know"-instances, even though the *entire* set of "don't know"-instances, of course, remains the same.

Our approach can be characterized as an errorless heuristic, which achieves exponentially low failure rate, capitalizing on an appropriate p-isomorphic encoding of the input. The class of languages for which it is possible forms a new complexity class, which we call *Roughly Polynomial Time*, abbreviated **RoughP**. The main result is that **RoughP** contains the family of paddable languages, which is a large class, including all known intuitively natural **NP**-complete problems.

2   Notations and Definitions

Set $\mathbb{N} = \{0, 1, 2, \ldots\}$. The length of a bit string x is denoted by $|x|$. The length of the empty string is 0. If a string x is of the form $x = uu$ for some $u \in \{0,1\}^*$, then x is called *symmetric*, otherwise it is *asymmetric*. A language $L \subseteq \{0,1\}^*$ is called *trivial* if $L = \emptyset$ or $L = \{0,1\}^*$, otherwise it is called *nontrivial*. We only consider decidable languages throughout the paper, without repeating this condition each time.

Definition 1 (p-isomorphic encoding). *A function $\alpha : \{0,1\}^* \mapsto \{0,1\}^*$ is called a polynomial time isomorphic (p-isomorphic) encoding, if it is a bijection, computable in polynomial time, and its inverse is also computable in polynomial time.*

Definition 2 (Sphere, α-sphere). *For any $n \in \mathbb{N}$, the set $S_n = \{x \in \{0,1\}^* \mid |x| = n\}$ is called the sphere of radius n. For a p-isomorphic encoding α, the set $S_n^{(\alpha)} = \alpha(S_n) = \{\alpha(x) \mid x \in S_n\}$ is called α-sphere (of radius n).*

The system of α-spheres will be used as an "accounting scheme" for the failure rate of the errorless heuristic. The standard metric is to count the failures on the set of n-bit strings. Our accounting scheme is somewhat more lenient, but still has quite natural properties: each α-sphere $S_n^{(\alpha)}$ has size 2^n, they form a partition of $\{0,1\}^*$, the mapping between n-bit strings and the elements of $S_n^{(\alpha)}$ is 1–1, α does not change string lengths more than polynomially, and the mapping is computable and invertible in polynomial time, making it just a feasible re-encoding of strings.

Now we can define the complexity class **RoughP**, the family of languages that are accepted in roughly polynomial time.

Definition 3 (RoughP). *Let $L \subseteq \{0,1\}^*$ be a language. We say that $L \in$ **RoughP**, if there exist a p-isomorphic encoding α, and a polynomial time algorithm $\mathcal{A} : \{0,1\}^* \mapsto \{accept, reject, \perp\}$, such that the following hold:*

(i) $\mathcal{A}$ *correctly decides* L, *as an errorless heuristic. That is, it never outputs a wrong decision: if* $\mathcal{A}$ *accepts a string* x, *then* $x \in L$ *always holds, and if* $\mathcal{A}$ *rejects* x, *then always* $x \notin L$.

(ii) *Besides accept/reject,* $\mathcal{A}$ *may output the special sign* $\perp$, *meaning "don't know" (failure). This can occur, however, only for an exponentially vanishing fraction of strings in* $S_n^{(\alpha)}$. *That is, there exist a constant* c *with* $0 \le c < 1$, *such that for every* $n \in \mathbb{N}$

$$\frac{|S_n^{(\alpha)} \cap \{x \mid \mathcal{A}(x) = \perp\}|}{|S_n^{(\alpha)}|} \le c^n. \tag{1}$$

Remark: It follows directly from the definition that $\mathbf{P} \subseteq \mathbf{RoughP}$, since for $L \in \mathbf{P}$ we can always choose for $\mathcal{A}$ the polynomial time algorithm that decides L, and use $\alpha(x) = x$.

A concept that will be important in our treatment is the *paddability* of a language. This notion originally gained significance from the role it played in connection with the well known Isomorphism Conjecture of Berman and Hartmanis [2]. The conjecture states that all **NP**-complete languages are polynomial time isomorphic (*p*-isomorphic, for short), see [2]. (Note that a *p*-isomorphism between languages is not the same as our *p*-isomorphic encoding in Definition 1, because the latter does not depend on a particular language.)

Informally, a language is paddable, if in any instance we can encode arbitrary additional information, without changing the membership of the instance in the language. Moreover, both the encoding and unique decoding can be carried out in polynomial time. To the author's knowledge, all practical/natural decision tasks (whether in **NP** or not) can be represented by paddable languages[2].

A further important fact is that all known languages that represent intuitively natural **NP**-complete problems are paddable. While there are constructions in **NP** that are conjectured to lead to non-paddable languages, they arise via diagonalization, and do not represent any natural problem. Among the equivalent formal definitions of paddability we use the following:

Definition 4 (Paddability). *A language* $L \subseteq \{0,1\}^*$ *is called* paddable, *if there exists a polynomial time computable padding function* pad $: \{0,1\}^* \times \{0,1\}^* \mapsto \{0,1\}^*$ *and a polynomial time computable decoding function* dec $: \{0,1\}^* \mapsto \{0,1\}^*$, *such that for every* $x, y \in \{0,1\}^*$ *the following hold:*

(i) $\text{pad}(x,y) \in L$ *if and only if* $x \in L$.
(ii) $\text{dec}(\text{pad}(x,y)) = y$.

---

[2] This does not mean that every language that represents a practical problem is necessarily paddable. For example, it is known that polynomially sparse (nonempty) languages are not paddable (see, e.g., [6], Theorem 7.15), yet they may still represent practical problems. We only say that, to our knowledge, for any practical/natural problem it is possible to construct a paddable representation, not excluding that there may be other, non-paddable representations, as well.

3   Main Result: All Paddable Languages Are in RoughP

As our main result, we prove that all paddable languages have polynomial-time errorless heuristics, with exponentially small failure rates on the system of α-spheres, for an appropriate p-isomorphic encoding α. In other words, they are all in **RoughP**.

Theorem 1. *If $L \subseteq \{0,1\}^*$ is a paddable language, then $L \in$ **RoughP**. Furthermore, for paddable languages, the constant c in* (ii) *of Definition 3 can be chosen as $c = 1/\sqrt{2}$.*

Proof. If L is trivial[3] then $L \in$ **P** $\subseteq$ **RoughP**, so it is enough to consider a nontrivial L. For any string $x = x_1 \ldots x_n \in \{0,1\}^*$, define $w(x)$ as the number of 1-bits in x, which we refer to as the *weight* of x.

Using the paddable language L, we define an auxiliary language $H \subseteq \{0,1\}^*$ by

$$H = \{xx \mid x \in L\} \cup \{x \mid w(x) \text{ is odd}\}. \tag{2}$$

To show that H has useful properties, let us also define a polynomial time computable auxiliary function $u : \{0,1\}^* \mapsto \{0,1\}^*$. Fix two strings $w_0 \notin L$, $w_1 \in L$ (they always exist for nontrivial L), and define u as follows:

$$u(z) = \begin{cases} x & \text{if } z = xx \text{ for } x \in \{0,1\}^* \\ w_0 & \text{if } z \text{ is asymmetric and } w(z) \text{ is even} \\ w_1 & \text{if } w(z) \text{ is odd} \end{cases}$$

Recall that a string z is called symmetric if $z = xx$ for some $x \in \{0,1\}^*$, otherwise z is asymmetric. Symmetry can be easily checked in polynomial time by comparing the two halves of the string (if it has even length, which is obviously necessary for symmetry). Now we prove some properties of H that we are going to use in the sequel.

(a) L has a $\leq_m^P$ (polynomial time many-one) reduction to H. Observe that $x \in L$ if and only if $xx \in H$. (Note that $w(xx)$ is always even, so $xx \in H$ can only occur through the first set on the right-hand side of (2).) Thus, the reduction can be implemented by the function $f : \{0,1\}^* \mapsto \{0,1\}^*$ defined by $f(x) = xx$, which is clearly computable in polynomial time.

(b) H has a $\leq_m^P$ reduction to L. It can be implemented by the function $g : \{0,1\}^* \mapsto \{0,1\}^*$ defined as $g(z) = u(z)$. To see that it is indeed a $\leq_m^P$ reduction, consider first $z \in H$. Then either $z = xx$ with $x \in L$, or $w(z)$ is odd. In the first case $u(z) = x \in L$, in the second case $u(z) = w_1 \in L$. Therefore, $z \in H$ implies $u(z) \in L$. Consider now $z \notin H$. In this case $w(z)$ must be even. Then there are two possibilities: (1) z is asymmetric. Since $w(z)$ is even, we have $u(z) = w_0 \notin L$. (2) $z = xx$ for some $x \in \{0,1\}^*$,

---

[3] Recall that L is called trivial if either $L = \emptyset$ or $L = \{0,1\}^*$. Observe that a trivial language formally satisfies Definition 4, via the functions $\text{pad}(x,y) = y$ and $\text{dec}(z) = z$.

but $x \notin L$. Then $u(z) = x \notin L$, so in either case we obtain that $z \notin H$ implies $u(z) \notin L$. Thus, noting the polynomial time computability of $u(z)$, we indeed get a $\leq_m^P$ reduction of H to L.

(c) H is paddable. Using that L is paddable by assumption, let $\mathrm{pad}(x, y)$ be a padding function for L, with decoding function $\mathrm{dec}(z)$. Then a padding function for H can be defined as

$$\mathrm{pad}'(z, y) = \mathrm{pad}(u(z), y) \, \mathrm{pad}(u(z), y). \tag{3}$$

To see that it satisfies Definition 4, take first $z \in H$. Then there are two possibilities:

(α) $z = xx$ for some $x \in L$, leading to $u(z) = x$. Then $\mathrm{pad}(u(z), y) = \mathrm{pad}(x, y) \in L$, due to $x \in L$, from which $\mathrm{pad}'(z, y) = \mathrm{pad}(x, y)\mathrm{pad}(x, y) \in H$ follows.

(β) $w(z)$ is odd, so $u(z) = w_1$. Then $\mathrm{pad}(u(z), y) = \mathrm{pad}(w_1, y) \in L$, due to $w_1 \in L$, resulting in $\mathrm{pad}'(z, y) = \mathrm{pad}(w_1, y)\mathrm{pad}(w_1, y) \in H$.

Now take $z \notin H$. Then there are again two possibilities:

(α) $z = xx$, but $x \notin L$. In this case $u(z) = x$, yielding $\mathrm{pad}'(z, y) = \mathrm{pad}(x, y)\mathrm{pad}(x, y)$. Since $\mathrm{pad}(x, y) \notin L$, due to $x \notin L$, and $w(\mathrm{pad}(x, y)\mathrm{pad}(x, y))$ is always even, therefore, $\mathrm{pad}'(z, y) \notin H$.

(β) $z \neq xx$ for any x, but $w(z)$ is even. Then we get $u(z) = w_0$, which gives $\mathrm{pad}'(z, y) = \mathrm{pad}(w_0, y)\mathrm{pad}(w_0, y)$. Since $\mathrm{pad}(w_0, y) \notin L$, due to $w_0 \notin L$, and $w(\mathrm{pad}(w_0, y)\mathrm{pad}(w_0, y))$ is always even, therefore, $\mathrm{pad}'(z, y) \notin H$.

Thus, we indeed have $\mathrm{pad}'(z, y) \in H$ if and only if $z \in H$. To get a decoding function dec' for H, define

$$\mathrm{dec}'(z) = \mathrm{dec}(u(z)). \tag{4}$$

We need to show that $\mathrm{dec}'(\mathrm{pad}'(v, y)) = y$ holds for any $v, y \in \{0, 1\}^*$. Observe that (3) and the definition of u imply

$$u(\mathrm{pad}'(v, y)) = \mathrm{pad}(u(v), y).$$

Using this, and (4), we get

$$\mathrm{dec}'(\mathrm{pad}'(v, y)) = \mathrm{dec}(\underbrace{u(\mathrm{pad}'(v, y))}_{\mathrm{pad}(u(v), y)}) = \mathrm{dec}(\mathrm{pad}(u(v), y)) = y,$$

where the last equality follows from (ii) in Definition 4. Thus, the function dec' indeed carries out correct decoding for pad'.

Now we know that both L and H are paddable. Furthermore, we have shown that they are both $\leq_m^P$ reducible to the other. Therefore, it follows from the well known and fundamental results of Berman and Hartmanis [2] that there is a p-isomorphism between H and L. That is, there exists a bijection $\varphi : \{0, 1\}^* \mapsto$

$\{0,1\}^*$, such that both φ and φ^{-1} are computable in polynomial time, and for every $x \in \{0,1\}^*$ it holds that $x \in L$ if and only if $\varphi(x) \in H$.

Let us define the p-isomorphic encoding α by $\alpha(x) = \varphi^{-1}(x)$, and define the algorithm $\mathcal{A}$ by

$$\mathcal{A}(x) = \begin{cases} \text{accept} & \text{if } w(\varphi(x)) \text{ is odd} \\ \text{reject} & \text{if } w(\varphi(x)) \text{ is even and } \varphi(x) \text{ is asymmetric} \\ \bot & \text{if } \varphi(x) \text{ is symmetric.} \end{cases} \qquad (5)$$

Next we show that this α and $\mathcal{A}$ together satisfy Definition 3:

- The function α is a p-isomorphic encoding: it is a bijection, plus both α and α^{-1} are computable in polynomial time, due to the same properties of φ.
- The algorithm $\mathcal{A}$ runs in polynomial time, as φ is computable in polynomial time, likewise the symmetry and the parity of the weight of any string can be checked in polynomial time.
- $\mathcal{A}$ is an errorless heuristic for L, that is, $\mathcal{A}$ correctly decides L, whenever $\mathcal{A}(x) \neq \bot$. Indeed, if $\mathcal{A}$ accepts, then $w(\varphi(x))$ is odd. This means, $\varphi(x) \in H$. Then, due to the properties of φ, it must hold that $x \in L$. Similarly, if $\mathcal{A}$ rejects, then $w(\varphi(x))$ is even and $\varphi(x)$ is asymmetric. This implies $\varphi(x) \notin H$, yielding $x \notin L$. Thus, condition (i) in Definition 3 is satisfied.
- Finally, it remains to prove condition (ii) in Definition 3. Let $F = \{z \mid \mathcal{A}(z) = \bot\}$ be the set where $\mathcal{A}$ fails. We need to prove that there is a constant $c < 1$, with

$$\frac{|S_n^{(\alpha)} \cap F|}{|S_n^{(\alpha)}|} \leq c^n.$$

From (5) we know that $\mathcal{A}(z) = \bot$ if and only if $\varphi(z)$ is symmetric. Let Y be the set of all symmetric strings in $\{0,1\}^*$, then $F = \{z \mid \varphi(z) \in Y\}$. Consider now the set $S_n^{(\alpha)} \cap F$. The α-sphere $S_n^{(\alpha)}$ contains all strings of the form $\alpha(x)$ with $|x| = n$. Among these, those strings z belong to F, for which $\varphi(z) \in Y$ also holds. Therefore, we can write

$$S_n^{(\alpha)} \cap F = \{z \mid z = \alpha(x), |x| = n, \varphi(z) \in Y\}.$$

Observe that if $z = \alpha(x)$, then $\varphi(z) = x$, since $\alpha = \varphi^{-1}$. This gives us

$$S_n^{(\alpha)} \cap F = \{z \mid z = \alpha(x), |x| = n, x \in Y\} = \{\alpha(x) \mid |x| = n, x \in Y\}.$$

The number of symmetric strings among all n-bit strings is $2^{n/2}$, if n is even, as the first half already determines a symmetric string. If n is odd, then their number is 0. This yields $|S_n^{(\alpha)} \cap F| \leq 2^{n/2}$. Taking into account that, due to the bijective property of α, we have $|S_n^{(\alpha)}| = |S_n| = 2^n$, the bound

$$\frac{|S_n^{(\alpha)} \cap F|}{|S_n^{(\alpha)}|} \leq \frac{2^{n/2}}{2^n} = \left(\frac{1}{\sqrt{2}}\right)^n$$

follows. Thus, with the choice of $c = 1/\sqrt{2} < 1$ we can indeed satisfy condition (ii) in Definition 3, completing the proof.

♠

3.1   Viewing the Proof as an Algorithm Design Technique

The proof of Theorem 1 actually provides a way to *efficiently construct* the p-isomorphic encoding α, and the algorithm $\mathcal{A}$. Once the p-isomorphism φ, and its inverse φ^{-1} are available, α is expressed as $\alpha = \varphi^{-1}$, and $\mathcal{A}$ is given by (5). In order to obtain φ and φ^{-1}, recall that we constructed the $\leq_m^P$ reductions f, g between L and H, as well as the padding/decoding function pair (pad', dec') for H, using the padding/decoding function pair (pad, dec) which is assumed available for L. Having the six polynomial time computable functions $f, g, \text{pad}, \text{dec}, \text{pad}', \text{dec}'$, we can then obtain the p-isomorphism φ and its inverse φ^{-1} via the method of Berman and Hartmanis [2] (see also the textbook description of Du and Ko [6], Theorem 7.14). The obtained functions φ, φ^{-1} also determine the encoding α, via $\alpha = \varphi^{-1}$. While the construction of φ, φ^{-1} is nontrivial, it can be carried out in polynomial time. With all this, our solution provides a rather general, novel **algorithm design technique** for approximating decision problems, which can be applied to any paddable language. Note that even though the expression (5) for the algorithm $\mathcal{A}$ may appear deceptively simple, in fact it can represent a rather complex polynomial time algorithm, since the function φ may be complicated.

4   Hardness Results

First we address the question: is it possible that *every* **NP**-complete problem has a polynomial-time errorless heuristic with vanishing failure rate? In this section we consider the standard sense for such heuristics, that is, the error rate is measured over the n-bit strings. Furthermore, we do not require here that the failure rate vanishes exponentially, it can tend to 0 arbitrarily. In this sense, can we expect the *universal approximability* of **NP**-complete problems?

Having this level of efficiency for *all* **NP**-complete problems appears quite unlikely. Interestingly, however, it is not known to conflict with the "standard" hypotheses of complexity theory, including **P≠NP, NP≠co-NP, E≠NE, EXP≠NEXP, NP≠PSPACE, NP≠EXP, NP⊈P/poly, P=BPP, PH** does not collapse, etc. There is, however, a (somewhat less standard) hypothesis, which already rules it out. It comes from the theory of resource bounded measure, for a survey see Lutz and Mayordomo [9]. In this theory a central conjecture is that the so called p-measure of **NP**, denoted by $\mu_p(\textbf{NP})$, is nonzero. Informally, this means that **NP**-languages within $\textbf{E} = \cup_{c>0}\text{DTIME}(2^{cn})$ do not constitute a negligible subset. The $\mu_p(\textbf{NP}) \neq 0$ conjecture can be viewed as a stronger form of the $\textbf{P} \neq \textbf{NP}$ conjecture, as $\mu_p(\textbf{NP}) \neq 0$ implies $\textbf{P} \neq \textbf{NP}$, but the reverse implication is not known.

Lemma 1. *If $\mu_p(\textbf{NP}) \neq 0$, then there is a language in* **NP** *which does not have a polynomial-time errorless heuristic with vanishing failure rate.*

Proof. An infinite and co-infinite language L is called **P**-bi-immune, if for every infinite $L_0 \in \textbf{P}$ it holds that $L_0 \nsubseteq L$ and $L_0 \nsubseteq \overline{L}$. In other words, neither L, nor

$\overline{L}$ can have an infinite subset in $\mathbf{P}$. It is known that the hypothesis $\mu_p(\mathbf{NP}) \neq 0$ implies the existence of a $\mathbf{P}$-bi-immune language in $\mathbf{NP}$, see Mayordomo [10]. Assuming $\mu_p(\mathbf{NP}) \neq 0$, pick a $\mathbf{P}$-bi-immune language $L \in \mathbf{NP}$. If every $\mathbf{NP}$-language has a polynomial-time errorless heuristic algorithm with vanishing failure rate, then L must have one, too, let $\mathcal{A}$ be this algorithm. Let A be the set on which $\mathcal{A}$ accepts. Then $A \in \mathbf{P}$, since $\mathcal{A}$ runs in polynomial time. Furthermore, since $\mathcal{A}$ is an errorless heuristic, it never accepts falsely, so $A \subseteq L$. Similarly, let B be the set where $\mathcal{A}$ rejects. Again, $B \in \mathbf{P}$, as $\mathcal{A}$ runs in polynomial time, and $B \subseteq \overline{L}$, due to that $\mathcal{A}$ never rejects falsely. By the vanishing failure rate requirement $A \cup B$ must be infinite. Therefore, at least one of A, B is infinite, so either L or $\overline{L}$ has an infinite subset in $\mathbf{P}$. Thus, L cannot be $\mathbf{P}$-bi-immune, a contradiction, proving the claim.

♠

Remarks

1. The proof shows that if $\mathbf{NP}$ contains a $\mathbf{P}$-bi-immune language, then Lemma 1 would hold unconditionally, without assuming $\mu_p(\mathbf{NP}) \neq 0$. While it is not known whether $\mathbf{NP}$ contains a $\mathbf{P}$-bi-immune language, nevertheless, there is some evidence which supports that it does: Hemaspaandra and Zimand [8] prove that relative to a random oracle $\mathbf{NP}$ contains a $\mathbf{P}$-bi-immune language, with probability 1.

2. The $\mathbf{P}$-bi-immune language in $\mathbf{NP}$, based on the $\mu_p(\mathbf{NP}) \neq 0$ hypothesis, does not arise, however, from a natural problem, in the intuitive sense of the word. To the author's knowledge, no hypothesis is known that would rule out the existence of polynomial-time errorless heuristics with vanishing failure rate for all *natural* $\mathbf{NP}$-complete problems. Instead of the informal (and slippery) notion of "natural" we could refer to paddable $\mathbf{NP}$-complete languages, since they contain all the known intuitively natural ones. Apparently, no hypothesis rules out that they are all approximable in the sense we consider here.

Another related question worth looking into is this: which is the smallest mainstream complexity class that provably does not have universal approximability in our sense, without assuming any unproven hypothesis? We can prove the following:

Lemma 2. *There is a language in $L \in \mathbf{E} = \cup_{c>0}\mathrm{DTIME}(2^{cn})$, such that L does not have a polynomial-time errorless heuristic with vanishing failure rate.*

Proof. We can re-use the proof of Lemma 1, with the only modification that $\mathbf{E}$ is known to unconditionally contain a $\mathbf{P}$-bi-immune language (see Balcàzar and Schöning [1]).

♠

Regarding our new class $\mathbf{RoughP}$, Lemma 1 implies that if $\mu_p(\mathbf{NP}) \neq 0$, then $\mathbf{NP} \not\subseteq \mathbf{RoughP}$. However, $\mu_p(\mathbf{NP}) \neq 0$ is not known. Without that, how hard is it to decide the $\mathbf{NP} \subseteq ?\,\mathbf{RoughP}$ question? Observe that while there are plenty of natural problems that are provably in $\mathbf{NP} - \mathbf{P}$, assuming the set is not empty, the situation is different with $\mathbf{NP} - \mathbf{RoughP}$. The reason is that any

$L \in \mathbf{NP} - \mathbf{RoughP}$ must be non-paddable, by Theorem 1, and, of course, be outside $\mathbf{P}$. Such languages in $\mathbf{NP}$ are in short supply. In fact, it is not known if $\mathbf{NP} - \mathbf{P}$ contains any non-paddable language at all, assuming only $\mathbf{P} \neq \mathbf{NP}$. The point is that deciding the $\mathbf{NP} \subseteq ? \mathbf{RoughP}$ question in either direction is likely to be hard, because in either case it resolves a long-standing, mainstream complexity class separation.

Lemma 3. *If* $\mathbf{NP} \nsubseteq \mathbf{RoughP}$, *then* $\mathbf{P} \neq \mathbf{NP}$. *If* $\mathbf{NP} \subseteq \mathbf{RoughP}$, *then* $\mathbf{NP} \neq$ $\mathbf{EXP}$, *where* $\mathbf{EXP} = \cup_{c>0}\mathrm{DTIME}(2^{n^c})$.

Proof. The first implication follows from $\mathbf{P} \subseteq \mathbf{RoughP}$. The second claim is implied by Lemma 2, along with $\mathbf{E} \subseteq \mathbf{EXP}$.

♠

Remark: Note that $\mathbf{NP} \subseteq \mathbf{RoughP}$ also implies $\mathbf{NP} \neq \mathbf{E}$, but that is not an open problem, as $\mathbf{NP} \neq \mathbf{E}$ has been known for a long time (see Book [5]). But $\mathbf{NP} \subseteq \mathbf{E}$ is not known, in contrast to $\mathbf{NP} \subseteq \mathbf{EXP}$.

5 Discussion

We have introduced the complexity class **RoughP**, and shown that it contains a large family of languages, including all known intuitively natural **NP**-complete problems. This means, they can all be approximated by efficient algorithms in the relaxed sense we have defined: by a polynomial time errorless heuristic with exponentially vanishing failure rate over the α-spheres.

It is natural to contemplate: how this relates to the standard metric, where we measure the failure rate of an errorless heuristic over n-bit strings? Our intuition is this: whenever the "don't know" instances congregate around the bottom of the α-spheres, i.e., they gravitate towards the shorter strings in the α-sphere, then we can expect high standard failure rate, as the failures that belong to some $S_N^{(\alpha)}$ with $N > n$ may fill up a large part of S_n. If, however, the failure instances are well spread within every α-sphere, then we may expect an exponentially small failure rate for the standard setting, similar to what we have over the α-spheres. Could we somehow force a near-uniform distribution of the failure instances over each α-sphere? We might perhaps capitalize on the fact that neither the algorithm $\mathcal{A}$ nor the p-isomorphic encoding α are determined uniquely by the language. We may also transform the original language, replacing $L(x)$ with $L(f(x))$, where f is some p-isomorphic encoding that we can choose. If $f(x)$ behaves in a pseudo-random way, then it may provide sufficient mixing to spread the failure instances appropriately.

Thus, it appears, there is room for improvement. We do not expect, however, that such attempts can *always* produce low failure rate with the standard metric for **NP**-complete languages, even though apparently nothing excludes this for the natural problems, represented by paddable languages. Nevertheless, we expect that for many specific problems it is still achievable, leading to good errorless heuristics for a number of tasks, for which no such schemes have been known before.

Let us also mention a further point: our approach may provide some contribution to explain the curious observation that fine-tuned algorithms often exhibit better performance in practice than what follows from their theoretical analysis. For example, carefully engineered modern SAT solvers routinely (and successfully!) attack industrial SAT instances with millions of variables, despite the conjectured exponential worst-case running time, as pointed out by Vardi [11]. We might ponder that they (unwittingly) implement a strategy that is equivalent (or close) to some fine-tuned **RoughP** algorithm. At least we already know that the latter must exist for the known natural **NP**-complete problems, including SAT.

As a final note, let us mention that a nontraditional aspect of our approach is that it builds on complexity theoretical concepts (padding, p-isomorphism) to derive a positive and constructive result in algorithm design. This is somewhat unusual, because algorithm design and complexity theory view the same world from complementary perspectives: algorithm design aims at creating efficient algorithms, while complexity theory explores the limitations of what can be achieved. In this sense, they are the *yin and yang* of the algorithms universe. Our approach tries to bring them together, in order to create a new, general purpose algorithm design technique for approximating decision problems, with a certain sense of efficiency, yet avoiding conflict with any known concept of intractability.

References

1. Balcàzar, J.L., Schöning, U.: Bi-immune sets for complexity classes. Math. Syst. Theor. **18**, 1–10 (1985)
2. Berman, L., Hartmanis, J.: On isomorphisms and density of NP and other complete sets. SIAM J. Comput. **6**(2), 305–322 (1977)
3. Bogdanov, A., Trevisan, L.: Average-Case Complexity. Found. Trends Theor. Comput. Sci. **2**(1), 1–106 (2006)
4. Bollobás, B.: Random Graphs. Cambridge University Press, Cambridge (2001)
5. Book, R.V.: On languages accepted in polynomial time. SIAM J. Comput. **1**(4), 281–287 (1972)
6. Du, D.-Z., Ko, K.-I.: Theory of Computational Complexity. Wiley, Hoboken (2000)
7. Hemaspaandra, L.A., Williams, R.: An atypical survey of typical-case heuristic algorithms. ACM SIGACT News **43**(4), 70–89 (2012). Complexity Theory Column 76
8. Hemaspaandra, L.A., Zimand, M.: Strong self-reducibility precludes strong immunity. Math. Syst. Theor. **29**(5), 535–548 (1996)
9. Lutz, J.H., Mayordomo, E.: Twelve problems in resource-bounded measure. In: Păun, G., Rozenberg, G., Salomaa, A. (eds.) Current Trends in Theoretical Computer Science: Entering the 21st Century, pp. 83–101. World Scientific (2001)
10. Mayordomo, E.: Almost every set in exponential time is P-bi-immune. Theor. Comput. Sci. **136**(2), 487–506 (1994)
11. Vardi, M.Y.: Boolean satisfiability: theory and engineering. Commun. ACM **57**(3), 5 (2014)

Community-Based Acceptance Probability Maximization for Target Users on Social Networks

Ruidong Yan[1], Yuqing Zhu[2], Deying Li[1(✉)], and Yongcai Wang[1]

[1] School of Information, Renmin University of China, Beijing 100872, China
{yanruidong,deyingli,ycw}@ruc.edu.cn
[2] Department of Computer Science, California State University at Los Angeles,
Los Angeles, CA 90032, USA
yuqing.zhu@calstatela.edu

Abstract. Social influence problems, such as *Influence Maximization* (IM), have been widely studied. But a key challenge remains: How does a company select a small size seed set such that the acceptance probability of target users is maximized? In this paper, we first propose the *Acceptance Probability Maximization* (APM) problem, i.e., selecting a small size seed set S such that the acceptance probability of target users T is maximized. Then we use classical *Independent Cascade* (IC) model as basic information diffusion model. Based on this model, we prove that APM is NP-hard and the objective function is monotone non-decreasing as well as submodular. Considering community structure of social networks, we transform APM to *Maximum Weight Hitting Set* (MWHS) problem. Next, we develop a pipage rounding algorithm whose approximation ratio is $(1 - 1/e)$. Finally, we evaluate our algorithms by simulations on real-life social networks. Experimental results validate the performance of the proposed algorithm.

Keywords: Social influence · Community structure
Seed selection · Submodularity · Approximate algorithm

1 Introduction

In recent years, with the rapid development of the internet and computer technology, some significant social networks have been widely integrated into our daily life, such as Facebook, Twitter and Google+. These online social networks have become significant platforms for disseminating useful content such as news, ideas, opinions, innovations, interests, etc. In viral market, *Influence Maximization* (IM) has been extensively studied. This research has been found useful in market recommendations through the powerful word-of-mouth effect in social networks. Specifically, a company launches a kind novel product and wants to market it by social network. Due to limited budget, it can only choose a small number of initial clients (seeds) to use it (by giving them free samples). The

© Springer Nature Switzerland AG 2018
S. Tang et al. (Eds.): AAIM 2018, LNCS 11343, pp. 293–305, 2018.
https://doi.org/10.1007/978-3-030-04618-7_24

company hopes that these initial clients like this product and recommend it to their friends on the social network. Similarly, their friends influence their friends of friends and so on. Finally, the company wants to maximize the number of clients who adopt the products.

However, in some scenarios, one may consider the maximizing acceptance probability of target users. Specifically, assume each user on social network has a potential value for a company. The company pay more attention to the users with higher potential value. We call these higher value users as *target users* (Selecting them as seeds requires very high cost since they are influential and authoritative). Intuitively, the company will benefit a great deal if it can maximize acceptance probability of target users. In this situation, the company aims at finding an optimal seed set within a budget such that the sum of acceptance probability of target users is maximized. We call this problem as *Acceptance Probability Maximization* (APM). It's obvious that IM problem is different from APM problem. The former selects a seed set from all nodes in network within a budget such that the expected number of nodes influenced by seed set through information diffusion is maximized. However, the latter selects a seed set from all nodes except target nodes within a budget such that the sum of acceptance probability of target users is maximized.

In fact, this problem is challenging. Intuitively, it should select global influential users or target users' neighbors as seeds. However, this intuitive choice may not be effective: (1) Target users are far away from global influential nodes. The influence from global influential nodes is less than it from local influential nodes that are close to target users. (2) Although target users' neighbors have highly influence on target users, it is impossible to choose all the neighbors of target users as seeds with restriction of small size seed set. Considering these two points, we should focus on local (community) influential nodes. Further, APM can be applied to most applications, such as personalized services, targeted advertising, targeted information dissemination, recommendation system, etc.

To the best of our knowledge, only a few studies focus on APM problem even though it plays an essential role in viral marketing. The similar studies have been done, such as [6,13]. Guo et al. [6] propose a problem to find the top-k most influential nodes to a given user. They develop a simple greedy algorithm. We expand their work and solve APM from different perspective. In [13], Yang et al. advocate recommendation support for active friending, where a user actively specifies a friending target. In other worlds, to maximize the probability that the friending target would accept an invitation from the source user. The difference between APM and previous works are: (1) Instead of [6,13], APM has multiple target users; (2) APM requires the acceptance probability instead of expected number of influenced nodes. We summarize main contributions as follows:

- We propose the *Acceptance Probability Maximization* (APM) problem and prove it's NP-hard. And we show that computing APM is $\#p - hard$.
- We prove objective function is monotone non-decreasing and submodular.

- Considering community structure of social networks, we transform APM to *Maximum Weight Hitting Set* (MWHS) problem. Then we propose a pipage rounding algorithm for APM and prove approximation ratio is $(1 - 1/e)$.
- We run the proposed algorithm and compare with other existing methods.

The rest of this paper is organized as follows. In Sect. 2, we introduce related work. In Sect. 3, influence diffusion model is presented. In Sect. 4, we state problem description. In Sect. 5, we show the properties of objective function. Algorithm is designed in Sect. 6. The experiment results are shown in Sect. 7. We draw our conclusions in Sect. 8.

2   Related Work

Kempe et al. [8] model viral marketing as a discrete optimization problem, which is named *Influence Maximization* (IM). They propose a greedy algorithm with $(1 - 1/e)$-performance ratio since the function is submodular under *Independent Cascade* (IC) or *Linear Threshold* (LT) model. Previous researches without target users, which cannot be directly transplanted to APM, such as [9]. In [9], Kuhnle et al. consider *Threshold Activation Problem* (TAP) which finds a minimum size set triggering expected activation of at a certain threshold. They exploit the bicriteria nature of solutions to TAP and control the running time by a parameter.

The related work involves the target users such as [2, 10, 11, 14]. In [14], Zhou et al. study a new problem: Give an activatable set A and a targeted set T, finding the k nodes in A with the maximal influence in T. They give a greedy algorithm with guarantee of $(1 - 1/e)$. In [10], Song et al. formalize the problem targeted influence maximization in social networks and adopt a login model where each user is associated with a login probability and he can be influenced by his neighbors only when he is online. Moreover, they develop a sampling based algorithm that returns a $(1 - 1/e - \varepsilon)$-approximate solution. In [11], Temitope et al. extend the fundamental *Influence Maximisation* (IM) problem with respect to a set of target users on a social network. In doing so, they formulate the *Minimal Influencer for Target Users* (MITU) problem and compare with state of the art algorithms. Unfortunately, they don't have any theoretical analysis. In [2], Chang et al. study a novel problem: Given a period of promotion time and a set of target users, each of which can be activated by its neighbors multiple times, they aim at maximizing the total acceptance frequency of these target users by initially selecting k most influential seeds. They propose a generalized diffusion model called the *Multiple Independence Cascade* (MIC) and a greedy algorithm for solving this problem.

3   Influence Diffusion Model

We briefly introduce influence diffusion model: *Independent Cascade* (IC) model. Given a social network $G = (V, E, w)$, where V represents node set, $E \subseteq V \times V$

represents edge set, and w_{uv} of edge (u, v) denotes the probability that node u can activate v successfully. We call a node as *active* if it adopts the products or information from other nodes, *inactive* otherwise. Influence propagation process unfolds discrete time steps t_i, $(i = 0, 1, \ldots)$. Initial seed set $S_{t_0} = S$. Let S_{t_i} denote *active* nodes in time step t_i, and each node u in S_{t_i} has single chance to activate each *inactive* neighbor v through its out-edge with probability w_{uv} at time step t_{i+1}. Repeat this process until no more new nodes can be activated. A node can only switch from *inactive* to *active*, but not in the reverse direction.

4   Preliminaries and Problem Description

4.1   Preliminaries

A set function f is monotone increasing if $f(A) \leq f(B)$ whenever $A \subseteq B$. Submodular functions have a natural diminishing returns property. If V is a finite set, a submodular function is a set function $f : 2^V \to \Re$, where 2^V denotes the power set of V, which satisfies the following condition: for every $A \subseteq B \subseteq V$ and $x \in V \backslash B$, $f(A \cup \{x\}) - f(A) \geq f(B \cup \{x\}) - f(B)$.

Further, we introduce some basic definitions for later discussion. *Set cover* [7]: Given a ground set $U = \{u_1, u_2, \ldots, u_n\}$ and a collection of subsets of $C = \{C_1, \ldots, C_m\}$. The set cover problem is to identify the smallest sub-collection C' from C such that C' covers all elements in U, i.e., $\bigcup_{C_i \in C'} C_i = U$.

s-t connectedness [12]: Given a directed graph G and arbitrary two nodes s and t. The *s-t connectedness* finds number of subgraphs of G in which there is a directed path from s to t.

4.2   Problem Description

Given a directed social network $G = (V, E, w)$, an information diffusion model $\mathcal{M}$, a target users set $T = \{T_1, T_2, \ldots, T_q\}$, and a positive integer budget b, where V denotes all users, $E \subseteq V \times V$ denotes the relationships between users, and w_{uv} of edge (u, v) means the probability that u activates v successfully. We assume that the acceptance probability of a node v is equal to the v's activation probability. We define the acceptance probability of a node $v \in V$ when given a seed set S under IC model as follow

$$Pr_{\mathcal{M}}(v, S) = \begin{cases} 1, & \text{if } v \in S \\ 0, & \text{if } N^{in}(v) = \emptyset \\ 1 - \displaystyle\prod_{u \in N^{in}(v)} (1 - Pr_{\mathcal{M}}(u, S)w_{uv}), & \text{otherwise.} \end{cases} \tag{1}$$

where $N^{in}(v)$ is the set of in-neighbors of v and $Pr_{\mathcal{M}}(u, S)w_{uv}$ represents the probability u successfully activates v under the diffusion model $\mathcal{M}$ (IC model). As we can clearly see the acceptance probability of a node v depends on the acceptance probability of its in-neighbors u. Then we define the acceptance probability for target users $v \in T$ from seed set S under the diffusion model as

$$Pr_{\mathcal{M}}(T, S) = \sum_{v \in T} Pr_{\mathcal{M}}(v, S). \tag{2}$$

Now, we can formally define the *Acceptance Probability Maximization* (APM). Given a social network $G = (V, E, w)$, a target users set T, an information diffusion model $\mathcal{M}$, and a positive integer budget b, APM aims to find a seed set S^* such that

$$S^* = \arg \max_{S \subseteq V \setminus T, |S|=b} Pr_{\mathcal{M}}(T, S). \tag{3}$$

In particular, we omit the subscript $\mathcal{M}$ if the context is clear.

5   Properties of APM

We show the properties of APM problem as following theorems.

Theorem 1. *APM problem is NP-hard under the IC model even if $|T| = 1$.*

Proof. We prove it with reduction from the set cover problem [7]. We construct a new network $G' = (V', E', w')$. V' includes three parts: (1) Create a node C_i for each C_i; (2) Create a node u_j for each u_j; (3) Create a target node T. E' is defined as follows. If C_i contains u_j, then add a directed edge (C_i, u_j) from node C_i to node u_j with $w'_{C_i u_j} = 1$. Moreover, for each node u_j, add a directed edge (u_j, T) from node u_j to node T with $w'_{u_j T} = p$. Obviously, the above transformation can be done in polynomial time.

We prove that there is a subset $C' \subseteq C$ covering all nodes in U in the set cover problem if and only if there is a solution with acceptance probability $1 - (1-p)^{|U|}$[1] when selecting $b = |C'|$ nodes as seeds. We first prove the sufficient condition. If there exists a subset C' covering all node in U, which obtains acceptance probability $1 - (1-p)^{|U|}$ for target nodes. We then prove the necessary condition. If there exists a seed set C' with $|C'| = b$ obtaining acceptance probability $1 - (1-p)^{|U|}$, then C' must covering all nodes in U. If the set cover problem is solvable, then APM problem is also solvable. As we all know, the former is NP-hard, therefore the latter is also NP-hard.

Theorem 2. *Given a seed set S and a target set T, computing acceptance probability from seed set S to target set T is $\#p - hard$ under the IC model.*

Proof. We prove this theorem with reduction from a classical $\#p - complete$ problem named *s-t connectedness* [12]. For simplicity, we let $T = \{z\}$. We assign the probability of each edge as 0.5 to guarantee each subgraph with equal probability. Therefore it's straightforward to see that *s-t connectedness* is equivalent to compute the path probability from s to t.

---

[1] Notice that $1 - (1-p)^{|U|}$ is maximum probability of T under the IC model.

Let $Pr(T, S, G)$ denote the acceptance probability of T from a given seed set S on G under the IC model. First, let $S = \{s\}$ and $w_{uv} = 0.5$ for all $(u, v) \in E$. Therefore $P_1 = Pr(T, S, G)$. Next, we add a new node z' and two directed edges (z', z) with $w_{z'z} = 0.5$ and (t, z') with $w_{tz'} = 1$, obtaining a new graph G'. Then, we compute $P_2 = Pr(T, S, G')$. Therefore, $P_2 - P_1 = (1 - P_1) \cdot Pr(t, s) \cdot w_{tz'} \cdot w_{z'z}$, which is related to the probability $Pr(t, s)$ that s is connected to t. As we all know, s-t connectedness is $\#p - complete$ and thus theorem follows immediately.

Theorem 3. *The objective function (3) is monotone non-decreasing and submodular under the IC model.*

Proof. Obviously, increasing the seed nodes does not reduce the objective function value, so we omit its proof. We show that $Pr(T, S)$ is submodular. Consider following two cases.

Case 1: If $A = B$, for an arbitrary node v, then $Pr(T, A \cup \{v\}) - Pr(T, A) = Pr(T, B \cup \{v\}) - Pr(T, B)$ always stands.

Case 2: If $A \subset B$, let $\triangle A_v = Pr(T, A \cup v) - Pr(T, A)$ denote the marginal influence on the target nodes that are not already in the union $\bigcup_{u \in A} Pr(T, u)$. Let $\triangle B_v = Pr(T, B \cup v) - Pr(T, B)$ denote the marginal influence on the target nodes that are not in the union $\bigcup_{u \in B} Pr(T, u)$. Obviously, $\triangle A_v$ is no less than $\triangle B_v$, that is, $Pr(T, A \cup v) - Pr(T, A) \geq Pr(T, B \cup v) - Pr(T, B)$ which $Pr(T, S)$ is submodular with respect to S.

6   Algorithm

From Theorem 2, computing the APM is $\#p - hard$. Therefore we need to find an approximate method to calculate it. Intuitively, computing APM in local structures, such as communities, allows efficient computation. Further, in each community, constructs a local tree structure [3] and approximates local influence diffusion to the target nodes.

Give a social network $G = (V, E, w)$, a community set $\mathcal{C} = \{\mathcal{C}_1, \mathcal{C}_2, \ldots, \mathcal{C}_m\}$ where $\bigcup_{1 \leq j \leq m} \mathcal{C}_j = V$, a target user set $T = \{T_1, T_2, \ldots, T_q\}$ and a positive integer budget b. For a path $Path(u, v) = <u = p_1, p_2, \ldots, p_l = v>$ from u to v in $\mathcal{C}_j$, we define the probability of this path as $\mathcal{P}(u, v) = \prod_{i=1}^{i=l-1} w_{p_i p_{i+1}}$. If u successfully activates v through path $Path(u, v)$, u must activate all the nodes along this path. Let $Path_{\mathcal{C}_j}(u, v)$ denote the set of all paths from u to v in $\mathcal{C}_j$.

Definition 1. *(Maximum Influence Path (MIP)). For a community $\mathcal{C}_j$, we define $MIP_{\mathcal{C}_j}(u, v)$ from u to v in $\mathcal{C}_j$ as*

$$MIP_{\mathcal{C}_j}(u, v) = \arg \max_{Path(u,v)} \{\mathcal{P}(u, v) | Path(u, v) \in Path_{\mathcal{C}_j}(u, v)\}. \quad (4)$$

Note that if we transform w_{uv} to $1/w_{uv}$ for each edge (u, v), $MIP_{\mathcal{C}_j}(u, v)$ is equivalent to the shortest path from u to v in $\mathcal{C}_j$. The shortest path problem has polynomial time algorithms, e.g., Floyd-Warshall and Dijkstra algorithms. For a target node $v \in T$, we create a tree structure which is the union of MIPs to v, to estimate the acceptance probability to v from other nodes. Moreover, we use a threshold θ to delete MIPs which have small probabilities.

Definition 2. *(Maximum Acceptance Probability Tree (MAPT)). For a threshold θ, the maximum acceptance probability tree of a target node $v \in T$ in C_j, $MAPT(v, \theta)$, is*

$$MAPT(v, \theta) = \bigcup_{u \in C_j, MIPC_j(u,v) \geq \theta} MIP_{C_j}(u, v). \tag{5}$$

In fact, we assume that the influences only propagate within communities and their propagation in these communities are independent of each other. With this assumption, we can calculate the acceptance probability that $v \in T$ is activated when given a seed set S exactly. Considering community structure that plays a vital role in propagation [1,4], we transform APM problem into *Maximum Weight Hitting Set* (MWHS) problem.

Definition 3. *(Maximum Weight Hitting Set (MWHS)). Given an element set V, a family of subsets $C \subseteq 2^V$, a weight function $w : C \mapsto \Re^+$, and a positive integer b. MWHS finds a subset $S \subseteq V$ and $|S| = b$ such that maximizes the total weight of subsets in C hit by S. (S hits C, which means $S \cap C \neq \emptyset$.)*

Let S denote seed set. We should do following two steps so that APM can be transformed into MWHS. (1) We say that seed set S hits community C_j if seed set $S \cap C_j \neq \emptyset$. (2) Let $w_j = w(C_j) = \sum_{v \in C_j, S_j \in C_j} Pr(v, S_j)$, where S_j is the set of seed nodes in community C_j and $Pr(v, S_j)$ is probability that S_j activates v successfully. Now, we formalize APM as following integer programming.

$$\max H(v) = \sum_{j=1}^{m} w_j \cdot \min\{1, \sum_{i \in C_j} v_i\}$$

$$s.t. \sum_{v_i \in V \setminus T} v_i = b \tag{6}$$

$$v_i \in \{0, 1\}, v_i \in V \setminus T, i = 1, \ldots, |V \setminus T|.$$

where $v_i = 1$ if $v_i \in S$ or $v_i = 0$ otherwise. We label all nodes except the target nodes from 1 to $|V \setminus T|$ and $i \in C_j$ denotes the node label belonging to the community C_j. Note that $v_i = 0$ or 1, $H(v) = \sum_{j=1}^{m} w_j \cdot \min\{1, \sum_{i \in C_j} v_i\}$ can rewrite as $F(v) = \sum_{j=1}^{m} w_j \cdot (1 - \prod_{i \in C_j} (1 - v_i))$. Therefore we have

$$\max F(v) = \sum_{j=1}^{m} w_j \cdot (1 - \prod_{i \in C_j} (1 - v_i))$$

$$s.t. \sum_{v_i \in V \setminus T} v_i = b \tag{7}$$

$$v_i \in \{0, 1\}, v_i \in V \setminus T, i = 1, \ldots, |V \setminus T|.$$

We consider the relaxed problem of (6), i.e., $0 \le v_i \le 1$. This relaxed problem can be found an optimal solution in polynomial time [5]. Based on this, we propose the pipage rounding algorithm for APM to obtain an integer solution.

Algorithm 1. Pipage Rounding Algorithm (PRA)

Input: $G = (V, E, w)$, a community set $\mathcal{C} = \{\mathcal{C}_1, \ldots, \mathcal{C}_m\}$, a target user set $T = \{T_1, \ldots, T_q\}$, a parameter θ, a inter budget b and an influence diffusion model $\mathcal{M}$.
Output: seed set S.
1: Find an optimal solution $\mathcal{S} = \{v_1, \ldots, v_b\}$ to relaxed problem;
2: $S \leftarrow \mathcal{S}$;
3: **for** each community $\mathcal{C}_l$ **do**
4:     **if** there exists a target node $v \in \mathcal{C}_l$ **then**
5:         Create $MAPT(v, \theta)$;
6:         **while** S has an non-integral component **do**
7:             Choose $0 < v_k, v_j < 1$ in $MAPT(v, \theta)$;
8:             Define $S(\varepsilon)$ by

$$v_i(\varepsilon) = \begin{cases} v_i, & \text{if } i \ne k, j, \\ v_j + \varepsilon, & \text{if } i = j, \\ v_k - \varepsilon, & \text{if } i = k; \end{cases}$$

9:             Let $\varepsilon_1 \leftarrow \min\{v_j, 1 - v_k\}$;
10:            Let $\varepsilon_2 \leftarrow \min\{1 - v_j, v_k\}$;
11:        **end while**
12:        **if** $F(S(-\varepsilon_1)) \ge F(S(\varepsilon_2))$ **then**
13:            $S \leftarrow S(-\varepsilon_1)$;
14:        **else**
15:            $S \leftarrow S(\varepsilon_2)$;
16:        **end if**
17:    **end if**
18: **end for**
19: **return** S.

Round one or two non-integer components of optimal solution to relaxed problem in each iteration, which does not cause the objective function value decreasing. We have following theorem.

Theorem 4. *The approximation ratio of Algorithm 1 is $(1 - 1/e)$.*

Proof (Proof of Theorem 4). Let $\mathcal{S}$ denote the optimal solution to relaxed problem and round the $\mathcal{S}$ to get an integer solution S_I for (6). Since $F(S(\varepsilon))$ is convex with respect to ε, $\max\{F(S(-\varepsilon_1)), F(S(\varepsilon_2))\} \ge F(\mathcal{S})$ if $\varepsilon_1, \varepsilon_2 > 0$. Thus, the value of $F(S)$ is non-decreasing in the loop of Algorithm 1. Therefore, $F(S_I) \ge F(\mathcal{S})$. We note that S_I has only integer components, and $F(S_I) = H(S_I)$. According to [5], it follows that $H(S_I) = F(S_I) \ge F(\mathcal{S}) \ge (1 - \frac{1}{e})H(\mathcal{S})$.

Let us analyze the complexity of the Algorithm 1. Finding an optimal solution to relaxed problem can be done in $O(|V|m)$ on G. The loop from line 3 to 18 at most runs $O(m)$ times. In each iteration, there are at most q target users. The inner loop runs at most b times. Therefore, the time complexity is $O(|V|m+mqb)$.

7   Experiments

In this section, we evaluate our algorithm on real-life networks. We first describe the datasets and experiment setup, and then show the results. Furthermore, we compare with other popular approaches.

7.1   Experiment Setup

Datasets: We use three real-life networks with various scale from (SNAP)[2]. Table 1 provides the details of these datasets. Further, 'CC' represents clustering coefficient and '#Community' represents the number of communities. Note that Amazon and Youtube are undirected networks. Therefore we transform these two undirected networks into directed networks. Specifically, for an undirected edge (u, v) on Amazon and Youtube networks, we randomly generate a directed edge (u, v) or (v, u) with probability of 0.5 respectively. According to [3], we let $\theta = 0.03$ in all experiments.

Table 1. The statistics of data sets

Dataset	#Node	#Edge	CC	#Community
E-mail	1K	25.6K	0.399	42
Amazon	334.8K	925.8K	0.396	75K
Youtube	1134.8K	2987.6K	0.080	8K

E-mail. This network is generated using email data from a large European research institution. Each node represents a researcher and each directed edge (u, v) means that u sent at least one email to v.

Amazon. It is based on *Customers Who Bought This Item Also Bought* feature of the Amazon website. Each node is a product. If a product u is frequently co-purchased with product v, thus there is an edge (u, v) between u and v.

Youtube. This network is a video-sharing social network. Each node is a user on network. Users form friendship if they share same videos.

---

[2] http://snap.stanford.edu/data.

Comparison Methods: To compare with existing methods, other methods are as comparison methods: Local Cascade Algorithm (LCA) [6] and Greedy Algorithm (GA) [14]. Our pipage rounding algorithm is abbreviated as PRA.

Random (RAN) means that it randomly selects seed nodes.

Local Cascade Algorithm (LCA) [6]. LCA constructs a local cascade community consists of only the shortest paths between each node and the target node, then restricts computations within the shortest path community.

Greedy Algorithm (GA) [14]. The influence spread of seed set S in the targeted set $T \subseteq V$ is the expected number of activated nodes in T by S. And the greedy algorithm iteratively selects a new seed v that maximizes the incremental change of function, to be included into the seed set S, until b seeds are selected.

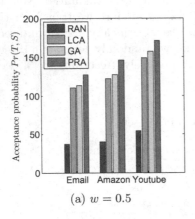

(a) $w = 0.5$

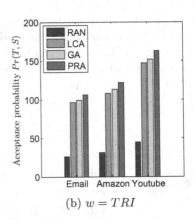

(b) $w = TRI$

Fig. 1. The total acceptance probability of target users under IC model: target users set size $|T| = 1000$ on Amazon and Youtube networks or $|T| = 500$ on E-mail network, $w = 0.5$ or $w = TRI, \theta = 0.03$ and $b = 30$.

7.2   Results

The Acceptance Probability of Target Users: We calculate the total acceptance probability of target users when $|T| = 1000$ or $|T| = 500$ and $b = 30$ on each network with different methods. Figure 1 illustrates the results. The horizontal axis represents the names of social networks. The vertical axis represents the total acceptance probability of target users. We compare RAN, PRA, LCA and GA where RAN means randomly selecting seed nodes. In both subfigures, total acceptance probabilities of target users show similar trends on each network. Specifically, on each network, total acceptance probabilities satisfy following relationship: RAN<LCA<GA<PRA. On the other hand, instead of utilizing Monte-Carlo simulation, it indicates MAPT is a more effective approximation to calculate the acceptance probability than other methods. In Fig. 1(a), we

let $w = 0.5$. PRA is 8.19%–13.01% more than GA, 12.87%–16.44% more than LCA, and 68.42%–72.60% more than RAN. In Fig. 1(b), we let $w = TRI^3$. PRA is 6.60%–7.38% more than GA, 9.43%–11.48% more than LCA, and 72.39%–75.47% more than RAN. Although GA has higher acceptance probability than LCA, it's too time consuming in experiments.

Seeds Size vs. Acceptance Probability of Target Users: In this part, for PRA, we analyze how the number of seeds affects the acceptance probability of target users when given a target user set in a fixed community. In fact, we randomly choose $|T| = 1000$ nodes on Amazon and Youtube networks as target users and choose a fixed community whose target users size is greater than 50. In particular, we randomly choose $|T| = 500$ on Email network and select a community whose target users size is greater than 40. Figure 2 shows the results. The horizontal and vertical axis indicate seeds size and acceptance probability of target users in the fixed community, respectively. Note that all methods (PRA, LCA, GA and RAN) show the property of diminishing marginal return. More precisely, acceptance probability sharply increases when seed size increases from $|S| = 1$ to $|S| = 5$. While it increases slowly from $|S| = 5$ to $|S| = 10$. Our PRA method is the best and RAN is the worst one because it has performance guarantee as we analyzed before. RAN randomly selecting seeds with high probability can not activate target users that leads to its worst.

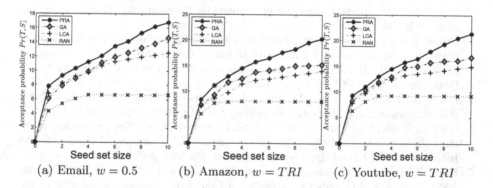

(a) Email, $w = 0.5$ (b) Amazon, $w = TRI$ (c) Youtube, $w = TRI$

Fig. 2. The relationship between total acceptance probability of target users and seeds size on each network under the IC model: $w = 0.5$ for Email network and $w = TRI$ for Amazon and Youtube networks, $|T| > 50$ in a fixed community for Amazon and Youtube networks, $|T| > 40$ in a fixed community for Email network, $\theta = 0.03$.

8 Conclusion

In this paper, we study a novel problem called *Acceptance Probability Maximization* (APM) problem that finds a small size seed set such that the acceptance

---

[3] We uniformly at random select a probability from $\{0.1, 0.3, 0.5\}$.

probability of target users is maximized. Based on IC model, we show APM is NP-hard and compute it is $\#p - hard$. And we prove objective function satisfies monotonicity and submodularity. Considering the community structure of social networks, we transform our APM problem into MWHS problem. We develop a pipage rounding algorithm which has a $(1-1/e)$ approximation ratio. In order to evaluate our proposed methods, extensive experiments have been conducted. The experiment results show that our method outperforms comparison approaches.

Acknowledgments. This work was supported in part by the National Natural Science Foundation of China Grant No. 11671400, 61672524. The Fundamental Research Funds for the Central University, and the Research Funds of Renmin University of China, 2015030273, and the Research Funds of Renmin University of China 16XNH116.

References

1. Algesheimer, R., Dholakia, U.M., Herrmann, A.: The social influence of brand community: evidence from European car clubs. J. Mark. **69**(3), 19–34 (2005)
2. Chang, C.W., Yeh, M.Y., Chuang, K.T.: On influence maximization to target users in the presence of multiple acceptances. In: 2015 IEEE/ACM International Conference on Advances in Social Networks Analysis and Mining (ASONAM), pp. 1592–1593. IEEE (2015)
3. Chen, W., Wang, C., Wang, Y.: Scalable influence maximization for prevalent viral marketing in large-scale social networks. In: Proceedings of the 16th ACM SIGKDD International Conference on Knowledge Discovery and Data Mining, pp. 1029–1038. ACM (2010)
4. Crandall, D., Cosley, D., Huttenlocher, D., Kleinberg, J., Suri, S.: Feedback effects between similarity and social influence in online communities. In: Proceedings of the 14th ACM SIGKDD International Conference on Knowledge Discovery and Data Mining, pp. 160–168. ACM (2008)
5. Du, D.Z., Ko, K.I., Hu, X.: Design and Analysis of Approximation Algorithms, vol. 62. Springer, New York (2011). https://doi.org/10.1007/978-1-4614-1701-9
6. Guo, J., Zhang, P., Zhou, C., Cao, Y., Guo, L.: Personalized influence maximization on social networks. In: Proceedings of the 22nd ACM International Conference on Information and Knowledge Management, pp. 199–208. ACM (2013)
7. Karp, R.: Reducibility among combinatorial problems. In: Miller, R.E., Thatcher, J.W., Bohlinger, J.D. (eds.) Complexity of Computer Computations, pp. 85–103. Springer, Boston (1972). https://doi.org/10.1007/978-1-4684-2001-2_9
8. Kempe, D., Kleinberg, J., Tardos, É.: Maximizing the spread of influence through a social network. In: Proceedings of the Ninth ACM SIGKDD International Conference on Knowledge Discovery and Data Mining, pp. 137–146. ACM (2003)
9. Kuhnle, A., Pan, T., Alim, M.A., Thai, M.T.: Scalable bicriteria algorithms for the threshold activation problem in online social networks. In: INFOCOM 2017-IEEE Conference on Computer Communications, pp. 1–9. IEEE (2017)
10. Song, C., Hsu, W., Lee, M.L.: Targeted influence maximization in social networks. In: Proceedings of the 25th ACM International on Conference on Information and Knowledge Management, pp. 1683–1692. ACM (2016)
11. Temitope, O.A.S., Ahmad, R., Mahmudin, M.: Influence maximization towards target users on social networks for information diffusion. J. Telecommun. Electron. Comput. Eng. (JTEC) **10**(1–10), 17–24 (2018)

12. Valiant, L.G.: The complexity of enumeration and reliability problems. SIAM J. Comput. **8**(3), 410–421 (1979)

13. Yang, D.N., Hung, H.J., Lee, W.C., Chen, W.: Maximizing acceptance probability for active friending in online social networks. In: Proceedings of the 19th ACM SIGKDD International Conference on Knowledge Discovery and Data Mining, pp. 713–721. ACM (2013)

14. Zhou, C., Guo, L.: A note on influence maximization in social networks from local to global and beyond. Procedia Comput. Sci. **30**, 81–87 (2014)

Knowledge Graph Embedding Based on Subgraph-Aware Proximity

Xiao Han[✉], Chunhong Zhang, Chenchen Guo, Tingting Sun, and Yang Ji

Key Laboratory of Universal Wireless Communications, Ministry of Education,
Beijing University of Posts and Telecommunications, Beijing, China
{hanxiao1007,zhangch,orangegcc,suntingting,jiyang}@bupt.edu.cn

Abstract. Knowledge graph (KG) embedding aims to project the original KG into a low-dimensional embedding vector space, so as to facilitate the completion of KGs and the application of KGs in other AI fields. Most existing models preserve certain proximity property of KGs in the embedding space, such as the first/second-order proximity and the sequence-aware higher-order proximity. However, the ubiquitous similarity relationship among different sequences has rarely been discussed. In this paper, we propose an unified framework to preserve the subgraph-aware proximity in the embedding space, holding that the sequences within a subgraph generally imply similar pattern. To analyze the impact of different composition of sequences on the subgraph-aware proximity, we classify the subgraphs into relation subgraph and complete subgraph based on the composition of their sequences. Accordingly, we provide three methods for KG sequence embedding module: (1) Simply adding the involved relations of the sequence in relation subgraph; (2) Recurrent neural networks for the sequences in complete subgraph; (3) Dilated RNN to match the special structure of KG sequences in complete subgraph. Empirically, we evaluate the proposed framework on the KG completion tasks of link prediction and entity classification. The results show that our framework performs better than the baselines by preserving the subgraph-aware proximity. Especially, exploring the special structure of KG sequences can further improve the performance.

Keywords: Knowledge graph completion · Embedding learning · Subgraph-aware proximity

1 Introduction

Knowledge graphs (KGs) are sub-collections of human knowledge, which are composed of entities as nodes and relations as edges. The facts in KGs are usually stored in triples of RDF form, i.e., (s, r, o) where s and o are the subject entity and object entity respectively, r is the relation between them. The abundant structured data contained in KGs can significantly boost the researches in diverse AI fields, including but not limited to decision-making, information retrieval and question answering [9]. However, despite enormous efforts in KG

© Springer Nature Switzerland AG 2018
S. Tang et al. (Eds.): AAIM 2018, LNCS 11343, pp. 306–318, 2018.
https://doi.org/10.1007/978-3-030-04618-7_25

maintenance, KGs still suffer from incompleteness. Instead of completing a KG based on symbol and logic, KG embedding models project the entities and relations of the KG into a low-dimensional embedding vector space, where knowledge reasoning can be conducted via algebraic operations.

In general, KG embedding models are supposed to preserve certain proximity property of the original KG in the embedding space. Particularly, since relations in KGs usually contain rich semantic meanings, the proximity definition is also different from it in general graphs, where edges only serve as topological links between nodes. To begin with, the first-order proximity, which is originally proposed to describe the pairwise similarity between two directly linked nodes in general graphs, has been modified in KGs. Specifically, it describes the similarity between the two entities in a triple based on the semantic meanings of the relation. For example, translation-based KG embedding models [3,12,14,21] compare the similarity between the translated subject entity $s+r$ and the object entity o. Neural network-based ones [2,15,18] view the relation as a weight factor in the comparison of the two entities. In addition, graph embedding models [19] and [20] use the second-order proximity to measure the similarity of two nodes according to their shared neighborhoods. Furthermore, [17] and [10] preserve higher order proximity by maximizing the probability of the query node given the past known ones along a single sequence, which can be view as **sequence-aware** higher-order proximity. However, the similarity relationship among different sequences has rarely been discussed.

In this paper, we propose an unified framework to preserve the **subgraph-aware** higher-order proximity in the embedding space, which describes the similarity among different sequences within the same subgraph. As Fig. 1 shows, our framework contains three major modules. In the subgraph pre-extraction module, we extract a subgraph for each triple in the KG, including the triple itself and a number of multi-hop sequences obtained by random walk. Since both relation sequence and complete sequence (relation & entity sequence) are used in previous work [7,13], we also classify the subgraphs into relation subgraph and complete subgraph according to the composition of their sequences. Accordingly, we provide three methods for the following KG sequence embedding module ψ, which reflect the different understanding of the KG sequence structure: (1) Adding the involved relations of the KG sequences in relation subgraph; (2) Using traditional RNN to compute the sequence embeddings in complete subgraph, considering that complete KG sequences contain complete semantic meanings just as sentence sequences in plain text; (3) Transforming RNN to dilated RNN, where a dilated recurrent skip connection is designed to match the special structure of complete KG sequences, i.e., a complete KG sequence is composed of alternating entities and relations and organized in an orderly form. Then, we compute the proximity scores in the embedding space. In our framework, the training objective is originally formulated based on the conditional probability of a triple given the corresponding multi-hop sequences within the same subgraph. It is further simplified as a hinge loss based on the fact that the proximity score of the real subgraph is higher than the corresponding negative one.

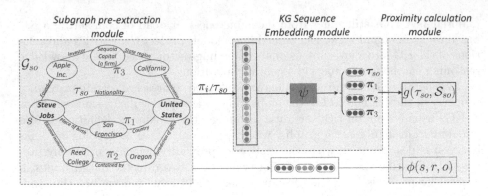

Fig. 1. The framework of KG embedding based on subgraph-aware proximity

We empirically evaluate the resulting embeddings of SA-KGE on two standard tasks of KG completion: link prediction and entity classification. The detailed comparison and analysis show that (i) The approaches under the framework of SA-KGE performs better than the baselines; (ii) The approaches based on complete subgraph are superior to the relation subgraph based one; (iii) Preserving the special structure of KG sequences is beneficial for KG embedding and further improves the performance of KG completion tasks.

The main contributions of this paper are summarized as follows:

- We propose an unified KG embedding framework to learn the subgraph-aware proximity of KGs, so as to preserve the ubiquitous inter-sequence similarity in the embedding space.
- We compare and analyze three methods for KG sequence embedding, which enables us get deep insight into the structure of KG sequences. Particularly, the orderly organizing form of KG sequences are supposed to be concerned.
- We empirically demonstrate that our framework are effective for KG completion tasks, especially when using complete subgraph and considering the special structure of KG sequences.

2   Related Work

The first-order proximity is originally used in general graphs [9], where the edges are homogeneous. However, considering KGs are multi-relational graphs, the relations between entities usually have different semantic meanings. As a result, the similarity between two directly linked entities should be measured based on the relation between them, which can be viewed as a modified first-order proximity. Both the translation-based embedding models [3,12,14,21,22] and the neural network-based models [2,15,18] fall into this category. In TransE [3], the first-order proximity is formulated as a distance measure between the translated subject entity $s + r$ and the object entity o. The following models

[12,14,21] project entities or relations into different spaces, so as to alleviate the problem that TransE cannot deal with complex relations. Meanwhile, [22] is another improvement with entity descriptions. The neural network-based models treat the relation as a weight factor in the similarity comparison between the subject and object entities. Specifically, [2] compares the similarity of the two entities after projecting the relation to them. [18] represents the relation as a tensor and computes how likely the two entities are after a neural network. In [15], a deep neural network is used to process the connection of the subject entity and the relation, whose outputs are compared with the object entity.

The second-order proximity is a supplement to the first-order proximity, which is used to obtain the similarity between two nodes with missing link. The graph embedding models [19,20] jointly use the above two proximity criterions. In addition, [10,17] extend skip-gram architecture to graphs, preserving higher-order proximity of nodes along a single sequence. However, the above related models rarely concern the similarity relationship among different sequences. In this paper, we focus on the subgraph-aware higher-order proximity that describes the inter-sequence proximity within each subgraph. It is worth noting that PTransE [13] uses the relation sequences to replace the direct relation of the score function, indicating that the relation sequences are similar to the direct relation. This is closely related to our framework with relation subgraph. Thus we make a comparison with PTransE in the experiment.

3 Methodology

In this section, we first give the formal definition of the subgraph-aware proximity. Then, we introduce the overall framework of SA-KGE, followed by the instantiation of the KG sequence embedding module and some other implementation details of the proposed framework.

Let us begin with some common notations. A KG is denoted as $\mathcal{G} = (\mathcal{E}, \mathcal{R})$, where $\mathcal{E}$ and $\mathcal{R}$ respectively represent the entity set and the relations set. Each fact in $\mathcal{G}$ is stored as a triple (s, r, o), where $s, o \in \mathcal{E}$ are the subject and object entities respectively, and $r \in \mathcal{R}$ is the relation between them. The proposed SA-KGE aims to preserve the subgraph-aware proximity of a KG into the embedding space by the following mapping: $e_i \rightarrow \mathbf{e}_i \in \mathbb{R}^d \ \forall e_i \in \mathcal{E}$ and $r_j \rightarrow \mathbf{r}_j \in \mathbb{R}^d \ \forall r_j \in \mathcal{R}$.

3.1 Definition of Subgraph-Aware Proximity

For a triple (s, r, o), there are a set of multi-hop sequences between the subject entity s and the object entity o, denoted as $\mathcal{S}_{so} = \{\pi_1, \cdots, \pi_i, \cdots, \pi_K\}$. The subgraph $\mathcal{G}_{so}$ is composed of the one-hop sequence τ_{so} and the multi-hop sequence set $\mathcal{S}_{so}$, i.e., $\mathcal{G}_{so} = \tau_{so} \cup \mathcal{S}_{so}$. The multi-hop sequences used in previous KG completion models include relation sequence [13] and relation & entity sequence [7]. Similarly, to analyze the impact of different composition of sequences on SA-KGE, we classify the subgraphs into **relation subgraph** and **complete subgraph** according to the composition of their sequences. In relation subgraph, the one-hop sequence $\tau_{so} = r$ and the multi-hop sequence $\pi_i = \{r_1, \cdots, r_j, \cdots, r_l\}$ with $r_j \in \mathcal{R}$ and l as the hop number. In addition to relation subgraph, we also discuss the complete subgraph whose sequences are composed of both relations and entities, i.e., the one-hop sequence $\tau_{so} = (s, r, o)$ and the multi-hop sequence $\pi_i = \{s, r_1, e_1, \cdots, r_j, e_j, \cdots, r_l, o\}$ with $r_j \in \mathcal{R}$ and $e_j \in \mathcal{E}$.

As the subgraph Fig. 1 shows, the multi-hop sequences generally indicate the similar semantic meanings of the one-hop sequence, no matter considering the relation subgraph or the complete subgraph. We formally define the above **subgraph-aware proximity** as follows:

Definition 1. *For a subgraph $\mathcal{G}_{so} = \tau_{so} \cup \mathcal{S}_{so}$ corresponding to the triple (s, r, o), the subgraph-aware proximity describes the similarity between the multi-hop sequences in the set $\mathcal{S}_{so}$ and the one hop sequence τ_{so}.*

3.2   Overall Framework of SA-KGE

In the framework of SA-KGE, we aim to learn the embeddings of the entities and relations in a KG, and preserve the subgraph-aware proximity in the learned embeddings. Since the multi-hop sequences usually reflect the inference pattern of the one-hop sequence, we can use the conditional probability of the one-hop sequence given the corresponding multi-hop sequences within the same subgraph to formulate the subgraph-aware proximity. Hence the objective of the framework is to find the optimal parameters by maximizing the following log-likelihood conditional probability

$$\hat{\theta} = \arg\max_{\theta} \sum_{\mathcal{G}_{so} \in \mathcal{G}} \log p\left(\tau_{so} | \mathcal{S}_{so}, \theta\right) \tag{1}$$

where θ denotes the parameters to be optimized, including the embeddings of entities and relations in the KG as well as the weight parameters of the framework. The conditional probability $p\left(\tau_{so} | \mathcal{S}_{so}, \theta\right)$ can be formulated over all the one-hop sequences of the subgraphs in the KG as follows

$$p\left(\tau_{so} | \mathcal{S}_{so}, \theta\right) = \frac{\exp\left(g\left(\tau_{so}, \mathcal{S}_{so}\right)\right)}{\sum_{\mathcal{G}_{xy} \in \mathcal{G}} \exp\left(g\left(\tau_{xy}, \mathcal{S}_{so}\right)\right)}. \tag{2}$$

where the subgraph-aware score function $g(\tau_{so}, \mathcal{S}_{so})$ measures the similarity between the one-hop sequence τ_{so} and the multi-hop sequences of $\mathcal{S}_{so}$ in the embedding space. τ_{xy} is the one-hop sequence of arbitrary subgraph $\mathcal{G}_{xy} \in \mathcal{G}$. Since the number of subgraphs of the KG can reach billions, it is intractable to compute the conditional probability $p\left(\tau_{so} | \mathcal{S}_{so}, \theta\right)$. To solve this problem, we resort to the hinge loss [6] which is widely used in previous KG embedding models. Thereby, we obtain the loss function for the subgraph $\mathcal{G}_{so}$

$$\mathcal{L}(\mathcal{G}_{so}) = \sum_{(\tau'_{so}, \mathcal{S}'_{so}) \in \Delta'_{\mathcal{G}_{so}}} \max\left[0, \gamma_1 - g\left(\tau_{so}, \mathcal{S}_{so}\right) + g\left(\tau'_{so}, \mathcal{S}'_{so}\right)\right] \tag{3}$$

where $\max[0, \cdot]$ indicates the hinge loss, $\Delta'_{\mathcal{G}_{so}}$ is the set of negative subgraphs with some components of $\mathcal{G}_{so}$ randomly replaced, and γ_1 is a predefined margin.

Furthermore, the framework is flexible to incorporate the first-order proximity that specializes in learning the similarity relationship within triples. Let $\phi(s, r, o)$ denote the first-order proximity score function, the loss function of the triple (s, r, o) is

$$\mathcal{L}(s, r, o) = \sum_{(s', r', o') \in \Delta'_{(s,r,o)}} \max\left[0, \gamma_2 - \phi(s, r, o) + \phi(s', r', o')\right] \qquad (4)$$

where $\Delta'_{(s,r,o)}$ indicates the set of negative triples with one of the three components randomly replaces. Finally, we get the loss function of the framework by combining Eqs. (3) and (4) over the KG

$$\mathcal{L} = \sum_{\mathcal{G}_{so} \in \mathcal{G}} \left(\alpha \cdot \mathcal{L}(\mathcal{G}_{so}) + \beta \cdot \mathcal{L}(s, r, o)\right) \qquad (5)$$

where α and β are adjustable weight coefficients of the two loss function, and (s, r, o) is the corresponding triple of the subgraph $\mathcal{G}_{so}$.

As Fig. 1 shows, the framework of SA-KGE includes three major modules. First, subgraph pre-extraction module extracts the multi-hop sequences for each triple of the KG by random walk. Next is the KG sequence embedding module, whose inputs are the initialized embeddings of all the components of each sequence, and outputs are the sequence embeddings. Finally, in the proximity calculation module, the score functions of the subgraph-aware proximity and the first-order proximity are computed respectively based on the sequence embeddings and the embeddings of each component of the triple. Next, we will introduce the specific methods for KG sequence embedding module and some other implementation details of the framework.

3.3   KG Sequence Embedding Module

In this section, we introduce three methods for representing KG sequences as embeddings, such that the proximity score $g(\tau_{so}, \mathcal{S}_{so})$ can be computed in the embedding space.

ADD for Relation Subgraph. In relation subgraph, the subgraph-aware proximity degenerates to the similarity between the direct relation r and each multi-hop sequence $\pi_i = \{r_1, \cdots, r_j, \cdots, r_l\}$ in $\mathcal{S}$. The embedding of the one-hop sequence is r, which is the embedding of the direct relation. Since the length of relation multi-hop sequences is only half of the corresponding complete ones and all the components are relations, we can obtain the embedding of $\pi_i \in \mathcal{S}$ by simply adding all the relations of the sequence as in Fig. 2(a)

$$\pi_i = \frac{1}{l} \sum_{j=1}^{l} r_j \qquad (6)$$

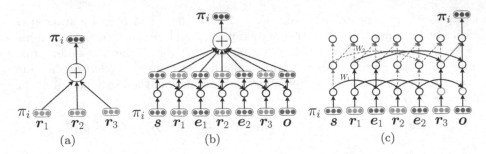

Fig. 2. The methods for KG sequence embedding module ψ. (a) ADD for relation subgraph; (b) RNN for complete subgraph; (c) DilatedRNN for complete subgraph.

RNN for Complete Subgraph. In complete subgraph, KG sequences usually imply complete semantic meanings just like sentence sequences in plain text. As a result, we can use the methods for sentence sequence embedding to deal with KG sequences, such as RNN which is widely used in linguistic model. As shown in Fig. 2(b), the hidden state of RNN is computed according to

$$\mathbf{h}_t = \delta(\mathbf{U}\mathbf{x}_t + \mathbf{W}\mathbf{h}_{t-1}) \tag{7}$$

where $\boldsymbol{x}_t$ is the embedding vector of the t-th component of the input sequence $\pi_i = \{s, r_1, e_1, \cdots, r_j, e_j, \cdots, r_l, o\}$, $\mathbf{U}$ and $\mathbf{W}$ denote weight matrixes, $\boldsymbol{h}_t$ and $\boldsymbol{h}_{t-1}$ are the hidden states of the steps t and $t-1$ respectively, δ is the activation function. The embedding of π_i is the average of the hidden states of each step

$$\pi_i = \frac{1}{T} \sum_{j=1}^{T} h_j \tag{8}$$

where T denotes the total number of components in π_i.

DilatedRNN for Complete Subgraph. Furthermore, we notice that KG sequences have some special properties that distinguish them from sentence sequences. Specifically, a KG sequence is organized in a orderly form of "*entity, relation, entity, relation, ..., entity*", which can be decomposed into an entity subsequence and a relation subsequence. Each subsequence has relatively independent dependencies. Considering the special structure of KG sequences, we incorporate dilated RNN [5] and design a dilated recurrent skip connection architectures to match the structure.

As show in Fig. 2(c), the dilated RNN in our paper contains two hidden layers and an output layer. The first hidden layer is designed to learn the independent dependencies by setting the dilation ratio $p^{(1)}$ to 2, such that the connections are within each subsequence, and there is no connection between them. Actually, we get the embedding results of the two subsequences at the last two timestamps. The hidden state in the first hidden layer is computed according to

$$h_t^{(1)} = \delta\left(U_1 x_t + W_1 h_{t-p^{(1)}}^{(1)}\right) \tag{9}$$

where x_t is the embedding vector of the t-th component of the input sequence $\pi_i = \{s, r_1, e_1, \cdots, r_j, e_j, \cdots, r_l, o\}$, U_1 and W_1 denote weight matrixes of the first hidden layer, $h_t^{(1)}$ and $h_{t-p^{(1)}}^{(1)}$ are the hidden states of the steps t and $t-p^{(1)}$ respectively, δ is the activation function.

The second hidden layer can be used to learn the dependencies of larger scale, whose dilation ratio $p^{(2)}$ is 4. We compute the hidden state according to

$$h_t^{(2)} = \delta \left(U_2 h_t^{(1)} + W_2 h_{t-p^{(2)}}^{(2)} \right) \tag{10}$$

where the notations are similar with Eq. (9). By now, the dependencies are still within each subsequence. To learn the interaction between relations and entities, we add connections between relation and entity timestamps at the output layer as in Fig. 2(c). The embedding of π_i is the output of the last step

$$\pi_i = \delta(W_r h_{T-1}^{(2)} + W_e h_T^{(2)}) \tag{11}$$

where W_r and W_e are the weight matrixes respectively corresponding to the relation and entity timestamps, $h_{T-1}^{(2)}$ and $h_T^{(2)}$ are respectively the hidden states of second hidden layer at $T-1$ and T steps, δ denotes the activation function.

3.4 Other Implementation Details

In this section, we give some implementation details of SA-KGE in our experiment. First, We compute the proximity score $g(\tau_{so}, \mathcal{S}_{so})$ with the following Log-SumExp function

$$g(\tau_{so}, \mathcal{S}_{so}) = \log \left(\sum_{i=1}^{K} \exp \left(f(\tau_{so}, \pi_i) \right) \right) \tag{12}$$

where $f(\tau_{so}, \pi_i)$ measures the similarity between the one-hop sequence τ_{so} and the multi-hop sequence π_i. All the similarity measure functions are applicable, such as distance similarity and cosine similarity. In our experiment, we use inner product to compute the similarity with the embeddings normalized

$$f(\tau_{so}, \pi_i) = \tau_{so}^{\mathsf{T}} \pi_i . \tag{13}$$

As for the negative subgraph set $\Delta_{\mathcal{G}_{so}}'$ in Eq. (3), it can be constructed by replacing arbitrary components of $\mathcal{G}_{so}$. To simplify, we choose to only replace the components of the one-hop sequence. Additionally, any score function of the KG embedding models based on the first-order proximity is feasible for $\phi(s, r, o)$. In our experiment, we use the following translation-based score function

$$\phi(s, r, o) = \|s + r - o\|_{L_n} \tag{14}$$

where $\|\cdot\|_{L_n}$ is the L_n-distance measure with n as 1 or 2.

4   Experiment

In this section, we evaluate SA-KGE on two standard KG completion tasks: Link Prediction and Entity Classification. The task of link prediction is conducted on FB15k, which is a subset of the typical large-scale knowledge base Freebase [1]. FB15k contains 14,951 entities and 1,345 relations. The dataset of entity classification is extracted from FB15k as in [22] and renamed as EN15k in [8]. EN15k contains the entity types with the frequency of the top 50 and remove the type of *common/topic* which every entity has. The related 13,445 entities are randomly split into 12,113 entities as the training set and 1,332 entities as the test set.

4.1   Link Prediction

Link prediction aims to complete the triple (s, r, o) when one of the components is missing, including two subtasks of entity prediction and relation prediction.

Protocols and Parameters. For a test triple, We fill up the missing position with each candidate entity in the entity set $\mathcal{E}$ or each relation in the relation set $\mathcal{R}$. The confidence score of each candidate triple is computed according to:

$$G(s, r, o) = g(\tau_{so}, \mathcal{S}_{so}) + \phi(s, r, o) \tag{15}$$

where $g(\tau_{so}, \mathcal{S}_{so})$ and $\phi(s, r, o)$ measures the subgraph-aware proximity and the first-order proximity respectively.

Then, we rank the confidence scores and record the ranking value of the correct candidate for each test triple. The evaluation metrics are the same as in [11,13]. For entity prediction, the evaluation metrics are Mean Rank of the correct candidates and the proportion of the correct candidates ranked in top 10 (Hits@10). For relation prediction, we report the results of Mean Rank and Hits@1, considering that the total number of relations is relatively small. Besides the above "Raw" results, we also report "Filter" results by filtering out all the valid candidate triples before ranking [3].

In the subtask of entity prediction, from Eq. (15) we know that the multi-hop sequences for each candidate triple are necessary for computing confidence score, which is impractical since we have to iterate all candidate entities of the KG. In practice, we use the re-rank method in [13] to simplify. In particular, we first use Eq. (14) to rank the candidate entities and select the top 500. Then, we re-rank the selected entities according to Eq. (15).

The hyper-parameters are determined by the Mean Rank of validation dataset, including learning rate λ, margins γ_1 and γ_2, embedding dimension d and the dissimilarity measure L_n in Eq. (14). The settings of the hyper-parameters for each KG sequence embedding method are: ADD: $\lambda = 0.005$, $\gamma_1 = 0.3$, $\gamma_2 = 0.5$, $d = 50$ and $L_n = L_2$; RNN: $\lambda = 0.001$, $\gamma_1 = 0.3$, $\gamma_2 = 0.3$, $d = 50$ and $L_n = L_2$; dilated-RNN: $\lambda = 0.001$, $\gamma_1 = 0.2$, $\gamma_2 = 0.3$, $d = 50$ and $L_n = L_2$. In addition, for fair comparison, we set the maximum hop number

Table 1. Entity prediction results.

Metric	Mean Rank		Hits@10(%)	
	Raw	Filter	Raw	Filter
SE [4]	273	162	28.8	39.8
SME (linear) [2]	274	154	30.7	40.8
SME (bilinear) [2]	284	158	31.3	41.3
TransE [3]	243	125	34.9	47.1
TransH [21]	212	87	45.7	64.4
TransR [14]	198	77	48.2	68.7
PTransE [13]	207	58	51.4	84.6
SA-KGE-ADD	210	52	51.5	**85.0**
SA-KGE-RNN	161	55	51.7	84.7
SA-KGE-dilatedRNN	**160**	**53**	**60.7**	84.9

l_{max} as 3 for both our framework and the baseline PTransE [13]. The activation function used in our experiment is sigmoid. The weight coefficients α and β in the loss function of Eq. (5) are both set to 1.

Result Analysis. The overall entity prediction results and detailed "Filter" results based on the mapping property of relations are respectively reported in Tables 1 and 2. The experiment results show that our framework performs better than other baselines on both Mean Rank and Hits@10. Moreover, SA-KGE-RNN and SA-KGE-dilatedRNN based on the complete subgraph improve the Mean Rank of "Raw" setting by a large margin, demonstrating that complete semantic meanings are essential for the similarity comparison of sequences. SA-KGE-dilatedRNN performs best since its architecture is capable to learn the special structure of KG sequences. Additionally, the performance of SA-KGE-ADD is close to PTransE, since they all focus on the similarity relationship between relation sequences.

Table 3 shows the relation prediction results, from where we observe that all the approaches, including our framework and PTransE, that consider the inter-sequence similarity within subgraphs provide relatively good performance. Furthermore, SA-KGE-RNN and SA-KGE-dilatedRNN based on the complete subgraph performs better than other approaches, which indicates that retaining the entities of subgraphs is also beneficial for relation prediction.

4.2 Entity Classification

Entity classification aims to predict the missing entity types, which is also a standard task of KG completion.

Table 2. Detailed results on FB15k by mapping properties of relation types. (%)

Tasks	Predicting head entities (Hits@10)				Predicting tail entities (Hits@10)			
	1-to-1	1-to-M	M-to-1	M-to-M	1-to-1	1-to-M	M-to-1	M-to-M
SE [4]	35.6	62.6	17.2	37.5	34.9	14.6	68.3	41.3
SME (linear) [2]	35.1	53.7	19.0	40.3	32.7	14.9	61.6	43.3
SME (bilinear) [2]	30.9	69.6	19.9	38.6	28.2	13.1	76.0	41.8
TransE [3]	43.7	65.7	18.2	47.2	43.7	19.7	66.7	50.0
TransH [21]	66.8	87.6	28.7	64.5	65.5	39.8	83.3	67.2
TransR [14]	78.8	89.2	34.1	69.2	79.2	37.4	**90.4**	72.1
PTransE [13]	90.1	92.0	58.7	86.1	90.7	70.7	87.5	88.7
SA-KGE-ADD	90.1	92.3	58.6	86.3	91.2	71.0	87.2	88.7
SA-KGE-RNN	**92.3**	91.5	61.1	86.5	**92.2**	73.9	88.1	**89.0**
SA-KGE-dilatedRNN	91.4	**93.0**	**61.3**	**86.7**	91.8	**74.0**	90.1	**89.0**

Table 3. Relation prediction results.

Metric	Mean Rank		Hits@1(%)	
	Raw	Filter	Raw	Filter
TransE [3]	2.8	2.5	65.1	84.3
PTransE [13]	1.8	1.4	68.5	94.0
SA-KGE-ADD	1.8	**1.3**	68.5	94.2
SA-KGE-RNN	**1.7**	**1.3**	**69.8**	93.8
SA-KGE-dilatedRNN	**1.7**	**1.3**	69.6	**94.3**

Table 4. Entity classification results.

Metric	FB15k
TransE [3]	87.9
DKRL [22]	**90.1**
PTransE [13]	86.7
node2vec [10]	63.2
SA-KGE-ADD	86.6
SA-KGE-RNN	88.5
SA-KGE-dRNN	89.3

Protocols and Parameters. Entity classification is essentially a multi-label classification problem, which can be decomposed to multiple binary classification tasks according to the one-versus-rest setting as in [16, 22]. Specifically, we utilize Logistic Regression [22] as classifier for fair comparison with baselines. In addition, we use the evaluation metric of Mean Average Precision (MAP), which is the mean of average precision over all entity types [11, 16].

Result Analysis. The results of entity classification is listed in Table 4. The results show that our framework achieves better results than TransE [3] based on the first-order proximity and node2vec [10] based on higher-order proximity along single sequence. This indicates that learning the inter-sequence proximity within subgraphs are significant for KG embedding. In addition, DKRL performs slightly better than SA-KGE-dilatedRNN. It may because DKRL utilizes entity descriptions which are highly related to entity types.

5   Conclusion and Future Work

In this paper, we propose an unified framework of SA-KGE to learn the subgraph-aware proximity, which describes similarity relationship among different sequences within a subgraph. Furthermore, the framework can be combined with the first-order proximity based models, so as to preserve more comprehensive property of KGs in the embedding space. By incorporating different KG sequence embedding methods, we show that framework is open to existing models. The experiment results of the two KG completion tasks show that the proposed framework can largely promote KG embedding by preserving the subgraph-aware proximity. We tend to use the complete subgraph since both relations and entities are indispensable for the exact semantic meanings of sequences. Moreover, the exploration of the special structure of KG sequences also improves the performance.

Acknowledgements. This work is supported by National Natural Science Foundation of China, 61602048, 61601046, 61520106007, BUPT-SICE Excellent Graduate Students Innovation Funds, 2016.

References

1. Bollacker, K., Evans, C., Paritosh, P., Sturge, T., Taylor, J.: Freebase: a collaboratively created graph database for structuring human knowledge. In: Proceedings of the 2008 ACM SIGMOD International Conference on Management of Data, pp. 1247–1250. ACM (2008)
2. Bordes, A., Glorot, X., Weston, J., Bengio, Y.: A semantic matching energy function for learning with multi-relational data. Mach. Learn. **94**(2), 233–259 (2014)
3. Bordes, A., Usunier, N., Garcia-Duran, A., Weston, J., Yakhnenko, O.: Translating embeddings for modeling multi-relational data. In: Advances in Neural Information Processing Systems, pp. 2787–2795 (2013)
4. Bordes, A., Weston, J., Collobert, R., Bengio, Y., et al.: Learning structured embeddings of knowledge bases. In: AAAI, vol. 6, p. 6 (2011)
5. Chang, S., et al.: Dilated recurrent neural networks. In: Advances in Neural Information Processing Systems, pp. 76–86 (2017)
6. Collobert, R., Weston, J.: A unified architecture for natural language processing: deep neural networks with multitask learning. In: Proceedings of the 25th International Conference on Machine Learning, pp. 160–167. ACM (2008)
7. Das, R., Neelakantan, A., Belanger, D., McCallum, A.: Chains of reasoning over entities, relations, and text using recurrent neural networks. arXiv preprint arXiv:1607.01426 (2016)
8. Fan, M., Zhou, Q., Zheng, T.F., Grishman, R.: Distributed representation learning for knowledge graphs with entity descriptions. Pattern Recognit. Lett. **93**, 31–37 (2016)
9. Goyal, P., Ferrara, E.: Graph embedding techniques, applications, and performance: a survey. arXiv preprint arXiv:1705.02801 (2017)
10. Grover, A., Leskovec, J.: node2vec: scalable feature learning for networks. In: Proceedings of the 22nd ACM SIGKDD International Conference on Knowledge Discovery and Data Mining, pp. 855–864. ACM (2016)

11. Han, X., Zhang, C., Guo, C., Ji, Y.: A generalization of recurrent neural networks for graph embedding. In: Phung, D., Tseng, V.S., Webb, G.I., Ho, B., Ganji, M., Rashidi, L. (eds.) PAKDD 2018. LNCS (LNAI), vol. 10938, pp. 247–259. Springer, Cham (2018). https://doi.org/10.1007/978-3-319-93037-4_20

12. Ji, G., He, S., Xu, L., Liu, K., Zhao, J.: Knowledge graph embedding via dynamic mapping matrix. In: ACL, vol. 1, pp. 687–696 (2015)

13. Lin, Y., Liu, Z., Luan, H., Sun, M., Rao, S., Liu, S.: Modeling relation paths for representation learning of knowledge bases. arXiv preprint arXiv:1506.00379 (2015)

14. Lin, Y., Liu, Z., Sun, M., Liu, Y., Zhu, X.: Learning entity and relation embeddings for knowledge graph completion. In: AAAI, pp. 2181–2187 (2015)

15. Liu, Q., et al.: Probabilistic reasoning via deep learning: neural association models. arXiv preprint arXiv:1603.07704 (2016)

16. Neelakantan, A., Chang, M.W.: Inferring missing entity type instances for knowledge base completion: new dataset and methods. arXiv preprint arXiv:1504.06658 (2015)

17. Perozzi, B., Al-Rfou, R., Skiena, S.: DeepWalk: online learning of social representations. In: Proceedings of the 20th ACM SIGKDD International Conference on Knowledge Discovery and Data Mining, pp. 701–710. ACM (2014)

18. Socher, R., Chen, D., Manning, C.D., Ng, A.: Reasoning with neural tensor networks for knowledge base completion. In: Advances in Neural Information Processing Systems, pp. 926–934 (2013)

19. Tang, J., Qu, M., Wang, M., Zhang, M., Yan, J., Mei, Q.: Line: large-scale information network embedding. In: Proceedings of the 24th International Conference on World Wide Web, pp. 1067–1077. International World Wide Web Conferences Steering Committee (2015)

20. Wang, D., Cui, P., Zhu, W.: Structural deep network embedding. In: Proceedings of the 22nd ACM SIGKDD International Conference on Knowledge Discovery and Data Mining, pp. 1225–1234. ACM (2016)

21. Wang, Z., Zhang, J., Feng, J., Chen, Z.: Knowledge graph embedding by translating on hyperplanes. In: AAAI, pp. 1112–1119 (2014)

22. Xie, R., Liu, Z., Jia, J., Luan, H., Sun, M.: Representation learning of knowledge graphs with entity descriptions. In: AAAI, pp. 2659–2665 (2016)

Author Index

Printed in the United States
By Bookmasters